RENEWALS 458-4574
DATE DUE

WITHDRAWN
UTSA Libraries

GOVERNING GLOBAL DESERTIFICATION

The Global Environmental Governance Series

Series Editors: John J. Kirton and Konrad von Moltke

Global Environmental Governance addresses the new generation of twenty-first century environmental problems and the challenges they pose for management and governance at the local, national, and global levels. Centred on the relationships among environmental change, economic forces, and political governance, the series explores the role of international institutions and instruments, national and sub-federal governments, private sector firms, scientists, and civil society, and provides a comprehensive body of progressive analyses on one of the world's most contentious international issues.

Also in the Series:

Sustainability, Civil Society, and International Governance
Edited by John J. Kirton and Peter I. Hajnal
ISBN 0 7546 3884 7

A World Environmental Organization
Edited by Frank Biermann and Stefan Bauer
ISBN 0 7546 3765 4

Hard Choices, Soft Law
Edited by John J. Kirton and Michael J. Trebilcock
ISBN 0 7546 0966 9

The Politics of Irrigation Reform
Peter Mollinga and Alex Bolding
ISBN 0 7546 3515 5

Agricultural Policy Reform
Wayne Moyer and Tim Josling
ISBN 0 7546 3050 1

Governing Global Biodiversity
Edited by Philippe Le Prestre
ISBN 0 7546 1744 0

Governing Global Desertification
Linking Environmental Degradation, Poverty and Participation

Edited by
PIERRE MARC JOHNSON
Heenan Blaikie, Canada

KAREL MAYRAND
Unisféra, Canada

MARC PAQUIN
Unisféra, Canada

With financial assistance from

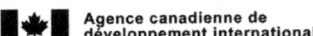

ASHGATE

© Pierre Marc Johnson, Karel Mayrand and Marc Paquin 2006

All rights reserved. No part of this publication may be reproduced, stored in a retrieval system or transmitted in any form or by any means, electronic, mechanical, photocopying, recording or otherwise without the prior permission of the publisher.

Pierre Marc Johnson, Karel Mayrand and Marc Paquin have asserted their moral right under the Copyright, Designs and Patents Act, 1988, to be identified as the editors of this work.

Published by
Ashgate Publishing Limited
Gower House
Croft Road
Aldershot
Hampshire GU11 3HR
England

Ashgate Publishing Company
Suite 420
101 Cherry Street
Burlington, VT 05401-4405
USA

Ashgate website: http://www.ashgate.com

British Library Cataloguing in Publication Data
Governing global desertification : linking environmental
   degradation, poverty and participation. - (Global
   environmental governance)
   1. Convention on Desertification (1994) 2. Desert reclamation
   - International co-operation 3. Community development -
   International cooperation 4. Desertification - Social
   aspects
   I. Johnson, Pierre-Marc II. Mayrand, Karel III. Paquin, Marc,
   1964-
   333.7'36153

Library of Congress Control Number: 2006921294

ISBN-10: 0 7546 4359 X

Printed and bound in Great Britain by MPG Books Ltd. Bodmin, Cornwall.

Contents

List of Tables	vii
List of Figures	ix
List of Contributors	xi
Preface	xiii
Acknowledgements	xv
List of Abbreviations and Acronyms	xvii

1. The United Nations Convention to Combat Desertification in Global Sustainable Development Governance
   Pierre Marc Johnson, Karel Mayrand, and Marc Paquin — 1

2. The Scientific Basis: Links between Land Degradation, Drought and Desertification
   Stefanie M. Herrmann and Charles F. Hutchinson — 11

3. Examining the Links between Poverty and Land Degradation: From Blaming the Poor toward Recognising the Rights of the Poor
   Sally-Anne Way — 27

4. Desertification and Migration
   Michelle Leighton — 43

5. Negotiating Desertification
   Adil Najam — 59

6. The United Nations and the Fight against Desertification: What Role for the UNCCD Secretariat?
   Steffen Bauer — 73

7. Civil Society's Role in Negotiating and Implementing the UNCCD
   Friederike Knabe — 89

8. Promoting Good Governance through the Implementation of the UNCCD
   Lene Poulsen and Masse Lo — 109

9. The Global Mechanism and UNCCD Financing: Constraints and Opportunities
   François Falloux, Susan Tressler, and Karel Mayrand — 131

10 Decentralisation and Sustainable Resources Management in West Africa: A Line of Action for Revisiting National Action Programmes
    Richard Pearce   147

11 Knowledge and the UNCCD: The Community Exchange and Training Programme
    Noel Oettlé   163

12 Agriculture, Trade, and Desertification: Implications for the UNCCD
    Karel Mayrand and Marc Paquin   177

13 Conclusion: The UNCCD at a Crossroad
    Pierre Marc Johnson, Karel Mayrand, and Marc Paquin   195

APPENDIX
United Nations Convention to Combat Desertification in those Countries Experiencing Serious Drought and/or Desertification, Particularly in Africa   205

Bibliography   *257*
Index   *287*

List of Tables

Table 5-1	UNCCD Timeline	60
Table 9-1	Opportunities and Constraints of the Global Mechanism	142
Table 11-1	Criteria Applied by SADC/RIOD for the Selection of Knowledge Exchange Projects	170

List of Figures

Figure 2-1	Distribution of Non-polar Arid Land	13
Figure 7-1	RIOD's Focal Point Structure 1997/98	94
Figure 11-1	Steps Followed in the Implementation of the Pilot Project	167

List of Contributors

Steffen Bauer is the co-ordinator of the MANUS Research Group (Managers of Global Change: Effectiveness and Learning of International Organizations) within the international Global Governance Project at the Environmental Policy Research Centre of Freie Universität Berlin and a doctoral candidate at the Freie Universität's Otto-Suhr-Institute for Political Science.

François Falloux is vice-president and member of the board of Eco-Carbone, of which he is also co-founder.

Stefanie M. Herrmann is a researcher in Arid Lands Resource Sciences at the University of Arizona.

Charles F. Hutchinson is director of the Office of Arid Lands Studies at the University of Arizona.

Pierre Marc Johnson is senior counsel with Heenan Blaikie and advisor to the Commission for Environmental Cooperation and the secretariat of the United Nations Convention to Combat Desertification.

Friederike Knabe is an independent consultant, working primarily with nongovernmental organisations (NGOs) and the Canadian International Development Agency (CIDA) on international environmental and development policies and strategies.

Michelle Leighton is assistant dean at John F. Kennedy University Law School.

Masse Lo is the Regional Director of ENDA — LEAD Francophone Africa, a not-for-profit organisation whose mission is to create, strengthen, and support networks of people and institutions promoting change toward sustainable development.

Karel Mayrand is director of research at Unisféra International Centre, an independent, not-for-profit research centre on sustainable development policy and law, and a lecturer in environment and international development issues at l'Université de Sherbrooke.

Adil Najam is associate professor of international negotiation and diplomacy at the Fletcher School of Law and Diplomacy at Tufts University.

Noel Oettlé is rural programme manager for the Environmental Monitoring Group (EMG) and is based in Nieuwoudtville, South Africa.

Marc Paquin is a lawyer specialising in international and comparative environmental and sustainable development law and policy, and co-founder and executive director of Unisféra International Centre, an independent not-for-profit research centre on sustainable development policy and law.

Richard Pearce is in charge of the desertification programme in the Bureau Regional pour l'Afrique de l'Ouest (BRAO) of the World Conservation Union (IUCN).

Lene Poulsen is an independent consultant in sustainable development processes with a focus on good governance and sustainable livelihood in rural areas.

Susan Tressler is a consultant specialising in project development and fundraising.

Sally-Anne Way is Senior Advisor to the United Nations Special Rapporteur on the Right to Food and heads the Research Unit on the Right to Food, based at the Graduate Institute of Development Studies of the University of Geneva.

Preface

This publication is the result of the combined work of experts and highly esteemed development practitioners with long-standing experience on desertification and drought-related matters. Some of the authors have, over the years, played a key role in the negotiation sessions of the United Nations Convention to Combat Desertification (UNCCD).

The book revisits many important dimensions of the UNCCD, including issues such as the nexus between land degradation and poverty, the role of civil society, the financing of activities to combat desertification, and the links between the action programmes and other national development strategies. In many chapters, we are reminded of the complexity of combating desertification and provided with sharp insights on the challenges ahead, now that the convention enters its implementation phase. At the same time, this volume highlights the importance of joining forces and strengthening our responses to address the adverse effects of drought and desertification fully and efficiently.

This publication is also a useful reference material, which contributes to a better knowledge of the importance of the convention, as well as opportunities and constraints associated with its implementation. Combating desertification through well-defined action programmes is a challenging task, especially in affected developing countries. The secretariat of the convention is committed to continuing to facilitate the implementation of the UNCCD jointly with interested partners, policy makers, academic, nongovernmental organisations (NGOs), and other relevant actors that deal with the convention daily.

Ten years after the entry into force of the UNCCD, the majority of countries affected by desertification have finalised their action programmes and moved the convention process from the planning phase to implementation. These countries, their regional or sub-regional organisations, and civil society should be encouraged to continue their efforts with resolve and determination. I would like to thank the authors of this book for sharing their visions, and thereby participating in the joint effort to bring forward the implementation of the UNCCD.

Hama Arba Diallo
Executive Secretary
United Nations Convention to Combat Desertification

Acknowledgements

The editors wish to acknowledge the generous financial contribution and continuous support of the Canadian International Development Agency (CIDA) without which this publication would not have been possible. Special thanks go to Joana Talafré and Nancy Hamzawi for their support early on this initiative, as well as to David Moloney, Paul Samson, and Charles Haines for their renewed commitment to this project at critical stages.

We wish to thank John Kirton, the series co-editor, for having given us the opportunity to contribute to the important field of desertification governance. We also express our gratitude to the contributing authors for their patience, commitment, and profoundly stimulating manuscripts. The diversity and richness of their views and perspectives significantly contribute to the literature on desertification and the United Nations Convention to Combat Desertification. We are grateful to Madeline Koch for her precious help in finalising and reviewing the manuscript, and wish to thank the four anonymous reviewers who contributed to improving this publication with their thoughtful and constructive comments.

The editors also wish to acknowledge the loss of two highly esteemed colleagues and friends involved in the preparation of this book: Konrad von Moltke, co-editor of the Global Environmental Governance series, who passed away in the spring of 2005, and Moussa Cissé, who disappeared during the fall of 2004. This book is dedicated to them and to their friends and families.

List of Abbreviations and Acronyms

ADB	Asian Development Bank
AfDB	African Development Bank
AFSED	Arab Fund for Social and Economic Development
AU	African Union
AHWG	Ad Hoc Working Group
CBD	Convention on Biological Diversity
CBO	community-based organisation
CDM	Clean Development Mechanism (of the United Nations Framework Convention on Climate Change)
CETP	Community Exchange and Training Programme (of the Global Mechanism)
CGIAR	Consultative Group on International Agricultural Research
CIDA	Canadian International Development Agency
CILSS	Permanent Inter-State Committee on Drought Control in the Sahel (Comité Permanent Inter États de Lutte Contre la Sécheresse au Sahel)
CONAGESE	Conseil National pour la Gestion de l'Environnement (of Burkina Faso)
CONEDD	Conseil National pour l'Environnement et le Développement Durable
COP	Conference of the Parties (to the United Nations Convention to Combat Desertification)
CRIC	Committee for the Review of the Implementation of the Convention (to Combat Desertification)
CST	Committee on Science and Technology (of the United Nations Convention to Combat Desertification)
DAC	Development Assistance Committee (of the Organisation for Economic Co-operation and Development)
DFID	United Kingdom Department for International Development
DGF	Development Grant Facility of the World Bank
ELCI	Environmental Liaison Centre International
EMG	Environmental Monitoring Group (South Africa)
ENDA	Environment and Development Action in the Third World
eniD	European Networking Initiative on Desertification
ENSO	El Niño Southern Oscillation
FAO	Food and Agriculture Organization
FDI	foreign direct investment
FIELD	Financial Information Engine on Land Degradation

G77	Group of 77 plus China (caucus of developing countries within the United Nations system)
GNI	gross national income
GEF	Global Environment Facility
GHG	greenhouse gas
GM	Global Mechanism (of the United Nations to Combat Desertification)
GNI	gross national income
GNP	gross national product
GTZ	Deutsche Gesellschaft für Technische Zusammenarbeit (German Technical Development Co-operation)
IADB	Inter-American Development Bank
ICTSD	International Centre for Trade and Sustainable Development
IDP	internally displaced person
IDRC	International Development Research Centre
IFAD	International Fund for Agricultural Development
IISD	International Institute for Sustainable Development
INCD	Intergovernmental Negotiating Committee on Desertification (of the United Nations Convention to Combat Desertification)
IMF	International Monetary Fund
IOM	International Organization for Migration
IsDB	Islamic Development Bank
ITCZ	Intertropical Convergence Zone
IUCN	World Conservation Union
LADA	Land Degradation Assessment for Drylands
LDC	least-developed country
MDGs	Millennium Development Goals
MEA	multilateral environmental agreement
NAFTA	North American Free Trade Agreement
NAO	North Atlantic Oscillation
NAP	National Action Programme (of the United Nations Convention to Combat Desertification)
NCB	national co-ordinating body (for implementing the United Nations Convention to Combat Desertification)
NEPAD	New Partnership for Africa's Development
NCCDs	National NGO Co-ordinating Committees on Desertification (established by the Réseau international des ONG pour la desertification [RIOD])
NDF	National Desertification Fund
NEAP	National Environmental Action Plan
NGO	nongovernmental organisation
OAU	Organization for African Unity
ODA	official development assistance
OECD	Organisation for Economic Co-operation and Development

OP-15	Operational Program No. 15 (of the Global Environmental Facility)
OPEC	Organization of Petroleum Exporting Countries
PACD	Plan of Action to Combat Desertification (adopted at the United Nations Conference on Desertification)
PDO	Pacific Decadal Oscillation
P/PET	ratio of precipitation to potential evapotranspiration
PrepCom	Preparatory Committee (for the United Nations Convention on Environment and Development)
PRSP	poverty reduction strategy paper
RAP	Regional Action Programme (of the United Nations Convention to Combat Desertification)
RCU	Regional Co-ordination Unit
RIOD	Réseau international des ONG pour la desertification (International Network on Desertification)
SADC	Southern African Development Community
SCS	Solidarité Canada Sahel
SRAP	Sub-regional Action Programme (of the United Nations Convention to Combat Desertification)
SST	sea surface temperature
TAG	Technical Advisory Group (of the Global Mechanism)
TPN	thematic programme networks
UNCCD	United Nations Convention to Combat Desertification in those Countries Experiencing Serious Drought and/or Desertification, Particularly in Africa
UNCOD	United Nations Conference on Desertification
UNCED	United Nations Conference on Environment and Development (the 1992 Earth Summit)
UNCTAD	United Nations Conference on Trade and Development
UNEP	United Nations Environment Programme
UNESCO	United Nations Educational, Scientific, and Cultural Organization
UNDP	United Nations Development Programme
UNFCCC	United Nations Framework Convention on Climate Change
UNGA	United Nations General Assembly
UNGASS	United Nations General Assembly Special Session
UNITAR	United Nations Institute for Training and Research
UNHCR	United Nations High Commissioner for Refugees
UNSO	United Nations Development Programme's Office to Combat Desertification and Drought
USAID	United States Agency for International Development
WFP	World Food Programme
WMO	World Meteorological Organization
WSSD	World Summit on Sustainable Development
WTO	World Trade Organization
WWF	World Wildlife Fund

Chapter 1

The United Nations Convention to Combat Desertification in Global Sustainable Development Governance

Pierre Marc Johnson, Karel Mayrand, and Marc Paquin

If one flew over central Niger in one of the Niger army's old Fokker aircraft in the winter of 2002, the sight was striking: arid, degraded lands stretched to the horizon in every direction. From the air, one could expect to find on the ground desperate, impoverished communities. But as the plane landed in Agadez, a vibrant, colourful and hospitable community revealed itself to the visitor. The contrast between the hostile living conditions and the vivid community was striking.

Even more striking was the sudden appearance of a palm tree plantation in the middle of nowhere north of Agadez. Thanks to the nearby presence of a coal-fuelled power plant, an electric pump–activated well had been drilled, allowing a local community to grow this plantation, which had become a new oasis in the middle of the drylands. Women were carrying water from the well to the trees and reforesting the area with sustainable techniques. By doing so, they were reversing the land degradation process and diversifying their sources of income.

The contrast between widespread rural poverty and environmental degradation, on the one hand, and the opportunities created on a small scale through community empowerment, access to groundwater, and sustainable land management, defines the ideals behind the United Nations Convention to Combat Desertification (UNCCD). This convention is not about fighting deserts, it is about reversing land degradation trends, improving living conditions and alleviating poverty in rural drylands. This dual focus on poverty and the environment makes it a sustainable development convention.[1]

Land Degradation, Poverty, and Agriculture: An Overview of the Issues

Drylands cover approximately 41 percent of the global landmass, comprise 44 percent of the world's cultivated lands, and are home to a third of the world's population, or 2 billion people (Millennium Ecosystems Assessment 2005, 1).[2] Estimates of rural dryland populations vary from 900 million to 1.2 billion and drylands population have among the highest population growth rates in the world with an average of 18.5 percent in the 1990s (Dobie 2001; Millennium Ecosystems Assessment 2005, 8). It is thought that desertification directly affects 1 to 6 percent of rural dwellers and that many more are threatened by this phenomenon (Millennium Ecosystems Assessment

2005, 7). According to the Millennium Ecosystem Assessment, 10 percent to 20 percent of the world's drylands — or 10 to 20 million square kilometres — are degraded (7). In Africa, approximately 500 million hectares of land are affected by land degradation, including two thirds of the region's productive agricultural land (United Nations Environment Programme [UNEP] 2003). It is estimated that between 5 and 6 million hectares are permanently lost to agriculture each year through human-induced soil degradation (GTZ 2005).

The causes of land degradation in the world include overgrazing (680 million hectares), deforestation (580 million hectares), agricultural mismanagement (550 million hectares), fuel wood over consumption (137 million hectares), and industry and urbanisation (19.5 million hectares) (UNEP 2003). As these figures show, land degradation is linked directly to agricultural practices. It is estimated that the annual economic costs of desertification worldwide exceed US$42 billion, including both output forgone and damage to natural resources, while the costs of combating desertification is estimated to be in the range of US$10 to 22 billion per year (Toulmin 1994).

There is a high correlation between extreme poverty, land degradation, and rural livelihoods in the world. This correlation is even more pronounced in rural drylands (see Chapter 3). Around the world, more than 1.2 billion people live in extreme poverty; 900 million of those people live in rural areas where their livelihoods depend on the consumption and sale of natural products (United Nations Development Programme [UNDP] 2001, 123). About two thirds of these rural poor live in ecologically vulnerable areas. Of these, a high proportion live in dryland areas, 250 million of whom are directly affected by desertification (Vitalis 2004, 2; Nelson 1997, cited in Scherr 1999).

There is a similar correspondence associating hunger, small-scale agriculture, and environmental vulnerability. Globally, some 852 million people suffer from hunger; of these, a large number are rural farmers (United Nations Millennium Project 2005, 3). The highest prevalence of hunger is found in remote areas where food production is affected by economic, environmental, or climatic shocks (3). This includes dryland areas where an estimated 44 percent of the world's malnourished children live (Sharma et al. 1996, cited in Scherr 1999, 70). According to the United Nations Millennium Project's Task Force on Hunger, about half the undernourished people in the world are small farmers, 20 percent are landless rural dwellers, 10 percent are pastoralists and fishers, and the remaining 20 percent are urban dwellers (UN Millennium Project 2005, 3–4).

These figures show that a significant share of the world's poor and undernourished are rural farmers suffering from a combination of socioeconomic marginalisation and the degradation of their resource base. Desertification is a major driver of poverty and hunger among these populations. Indeed, desertification is strongly correlated to poverty through its relation with declining agricultural yields.

Agricultural productivity is one of the most important determinants of economic growth and poverty in the developing world (Department for International Development [DFID] 2004a, 3). Indeed, poverty alleviation has been directly correlated with improvements in agricultural productivity throughout the world (DFID 2004b). Recent research shows that a 1 percent increase in agricultural yields reduces the proportion of people living on less than US$1 per day by between 0.6 and 1.2 percent (DFID 2003, 1). Another recent study covering 58 developing countries concluded that a 10 percent increase in agricultural productivity is associated with a 6 percent reduction in the proportion of people living with less than US$1 a day (Lin et al. 2001, cited in DFID 2002, 4). This correlation is even higher for sub-Saharan Africa. Moreover, considering that dryland communities often compensate their low monetary income level with free access to ecological goods and services, the deterioration of their resources base further contributes to their impoverishment, even when this is not reflected in income statistics.

Under such conditions, poverty alleviation strategies must involve actions to combat desertification and favour the development of sustainable land management practices that improve the livelihoods of poor rural peoples. By preventing land degradation and improving agricultural practices, actions to combat desertification can lead to increased agricultural productivity, thereby directly raising incomes and food security among rural dryland communities.

The UNCCD in Global Sustainable Development Governance

Since the beginning of this decade, global sustainable development governance is greatly influenced by the framework of the Millennium Development Goals (MDGs), which establish a set of objectives and targets that have drawn almost universal adhesion among states and international institutions. Together, these goals set a new strategic framework to address development issues, prioritise action, and improve coherence in international activity. Given the diversity of the MDGs and their interrelations, their attainment will require the coherent mobilisation of international policies and financial resources.

Among these goals are the two interrelated and cross-cutting goals of poverty alleviation and environmental sustainability. The first MDG sets the objective to halve by 2015 the number of people living on less than US$1 a day (Target 1) and of those suffering from hunger (Target 2).[3] The seventh MDG sets the objective to integrate the principles of sustainable development into country policies and programmes and to reverse the loss of environmental resources (Target 9).

With its dual focus on poverty and land degradation in rural drylands, the UNCCD appears to be the instrument of choice in the international governance system to tackle the challenges raised by the attainment of the MDGs.[4] Accordingly, the World Summit on Sustainable Development (WSSD) recognised the UNCCD as 'one of the tools for poverty eradication' in the world and as being in line with the MDGs (United Nations 2002, para. 7[1]).

The convention was also listed in the UN Secretary General's Millennium Report — along with the United Nations Framework Convention on Climate Change (UNFCCC) and the Convention on Biological Diversity (CBD) — among a core group of treaties that represent the UN's key objectives. It is also one of the three sustainable development companion conventions born out of the UN Conference on Environment and Development (UNCED) at Rio, the only one that stems from a direct recommendation of Agenda 21 (UNCED 1992, ch. 12(40)), and the only instrument in the land conventions cluster within the classification developed by UNEP.

The convention itself seeks to co-ordinate its activities with other instruments of environmental governance. According to Article 8, the UNCCD should encourage co-ordination with other multilateral environmental agreements (MEAs), particularly the United Nations Framework Convention on Climate Change (UNFCCC) and the CBD. This co-ordination can extend to the conduct of joint programmes, particularly in the fields of research, training, systematic observation, and information collection and exchange.

Accordingly, the UNCCD has developed a work programme with the CBD and is working with the UNFCCC to develop joint objectives. A memorandum of understanding has been signed with the Ramsar Convention in 1998 and the UNCCD has collaborated on multiple initiatives, including with UNEP on the Land Degradation Assessment for Drylands (LADA) and the Millennium Ecosystem Assessment. Other arrangements for co-operation have also been concluded with the UNDP, the UN Educational, Scientific, and Cultural Organization (UNESCO), and the World Meteorological Organization (WMO).

While the UNCCD has been effectively integrated among global environmental governance instruments, its integration among developmental instruments such as the New Partnership for Africa's Development (NEPAD) and poverty reduction strategy papers (PRSPs) as an instrument to fight poverty and favour rural development has not been concretised (see Chapter 10). Although the UNCCD has a dual focus on development and environment, it is still mostly perceived as an environmental instrument rather than as a sustainable development one that could add value to development governance. As a consequence, it remains a marginal instrument in rural development and agricultural policies at the national and international levels.

An Overview of the UNCCD Approach

The signature of the UNCCD in 1994 introduced an innovative approach to combat desertification that focussed on both natural and socioeconomic processes as well as on participation. This innovative approach was strongly inspired by the concept of sustainable development and by new insights on the linkages among desertification, environmental degradation, and poverty. It was also the result of a compromise between developed and developing countries and the product of intense lobbying on the part of African countries (see Chapter 5).

The UNCCD defines desertification as 'land degradation in arid, semi-arid and dry sub-humid areas resulting from various factors, including climatic variations and human activities' (UNCCD 1994, art. 1[a]).[5] Land degradation itself is further defined as a 'reduction or loss, in arid, semi-arid and dry sub-humid areas, of the biological or economic productivity and complexity of rainfed cropland, irrigated cropland, or range, pasture, forest and woodlands resulting from land uses or from a process or combination of processes, including processes arising from human activities and habitation patterns, such as: (i) soil erosion caused by wind and/or water; (ii) deterioration of the physical, chemical and biological or economic properties of soil; and (iii) long-term loss of natural vegetation' (art. 1[f]).

These definitions combine both the anthropogenic and biophysical processes that are at play in the land degradation process. As J.F. Reynolds and D.M. Stafford Smith state: 'Land degradation in drylands involves complex interactions between biophysical factors (the meteorological and ecological dimensions of desertification that include sensitive soils, extreme rain events etc.) and socioeconomic factors (the human dimensions of desertification that include overgrazing, deforestation, intensive use of soils and water resources etc.)' (Reynolds and Stafford Smith 2002, 8). The dual focus on human and natural causes of desertification constitutes the conceptual underpinning of the UNCCD: addressing both the socioeconomic and environmental interactions at play in the desertification process.

The convention preamble explicitly recognises that 'desertification and drought affect sustainable development through their interrelationships with important social problems such as poverty, poor health and nutrition, lack of food security, and those arising from migration, displacement of persons and demographic dynamics' (UNCCD 1994, prologue). In line with this recognition, the UNCCD commits parties to 'adopt an integrated approach addressing the physical, biological and socio-economic aspects of the processes of desertification and drought' (art. 4.2 [a]), and to 'integrate strategies for poverty eradication into efforts to combat desertification and mitigate the effects of drought' (art. 4.2[c]). The UNCCD also commits affected country parties to 'address the underlying causes of desertification and pay special attention to the socio-economic factors contributing to desertification processes' (art. 5[c]).

In addition to this dual focus on poverty and environmental degradation, several attributes of this convention make it an innovative and strategic instrument in the environment and development governance systems. The UNCCD provides a framework for community-based, participatory approaches in the elaboration and implementation of activities to combat desertification. It is committed to the improvement of living conditions, 'in particular at the community level' (UNCCD 1994, art. 2.2), and contains several provisions supporting the participation of local communities, community-based organisations (CBOs), nongovernmental organisations (NGOs), and women in the design and implementation of activities to combat desertification.

Moreover, the National, Regional, and Sub-Regional Action Programmes (NAPs, RAPs, SRAPs respectively), which constitute the convention's primary implementation tool, must 'provide for effective participation at the local, national and regional levels of non-governmental organizations and local populations, both women and men, particularly resource users, including farmers and pastoralists and their representative organizations, in policy planning, decision-making, and implementation and review of national action programmes' (UNCCD 1994, art. 10.2[f]).

Through the action programmes, the UNCCD establishes a set of strategic instruments to combat desertification. NAPs are highly integrative and participatory instruments that support broader sustainable development objectives, provided they are integrated into development planning. They must 'promote policies and strengthen institutional frameworks which develop cooperation and coordination, in a spirit of partnership, between the donor community, governments at all levels, local populations and community groups, and facilitate access by local populations to appropriate information and technology' (UNCCD 1994, art. 10.2[e]).

The UNCCD also creates a governance framework similar in many ways to that of its two sister conventions. The convention is governed by the Conference of the Parties (COP) that meets every two years and has a scientific body, the intergovernmental Committee on Science and Technology (CST). It has also created the Committee for the Review of the Implementation of the Convention (CRIC), which met for the first time in 2002. The UNCCD's secretariat, which is based in Bonn, plays a very proactive role within the convention system (see Chapter 6).

The UNCCD provides a framework to co-ordinate and pool financial resources from multiple sources, with the aim of maximising impacts on dryland communities. In its approach to financing, it supports 'innovative methods and incentives for mobilizing and channelling resources, including those of foundations, non-governmental organizations and other private sector entities, particularly debt swaps and other innovative means which increase financing by reducing the external debt burden of affected developing country Parties, particularly those in Africa' (UNCCD 1994, art. 20.2[d]).

In addition, the mobilisation of financial resources under the convention is based on the 'full use and continued qualitative improvement of all national, bilateral and multilateral funding sources and mechanisms, using consortia, joint programmes and parallel financing' (UNCCD 1994, art. 20.4). In practice, this approach operates through partnership agreements among donors, affected countries, and dryland communities, in which one donor may play a *chef de file* role, or consortium leader and resources mobiliser.

Overview of the Structure of This Book

This volume seeks to provide an analysis of some of the complex issues related to sustainable development governance in the drylands and the role of the UNCCD in that context. It combines the expertise of international governance scholars with that of field practitioners from various regions of the world. This volume was conceived as an opportunity to reflect the diversity of perspectives on desertification, from theoretical approaches to field work, and from global governance to local resources management. It also purports to outline the contribution this convention should make to sustainable development in rural drylands, were governments of member countries of the Organisation for Economic Co-operation and Development (OECD), as well as the governments of affected countries, giving it the consideration it deserves within global sustainable development governance.

The first three chapters review the issues that are addressed by the UNCCD. Together, they lay out the key interactions among natural processes, environmental degradation, and socioeconomic development that constitute the problem of desertification as it was articulated in international sustainable development governance. In Chapter 2, Stefanie M. Herrmann and Charles F. Hutchinson define key concepts and present the natural processes at play in land degradation and desertification. This chapter also analyses some of the weaknesses in the desertification concept as defined in the UNCCD. In Chapter 3, Sally-Anne Way examines the linkages between poverty and desertification and addresses the broader economic and political marginalisation processes at play in their development. This chapter presents the evolution in thinking about the socioeconomic causes of desertification. Chapter 4 by Michelle Leighton presents the linkages between poverty, land degradation, and migration and shows evidence of a nexus between desertification and migration in several regions of the world.

Chapters 5 to 7 provide an historical perspective on the negotiation and early implementation of the UNCCD with a specific focus on different actors in the convention system. In Chapter 5, Adil Najam offers a historical perspective on the negotiation of the UNCCD, focussing on the internal dynamics and negotiation strategies of the G77 and China. Steffen Bauer takes a closer look in Chapter 6 at the proactive role played by the UNCCD secretariat in the negotiation and implementation phases of the convention, and in Chapter 7 Friederike Knabe analyses the role played by nongovernmental actors in different periods.

Chapters 8 and 9 present two aspects of UNCCD implementation in its first decade — good governance and financing — contrasting the strengths of its provisions with the challenges of their concrete implementation. In Chapter 8, Lene Poulsen and Masse Lo look at the concept of good governance and how it is reflected in provisions of the UNCCD. They also consider gender issues in the good governance framework and how gender equality is supported by the UNCCD. Chapter 9, by François Falloux, Susan Tressler, and Karel Mayrand, analyses the financing structure of the UNCCD and outlines the challenges and opportunities in expanding the financial base of the convention. This chapter assesses the role of the Global Mechanism (GM), the Global Environmental Facility (GEF), and other mechanisms in leveraging funds for the implementation of the convention.

Chapters 10 and 11 provide regional perspectives on the issues of the implementation of NAPs and knowledge exchanges. In Chapter 10, Richard Pearce looks at the articulation of NAPs in domestic and international development planning in West Africa, with a specific focus on the decentralisation processes that have taken place in the region in recent years. This chapter highlights some of the NAPs' key weaknesses and makes recommendations to enhance their status and effectiveness in development planning. Noel Oettlé in Chapter 11 focusses on Southern Africa and addresses science and knowledge issues from a community-based perspective by looking at the Community Exchange Training Programmes, which favour knowledge exchanges and capacity building at the community level.

In Chapter 12, Karel Mayrand and Marc Paquin situate the UNCCD within the larger framework of economic globalisation by analysing the connections among agricultural trade liberalisation, poverty, and desertification. They analyse the effects of agricultural trade liberalisation in the context of the Doha Round launched under the auspices of the World Trade Organisation (WTO) on land degradation and rural livelihoods in the drylands, and describe how the UNCCD could contribute attenuating these impacts.

But first, let us return to Niger in the fall of 2005, this time near the city of Zinder in the southern part of the country, where one can see the face of absolute poverty and despair. At this moment, 2.65 million Nigeriens were suffering from malnutrition, and as many as 150 000 children were suffering from severe starvation in the worst 'hunger season' to hit West Africa in years (World Food Programme 2005; Gosline 2005). According to the World Food Programme (WFP): 'Niger's food crisis is complex: weather patterns, food production, markets, technology, sanitation, healthcare, education, child rearing practices and poverty all play their roles'.

As described in the opening paragraphs above, targeted actions on a small scale can sometimes make the difference between a desperate situation and a thriving community. Dryland communities have great resources that can be put to fighting poverty and desertification, provided they are properly empowered and supported by national and international development partners. This is the unfulfilled promise of the UNCCD. This book offers a better understanding of the challenges of dryland development and of the role of the UNCCD in addressing this challenge. It is dedicated to dryland communities, for their courage, vibrant inspiration, and unfulfilled promise.

Notes

1   According to the Brundtland Commission: 'sustainable development is development that meets the needs of the present without compromising the ability of future generations to meet their own needs. It contains within it two key concepts: the concept of "needs", in particular the essential needs of the world's poor, to which overriding priority should be given; and the idea of limitations imposed by the state of technology and social organization on the environment's ability to meet present and future needs, (World Commission on Environment and Development 1987).
2   Drylands are the arid, semi-arid, and dry sub-humid areas (other than polar and sub-polar regions) in which the ratio of annual rainfall to potential evapotranspiration falls within the range between 0.05 and 0.65.
3   These targets are set on a 1990 timeline (see World Bank Group 2004).
4   The links between the MDGs and the UNCCD are analysed in greater detail in chapters 3, 8, and 12.
5   The operational usefulness of this definition is still debated in the scientific and development communities (see Chapter 2).

References

Department for International Development (United Kingdom) (2002). 'Better Livelihoods for Poor People: The Role of Agriculture'. Consultation Document. <www.dfid.gov.uk/pubs/files/agricultureconsult.pdf> (December 2005).

Department for International Development (United Kingdom) (2003). 'Agriculture and Poverty Reduction: Unlocking the Potential'. DFID Policy Paper, December. <www.dfid.gov.uk/pubs/files/agripovertyreduction.pdf> (December 2005).

Department for International Development (United Kingdom) (2004a). 'Agriculture, Growth and Poverty Reduction'. Working Paper 1, October. <dfid-agriculture-consultation.nri.org/summaries/wp1.pdf> (December 2005).

Department for International Development (United Kingdom) (2004b). 'Making Agricultural Markets Work for the Poor'. Working Paper 2, September. <dfid-agriculture-consultation.nri.org/summaries/dfidwp2.pdf> (December 2005).

Dobie, Philip (2001). 'Poverty and the Drylands'. United Nations Development Programme. <www.undp.org/seed/unso/gdp/docs/Poverty-and-the-Drylands.doc> (December 2005).

Gosline, Anna (2005). 'Early Warnings of Niger Famine Disregarded'. *New Scientist* 25 July. <www.newscientist.com/article.ns?id=dn7725> (December 2005).

GTZ (2005). 'Fact Sheet Desertification: Africa'. Convention Project to Combat Desertification. <www2.gtz.de/desert/download/basicfactsheets/factsheet_africa.pdf> (December 2005).

Lin, Lin, Virginia McKenzie, Jenifer Piesse, et al. (2001). 'Agricultural Productivity and Poverty in Developing Countries'. Extension to DFID Report No. 7946. Department for International Development (United Kingdom), London.

Millennium Ecosystems Assessment (2005). 'Ecosystems and Human Well-Being: Desertification Synthesis'. World Resources Institute, Washington DC. <www.maweb.org//proxy/document.355.aspx> (December 2005).

Nelson, Mike (1997). 'Report of the Study on CGIAR Research Priorities for Marginal Lands'. TAC Working Document, March. Technical Advisory Committee Secretariat, Food and Agriculture Organization.

Reynolds, J.F. and D.M. Stafford Smith (2002). 'Do Humans Cause Deserts?' In J.F. Reynolds and D.M. Stafford Smith, eds., *Global Desertification: Do Humans Cause Deserts?* Dahlem University Press, Berlin.

Scherr, Sara (1999). 'Dryland Degradation and Poverty'. pp. 69–77, in 'Drylands, Poverty, and Development', Proceedings of the World Bank Round Table, 15–16 June, Washington DC.

Sharma, Manohar, Marito Garcia, Aamir Quershi, et al. (1996). 'Overcoming Malnutrition: Is There an Ecoregional Dimension?' Food, Agriculture, and the Environment Discussion Paper 10. International Food Policy Research Institute, Washington DC.

Toulmin, Camilla (1994). 'Combating Desertification: Encouraging Local Action within a Global Framework'. In Fridtjof Nansen-stiftelsen på Polhøgda, ed., *Green Globe Yearbook of International Co-operation on Environment and Development*, pp. 79–88. Oxford University Press, Oxford.

United Nations (2002). 'Plan of Implementation of the World Summit on Sustainable Development'. Johannesburg. <www.un.org/esa/sustdev/documents/WSSD_POI_PD/English/WSSD_PlanImpl.pdf> (December 2005).

United Nations Conference on Environment and Development (1992). 'Agenda 21'. 14 June. Rio. <www.un.org/esa/sustdev/documents/agenda21/english/agenda21toc.htm> (December 2005).

United Nations Convention to Combat Desertification (1994). 'Text of the United Nations Convention to Combat Desertification'. <www.unccd.int/convention/text/convention.php> (December 2005).

United Nations Development Programme (2001). 'Human Development Report 2003: Millennium Development Goals: A Compact among Nations to End Poverty'. United Nations Development Programme, New York. <hdr.undp.org/reports/global/2003> (December 2005).

United Nations Environmental Programme (2003). 'Global Environmental Outlook 3'. <www.unep.org/geo/geo3/english> (December 2005).

United Nations Millennium Project (2005). 'Halving Hunger: It Can Be Done'. New York. <www.unmillenniumproject.org/reports/tf_hunger.htm> (December 2005).

Vitalis, Vangelis (2004). 'Trade, Agriculture, the Environment, and Development: Reaping the Benefits of a Win-Win-Win'. Paper presented at a Strategic Dialogue on Agriculture, Trade Negotiations, Poverty, and Sustainability, 14–16 July. Windsor UK. <www.ictsd.org/dlogue/2004-07-14/Vangelis.pdf> (December 2005).

World Bank Group (2004). 'About the Goals'. <www.developmentgoals.org> (December 2005).

World Commission on Environment and Development (1987). *Our Common Future* (Brundtland Report). Oxford University Press, Oxford.

Chapter 2

The Scientific Basis: Links between Land Degradation, Drought, and Desertification

Stefanie M. Herrmann and Charles F. Hutchinson

Land degradation, drought, and desertification have emerged as major problems affecting dryland environments. According to 'official' definitions crafted by the United Nations Convention to Combat Desertification (UNCCD), desertification is 'land degradation in arid, semi-arid and dry sub-humid areas resulting from various factors, including climatic variations and human activities', and land degradation is defined as 'reduction or loss ... of the biological or economic productivity and complexity of rainfed cropland, irrigated cropland, or range, pasture, forest and woodlands resulting from land uses or from a process or combination of processes, including processes arising from human activities and habitation patterns' (UNCCD 1994, art. 1[a], 1[f]). The definitions of these terms, as well as their alleged causes, interrelationships, and implications, have fuelled a great deal of controversy among scientists, land managers, and policy makers.

While concerns about desiccation and environmental deterioration in the world's drylands date back to colonial times in West Africa and the massive soil erosion during the 1930s that became known as the 'Dust Bowl' in the American Great Plains, more systematic scientific inquiry into dryland environments only began with the establishment in 1951 of the Arid Zone Research Programme of the United Nations Education, Science, and Cultural Organization (UNESCO).[1] That makes dryland research a relatively young field in science and a frontier of knowledge.

Interest in drylands, however, is vital to a range of countries, notably those on the fringe of the Sahara desert, which are classified — wholly or in large part — as arid and semi-arid. The economic future of these countries, most of which are judged to be 'developing countries', depends on the adaptation of their development strategies to the stresses imposed by their arid environments. Moreover, they are home to a significant, and rapidly growing, portion of the world population: estimates range from some 15 percent to nearly 40 percent, depending on the definition of drylands and demographic data used (Clarke and Noin 1997; White and Nackoney 2003).

The large populations affected by an unusually long series of drought years that hit the African Sahel region from the late 1960s well into the 1980s, and the resulting human suffering, brought a worldwide upsurge in scientific research on drylands, which culminated in the United Nations Conference on Desertification (UNCOD) in

Nairobi in 1977. Although much research since then has been aimed at understanding the mechanisms behind desertification and its links to drought and human land use, the scientific basis for recognition of desertification as an irreversible process remains rather weak. While the occurrence and severity of local-scale land degradation in many drylands is not disputed, there is insufficient scientific evidence of large-scale permanent desertification (Prince, Brown de Colstoun, and Kravitz 1998; Tucker and Nicholson 1999; Eklundh and Olsson 2003). For lack of a more precise concept, however, the term 'desertification' is still widely used in scientific or popular circles to refer to a multitude of land-cover change phenomena, despite its acknowledged limitations and lack of clarity about the fundamental nature of the problem and the lack of measurable criteria.

The Physical Characteristics of Drylands

Drylands are regions that are affected routinely by moisture deficits. These regions are marked not only by low precipitation totals, but also by extreme interannual variability and extended drought. Water availability is the key to understanding dryland environments, yet absolute amounts of precipitation tell little about dynamic processes operating in any particular region. Different disciplines involved in the study of drylands also rely on different indicators to define them (such as botanic and climatic indicators). Therefore, finding an unambiguous and commonly agreed-on definition and classification of drylands is hard to achieve (Beaumont 1993).

The work of Peveril Meigs (1953), commissioned by UNESCO, still forms the basis of most definitions of arid conditions to date. He used a moisture index that relates precipitation, the main input of water into an ecosystem, to evapotranspiration, its main output, to map the distribution of arid lands worldwide (see Figure 2-1). According to the index, drylands are those areas in which potential evaporative demand exceeds total rainfall.

Geographically, the world's drylands are found in four areas: 1) beneath the belt of subtropical atmospheric high-pressure zones marked by the Tropics of Cancer and Capricorn (trade wind deserts), 2) in interior drainage basins of the mid-latitudes (continental deserts), 3) on the lee sides of mountain ranges (rain shadow deserts), and 4) on the western edges of continents affected by cold ocean currents (coastal deserts) (Mainguet 1999). They occupy approximately one third of the world's land surface and include a whole range of conditions from extremely arid to semi-arid.

While causes of aridity are complex and differ from one region to another, all these drylands share the experience of recurring drought and thus fluctuating productivity of the resource base, although these experiences vary in their extents and implications. Because extremely arid regions can support so few people, the more serious problems of land degradation are generally reported from the semi-arid where the effects of drought are more profound due to a higher concentration of population (Dregne 1983).

Ecologically, arid and semi-arid environments, unlike their more humid counterparts, function primarily as non-equilibrium systems. Equilibrium systems are associated with the concepts of succession and climax vegetation and served as the ecological paradigm for much of the 20th century (Clements 1916; Dyksteruis 1949). In contrast, non-equilibrium systems are governed by variability and unpredictability and have offered an alternative to equilibrium models for the past quarter century. Non-equilibrium systems possess a limited capacity for internal regulation and greater potential for transit among multiple equilibrium points (Briske, Fuhlendorf, and Smeins 2003). Vegetation dynamics are driven by stochastic climatic events, which result in discontinuous and possibly non-reversible changes.[2] Ecological thresholds, once crossed, offer transition into a number of other possible equilibrium states, distinguished on the basis of community features, plant growth forms, or soil properties (Holling 1973; Westoby, Walker, and Noy-Meir 1989). In fact, equilibrium and non-equilibrium dynamics are not mutually exclusive, but represent two ends of a continuum within which most systems fall (Wiens 1984). Depending on spatial and temporal scales, drylands can exhibit both equilibrium and non-equilibrium characteristics, and shifts over time between equilibrium and non-equilibrium dynamics are not uncommon (Illius and O'Connor 1999).

Due to periodic extreme climatic conditions, drylands were once assumed to be especially fragile in the face of disturbances (White 1956). However, the view that their evolutionary response to variability increased the resilience of dryland species (Walker et al. 1981; Holling 1973) has become widely accepted (Mortimore 1988). This implies that drylands, although easily perturbed from an initial state of equilibrium, might have a high potential to recover from natural disturbances and might also respond to rehabilitation efforts better than initially believed.

Figure 2-1   Distribution of Non-polar Arid Land

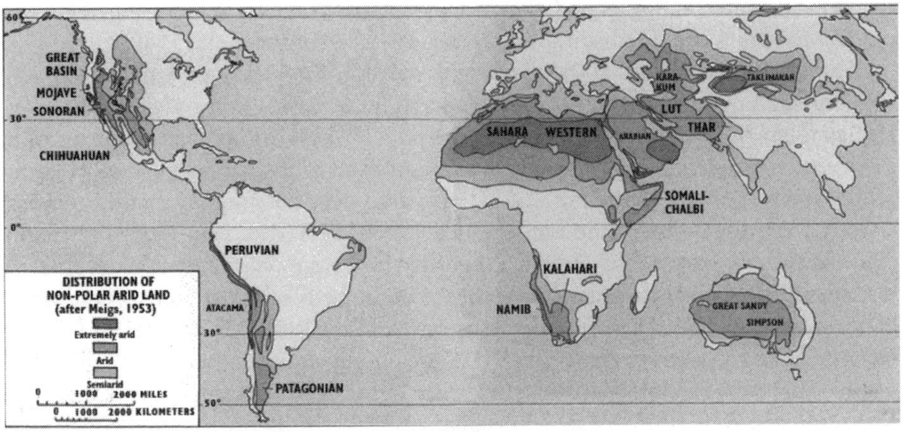

Source: United States Geological Survey (1997).

Aridity and Drought

Aridity and drought, both natural features of drylands, need to be differentiated from one another, even though in practice they are often confused. While similar in their appearance, both operate on different time scales. Aridity is a long term climatic phenomenon and a defining physical characteristic of drylands. Drought is an episodic feature, which can affect any environment, but is also a frequent and defining characteristic of drylands.

Nature and Causes of Aridity

Aridity is the defining physical characteristic of drylands. It entails not only a permanent rainfall deficit, which is connected also to other climatic phenomena such as strong insolation, elevated temperatures, and high evapotranspiration, but is also manifest in a sparse and discontinuous vegetation cover and poorly evolved soils, as soils develop slowly in the absence of water.[3] In addition to a general rainfall deficit, aridity is associated with a high variability and unpredictability of precipitation (Mainguet 1999). Fauna and flora respond to these conditions with a 'pulse-reserve' strategy, which entails pulses of activity triggered by rainfall events, such as plant growth (Noy-Meir 1973). A portion of each pulse is held in reserve in the form of seeds or energy stores in roots and thus contributes to sustainability during the dry season (Whitford 2002).

Precipitation totals alone do not suffice to define aridity because their significance can only be established relative to temperature and potential evapotranspiration. Over the years, a number of approaches to measure aridity have been put forward (Beaumont 1993). Classical approaches focus on climatic elements and their relationships to vegetation zones; index approaches apply standard formula to define boundaries of the arid zone in terms of annual precipitation and temperature (Köppen 1931); and water and energy balance concepts portray the relationship between precipitation and evapotranspiration (Penman 1948; Thornthwaite 1948).

There is a gradation of aridity from hyper-arid to sub-humid regions, which may be disrupted to varying degrees by problems of land degradation and desertification. Drylands can be categorised either along the lines of these aridity measurements or in a descriptive way with respect to dominant vegetation forms. On a simple bioclimatic aridity index, defined as the ratio of precipitation over potential evapotranspiration (P/PET), the hyper-arid zone corresponds to a P/PET less than 0.03 and marks extreme deserts bare of vegetation except for some annual plants that respond to infrequent rains, and drought-tolerant shrubs in the beds of dry streams where water collects after storms. The arid zone is delineated by P/PET from 0.03 to 0.2 and consists of barren areas or areas covered by sparse vegetation of perennial and annual plants. This zone supports pastoral nomadism but rainfed agriculture is not possible. The semi-arid zone (P/PET 0.2–0.5) is distinguished by open vegetation cover, with perennial species dominating. Extensive livestock grazing can be practiced. In the sub-humid zone (P/PET 0.5–0.75) savannah, with or without trees, and dry forest prevail and permanent rainfed agriculture with crops adapted to seasonal drought can be practiced (UNESCO 1979).

Aridity is associated with a variety of factors, all resulting in blocking moist air masses from reaching an area or inhibiting atmospheric moisture from condensing and falling as precipitation. Descending and divergent air on the downward limb of the tropical Hadley circulation leads to permanent arid conditions paired with high insolation along a belt of subtropical anticyclones (Sahara, Arabian deserts, Thar). In the mid-latitudes, aridity occurs on the lee side of north-south–running mountain ranges, which act as barriers for moist oceanic air masses, such as the Sierra Nevada in North America, and in the interior of continents so remote from oceans that air masses reaching there have lost much of their moisture, such as in the Central Asian deserts. Anthropogenic aridisation has been discussed in the context of desertification; however, only a very small portion of today's drylands can be considered truly 'man-made', for example the Aralkum desert surrounding the remainder of the Aral Sea.

Rainfall Variability

Rainfall variability in drylands occurs on different time scales, from seasonal to multi-decadal. As a rule, the lower the annual rainfall totals at a particular location, the higher the variability.

Today's understanding of rainfall variability in drylands, particularly with respect to the persistent droughts in the Sahel, has been modified in past decades by advances in climate observation, monitoring technology, and mathematical modelling. Both temporal and spatial variability in rainfall have come to be understood as a normal part of dryland conditions, such that isohyets (lines joining points of equal precipitation on a map) showing annual averages have little meaning for drylands, because they suggest climate as an equilibrium condition (Glantz 1987; Hulme 2001).

While the factors driving seasonal variability of rainfall are found in global and regional circulation features linked to the yearly revolution of the Earth around the sun and the tilt of the Earth's axis, the causes of inter-annual and decadal scale variabilities are complex and can be explained by changing configurations of forcings and feedback mechanisms, which, increasingly seem to be induced by atmospheric and ocean circulation dynamics (Barry and Chorley 1998).

Sea surface temperature (SST) anomalies related to the El Niño Southern Oscillation (ENSO), the North Atlantic Oscillation (NAO) or the Pacific Decadal Oscillation (PDO), provide explanations for rainfall variability at different time scales. ENSO cycles, for example, a combination of shifting pressure differences and varying ocean temperatures between the eastern and the western Pacific that occur every few years, not only have consequences for the Pacific region — where they are directly correlated with rainfall variability in Chile's Atacama desert — but also have repercussions in other parts of the world via large-scale teleconnections. Sharon Nicholson (2001) described the relationships between ENSO and rainfall variability in the arid and semi-arid regions of West, East, and southern Africa. The PDO is a similar mechanism, but operates at longer time scales of 20 to 30 years and its fingerprints are less visible in West Africa. Matthew Barlow, Heidi Cullen, and Bradfield Lyon (2002) analysed the influence of ENSO on recent droughts in Central and southwest Asia and found a possible link between the prolonged duration of a recent ENSO high phase and an exceptionally harsh drought. Ranga Myneni, Sietse Los, and Jim Tucker (1996) found effects of ENSO-cycle SST anomalies on temporal rainfall patterns in Africa, Australia, and South America.

The explanation of climate variability from large-scale modes of atmospheric pressure patterns and sea surface temperatures contrasts with earlier interpretations, which attributed rainfall variability to more regional factors, such as a latitudinal displacement of the Intertropical Convergence Zone (ITCZ) over the Sahel (Nicholson 2001).

Drought

Drought, in contrast to aridity, is an episodic short-term phenomenon and as such must be clearly differentiated from aridity. Defined as a period of below-average rainfall, it is not restricted to any particular environment, but is particularly associated with drylands, since they tend to experience more variable climate conditions. Although a natural hazard by definition, drought is not only a physical phenomenon but also has an important social component in that its impacts are influenced by drought vulnerability of particular societies or social groups. Therefore, drought is considered by many to be the most complex of all natural hazards, the effects of which often accumulate slowly over a considerable period of time and persist longer than the individual event (Hagman 1985).

Donald Wilhite and Michael Glantz (1985) distinguish four kinds of droughts — or disciplinary perspectives — with increasing severity: meteorological, hydrological, agricultural, and socioeconomic. Meteorological droughts are solely defined on the basis of precipitation departures from average. Hydrological droughts are associated with the effects of precipitation shortfalls on surface and subsurface water supply rather than precipitation shortfalls themselves, and usually lag behind meteorological droughts. The definition of agricultural droughts focusses on the crop-specific impacts of reduced water supply for agriculture. Last, the term 'socioeconomic drought' refers to the cumulative impacts of meteorological, hydrological, and agricultural droughts on the functioning of the socioeconomic system, for example the supply and demand of some economic goods and services (Wilhite 2000).

On the ground, droughts manifest themselves in vegetation stress and ultimately loss of green vegetation cover, decreases in streamflow, and the drying out and cracking of soil surfaces. Overall ecosystem productivity declines in times of drought. Although longer-term fluctuations of rainfall are seen as a normal part of climate rather than an aberration, the occurrence of prolonged droughts over large areas that caused severe and long-lasting impacts on the ground, such as the great Sahelian droughts, might indicate a regional or even global desiccation trend linked to global climate change. At least in the 20th century, the magnitude and duration of this desiccation has been unprecedented (Hulme et al. 2001). However, how unique that desiccation has been in the recent human history cannot be established with certainty, because the meteorological record from this area is of relatively short duration (Nicholson 2001).

On one hand, the wet decades of the 1920s, '30s, and '50s might be more exceptional than the dry decades of the post 1960s (Hulme 2001). On the other hand, it has been suggested that the great Sahelian droughts, in particular, require explanations other than natural variability, and that human-induced land degradation might have played a role in exacerbating them — if not causing them — because of their long duration and intensity (see Charney, Stone, and Quirk 1975).

Land Degradation and Desertification

Much like aridity and drought, the often-linked pair of terms 'desertification' and 'land degradation' describe phenomena that can be similar in appearance, but are different in scope and time scale. While desertification is restricted to drylands, land degradation can affect any environment. Desertification implies irreversibility on long time scales, whereas land degradation has also been used to describe short-term processes.

Land degradation and desertification are complex phenomena that can be attributed to a number of causes. The contributions of natural, notably climatic, and human factors are difficult to disentangle, and blame has alternately been assigned to anthropogenic and natural processes (see Chapter 3). A vast literature and the existence of more than a hundred definitions of desertification reflect — or perhaps promote — some of the confusion and controversy surrounding the concept of desertification (Glantz and Orlovsky 1983).

The official UNCCD definitions of desertification and land degradation are so broad in scope that virtually any change in land cover conditions, whatever the cause or time period, may be interpreted as land degradation or, in the particular case of dryland environments, desertification. As a result, estimates of the extent of desertification range around one third of the world's land surface, which corresponds approximately to the proportion of arid and semi-arid lands (see Dregne 1983). This obviously debatable figure should be seen as describing an area potentially at risk rather than the area actually affected by desertification. Given their already low biological potential, (hyper-)arid desert cores are less prone to desertification than the semi-arid and sub-humid environments — a fact that should put to rest the image of an advancing desert tide. Rather, land degradation starts as small localised pockets in the semi-arid zone.

In view of the importance of non-equilibrium dynamics in arid and semi-arid ecosystems, the concept of degradation appears in a different light. In non-equilibrium thinking, moving from one state to another does not necessarily imply degradation. Rather, different states might hold different opportunities for different uses. Reduction of productivity, especially economic productivity, is relative and depends on specific evaluations and objectives, that is, how this productivity is going to be used and to what end. For example, what constitutes a loss of productivity for the peasant farmer might actually be a gain for the livestock herder.

Impacts of Human Activity on Drylands

Especially in popular understanding, land degradation and desertification have commonly been ascribed to adverse impacts of human activity in drylands (see Mainguet 1991; Mensching 1990; Clarke and Noin 1997; Le Houerou 2002). Consequently, relationships between people and environment have figured prominently in the research agenda, especially after UNCOD.

Indeed, the biophysical background of drylands — recurrent droughts and slow rate of soil development — poses some constraints to human land use, which have not always been adequately considered in development schemes. The inherent variability and unpredictability of these ecosystems make them operate as non-equilibrium systems that translate into fluctuating productivity and resource availability. Flexible planning and increased preparedness for droughts and other uncertainties are required, if resources are to be managed sustainably. Traditional land use systems, which have evolved under these constraints, are adapted to cope with the uncertainties imposed by the environment (for example, shared risk distribution through social relations and opportunistic movement of livestock) (Bruins and Berliner 1998). Nevertheless, these land use systems have often been unfairly criticised as inappropriate and damaging to their environment (Hardin 1968).

Generally, the alleged human causes of land degradation and desertification are subsumed under the headings of overgrazing, overcultivation, and deforestation, often portrayed as almost inevitable consequences of population growth and poverty (Thomas and Middleton 1994).

Overgrazing refers to the presence of too many animals on the land so that vegetation cover is locally decreased, permitting the encroachment of noxious and unpalatable grasses and shrubs into pastures, compacting soils by animal trampling, and enhancing the susceptibility to wind and water erosion. Traditional nomadic pastoralism rarely results in overgrazing; however, problems arise when mobility is restricted or when subsistence livestock keeping is transformed into commercial production. While declines in vegetation cover may recover quickly once the pressure is removed, secondary effects on the soil resource can have long-term implications and can push the system to a new equilibrium state.

Overcultivation typically consists of a shortening of fallow periods to the point that soil moisture and nutrients are depleted, often accompanied by the expansion of cultivation into marginal lands as lower rates of return require that larger areas be put into production to compensate. Consequences are a loss of biodiversity, accelerated soil erosion, and deterioration of physical and chemical soil properties. Moreover, with farmers moving into marginal lands that do not permanently support rainfed agriculture, pastoralists often find themselves excluded from some of their best grazing grounds and are forced to retreat to even more marginal lands.

Deforestation, driven by demands of fuelwood — the major source of energy in many drylands — and the clearing of land for cultivation also increase exposure and vulnerability to soil erosion, as root systems are removed and natural wind barriers broken. Even more damage is done when clearing is done by mechanized equipment, which compacts the soil. Indirectly, deforestation might also increase the risk of desertification by contributing to the increase in atmospheric carbon dioxide while reducing the terrestrial carbon dioxide sink (see Parry 1996; Darkoh 1998).

Labelling these land uses as causes of land degradation, however, stems from subjective judgements and is often based on misunderstanding of indigenous land use systems on the part of western scientists. As a result of their long and intimate contact with their environment, most indigenous people have developed strategies to cope with uncertainties that make economic and ecological sense (see Mortimore 1988; Scoones 1994). Unless droughts hit with unusual severity, these strategies generally help to sustain rather than destroy their resource base (Broad 1994). However, socioeconomic factors such as poverty, marginality to local power structures, and uncertain or unfavourable land tenure relationships often undermine these strategies (see Chapter 3).

Although degraded 'sacrifice zones' can frequently be found in the vicinity of settlements and boreholes, regional-scale human-induced degradation is far less common than previously assumed. Thus, David Niemeijer and Valentina Mazzucato (2002) find no conclusive evidence of widespread soil degradation in the West African Sahel. Furthermore, Mary Tiffen and Michael Mortimore (2002), echoing Ester Boserup (1965), refute the notion of desertification and show with examples taken from case studies that a growing population can actually contribute to improved soil fertility, tree cover, and water conservation.

No simple relationship between human activity and desertification/land degradation can be established. On short time scales, both natural and anthropogenic stresses can produce similar effects on dryland ecosystems, which make the relative contributions of drought- and human-induced land degradation almost impossible to separate (Nicholson, Tucker, and Ba 1998). Moreover, natural and anthropogenic factors are interrelated in complex ways, adding to the difficulty of disentangling them. Only the irreversibility of desertification can be used to distinguish it from the relatively short-term effects of drought, which requires long time-scale observation to arrive at valid conclusions. Especially in drylands, which often occupy marginal positions with respect to national economic and political centres, reliable long-term records on climate and land use are scarce for many countries, so that the tension between anthropogenic and natural causes in the desertification concept remains largely unresolved.

Multiple Linkages and Feedback Mechanisms

Cause and effect relationships among drought, human activities, and desertification are neither linear nor simple, but are characterised by multiple linkages and feedback mechanisms. While impacts of desertification and land degradation are mostly of local extent, processes that lead to them can be regional or global in nature. Often, a range of factors, both climatic and human, must combine to trigger a process of degradation or to make a land degradation process cross the threshold to quasi-irreversible desertification. To date, these thresholds remain poorly specified, and further empirical research and the availability of long-term data are needed to accurately establish them for specific dryland locations.

Albedo Hypothesis

Joseph Otterman (1974) and Julie Charney, Peter Stone, and William Quirk (1975) were the first to formulate hypotheses that the modification of land cover characteristics in dryland regions might have climatic effects by altering land surface albedo (the fraction of incident radiation reflected by the land surface). Their hypotheses argued that desertification actually contributes to drought via a positive feedback mechanism, and not vice versa. Otterman cited the example of the Sinai-Negev region, where the denudation of bright, sandy soil by grazing on the Egyptian side increased albedo and decreased surface temperature compared to the more densely vegetated Negev side. Charney and her colleagues used a global circulation model to show a positive feedback mechanism between a decrease in plant cover and corresponding decrease in precipitation via increasing albedo, radiative cooling of the air column above, and, thereby, an enhancement of large-scale atmospheric subsidence and desiccation.

The albedo hypotheses received broad interest and substantial effort has been made to examine the sensitivity of regional rainfall to large-scale changes in land cover through climate modelling experiments, the results of which support Charney's basic hypothesis that sufficient changes in albedo can, at least potentially, produce droughts. However, satellite measurements of actual sub-Saharan albedo show no evidence for the persistent increase in albedo necessary to produce significant differences in rainfall (Folland et al. 1991; Hulme 2001). Observed changes in albedo due to changes in land surface characteristics have been localised in extent and often short in duration, in contrast to the widespread and sustained changes assumed in the modelling studies. Despite the absence of supporting empirical evidence for the Charney hypothesis, modelling studies are valuable as simulation experiments for understanding the interrelationships between land surface and atmospheric processes (Hulme and Kelly 1993).

Role of Atmospheric Dust

The role of dust in affecting precipitation is also controversial. Contrary to what some theoretical models predict (Yin et al. 2000), Daniel Rosenfeld, Yinon Rudich, and Ronen Lahav (2001) suggest that mineral dust in the atmosphere actually reduces precipitation efficiency of clouds due to the coalescence-suppressing effects of large concentrations of dust particles. In addition, dust can inhibit the formation of convective clouds because of radiative cooling and increased atmospheric subsidence (the descending motion of air in the atmosphere occurring over a rather broad area). That would mean that higher dust storm frequency might be the cause rather than the result of the decreased rainfall. Thus, dust emissions from anthropogenic sources might provide a mechanism for initiating a desertification feedback cycle. On the other hand, the dust loading over the Sahel has clearly followed trends in precipitation, rather than preceded them (Nicholson, Tucker, and Ba 1998).

The impact of dust on warming is a complex issue. Because it modifies both the incoming shortwave solar radiation and outgoing longwave radiation, either a cooling or heating effect can occur, depending on cloud cover and the albedo of the underlying surface (Nicholson 2001). In addition to its potential impacts on climate, long-range dust transport and deposition influences global biogeochemical fluxes (Schlesinger et al. 1990).

Global Warming and Desertification

Over the period of instrumental records, no global long-term trend in rainfall was established, but a carbon dioxide–induced global temperature increase of 0.5 degrees has been observed over the past hundred years, which has potential repercussions on desertification.

On the one hand, a carbon dioxide–enriched atmosphere is hypothesised to have positive effects on plant growth ('carbon fertilisation'), as it can boost photosynthesis and lead to improved water use efficiencies (Kimball et al. 1993); however, deficiency in soil nutrients could potentially offset positive effects in the long run (Schlesinger 2004). On the other hand, an increase in temperature in drylands would increase potential evapotranspiration. Given no substantial changes in rainfall, this would lead to desiccation and, in combination with unchanged human-use pressure under drier conditions, could ultimately cause desertification. In return, desertification would intensify global warming through the release of carbon dioxide from cleared vegetation and reduction of carbon sequestration potential of the degraded land.

Whether the carbon dioxide–induced global warming trend would ultimately result in a negative feedback cycle, leading to higher productivity and more efficient water use, or in a positive feedback cycle of increasing aridity and desertification, is difficult to predict (Le Houerou 1996).

Conclusion

Land degradation, drought, and desertification are linked through complex cause-and-effect relationships, which may be widely quoted but only partially understood. The realisation, which has evolved over the past 30 years, that the climatic and ecological functioning of drylands is fundamentally different from the functioning of their more mesic counterparts has only slowly been integrated into the discussion about desertification and land degradation, and has rendered some established concepts obsolete.

Although it remains uncontested that land degradation and desertification are interrelated with both drought and human land-use practices, the understanding of the nature and relative magnitude of these relationships, which are far more complex than previously assumed, is still rather weak (see Chapter 3). Marginal areas in terms of global economy, the drylands of the world are also under-researched.

It can be argued that the upsurge in interest in desertification following UNCOD was driven more by politics than science and that its impacts will sort out accordingly. The institutionalisation of the desertification issue has offered an opportunity for many dryland nations to garner international attention and support for combating desertification (namely, attracting economic development) that would not have happened otherwise. However, uncertainties about the scientific basis of understanding of the processes involved — particularly in the near future — make it difficult to define a clear course of action, paving the way to multiple interpretations and policy orientations.

Notes

1. Desiccation refers to the process of drying or to a state of extreme dehydration.
2. Stochastic refers to a random variable, or to patterns resulting from random effects.
3. Evapotranspiration refers to the loss of water from the soil both by evaporation and by transpiration from plants.

References

Barlow, Mathew, Heidi Cullen, and Bradfield Lyon (2002). 'Drought in Central and Southwest Asia: La Nina, the Warm Pool, and Indian Ocean Precipitation'. *Journal of Climate* vol. 15, no. 7, pp. 697–700.

Barry, Roger G. and Richard J. Chorley (1998). *Atmosphere, Weather, and Climate*. 7th ed. Routledge, London.

Beaumont, Peter (1993). *Drylands: Environmental Management and Development*. Routledge, London.

Boserup, Ester (1965). *The Conditions of Agricultural Growth: The Economics of Agrarian Change under Population Pressure*. Aldine Publishing, Chicago.

Briske, David D., Samuel D. Fuhlendorf, and Fred E. Smeins (2003). 'Vegetation Dynamics on Rangelands: A Critique of the Current Paradigm'. *Journal of Applied Ecology* vol. 40, no. 4, pp. 601–614.

Broad, Robin (1994). 'The Poor and the Environment: Friends or Foes?' *World Development* vol. 22, no. 6, pp. 811–822.

Bruins, Hendrick J. and Pedro R. Berliner (1998). 'Bioclimatic Aridity, Climatic Variability, Drought, and Desertification: Definitions and Management Options'. In H.J. Bruins and H. Lithwick, eds., *The Arid Frontier: Interactive Management of Environment and Development*, pp. 97–116. Kluwer Academic Publishers, Dordrecht.

Charney, Julie, Peter H. Stone, and William J. Quirk (1975). 'Drought in the Sahara: A Biogeophysical Feedback Mechanism'. *Science* vol. 187, no. 4175, pp. 434–435.

Clarke, John Innes and Daniel Noin (1997). *Population and Environment in Arid Regions.* Parthenon Publishing, New York.
Clements, Frederic E. (1916). *Plant Succession: An Analysis of the Development of Vegetation.* Carnegie Institution of Washington, Washington DC.
Darkoh, M.B.K. (1998). 'The Nature, Causes, and Consequences of Desertification in the Drylands of Africa'. *Land Degradation and Devopment* vol. 9, no. 1, pp. 1–20.
Dregne, Harold E. (1983). *Desertification of Arid Lands.* Harwood Academic Publishers, Chur, Switzerland.
Dyksteruis, E.J. (1949). 'Condition and Management of Range Land Based on Quantitative Ecology'. *Journal of Range Management* vol. 2, pp. 104–115.
Eklundh, Lars and Lennart Olsson (2003). 'Vegetation Index Trends for the African Sahel 1982–1999'. *Geophysical Research Letters* vol. 30, no. 8, pp. 13-11–13-14.
Folland, C., J. Owen, M.N. Ward, et al. (1991). 'Prediction of Seasonal Rainfall in the Sahel Region Using Empirical and Dynamical Methods'. *Journal of Forecasting* vol. 10, no. 1-2, pp. 21–56.
Glantz, Michael H. and Nicolai Orlovsky (1983). 'Desertification: A Review of the Concept'. *Desertification Control Bulletin* vol. 9, pp. 15–22.
Glantz, Michael H. (1987). 'Drought and Economic Development in Sub-Saharan Africa'. In M.H. Glantz, ed., *Drought and Hunger in Africa: Denying Famine a Future.* Cambridge University Press, Cambridge.
Hagman, Gunnar (1985). *Prevention Better than Cure: Report on Human and Environmental Disasters in the Third World.* 3rd ed. Red Cross, Stockholm.
Hardin, Garrett (1968). 'Tragedy of the Commons'. *Science* vol. 162, no. 3859, pp. 1243–1248.
Holling, C.S. (1973). 'Resilience and Stability of Ecological Systems'. *Annual Review of Ecology and Systematics* vol. 4, pp. 1–23.
Hulme, Mike and Mick Kelly (1993). 'Exploring the Links between Desertification and Climate Change'. *Environment* vol. 35, no. 6, pp. 6–11, 39–45.
Hulme, Mike (2001). 'Climatic Perspectives on Sahelian Desiccation: 1973–1998'. *Global Environmental Change* vol. 11, pp. 19–29.
Hulme, Mike, Ruth Doherty, Todd Ngara, et al. (2001). 'African Climate Change: 1900–2100'. *Climate Research* vol. 17, no. 2, pp. 145–168.
Illius, A.W. and T.G. O'Connor (1999). 'On the Relevance of Nonequilibrium Concepts to Arid and Semiarid Grazing Systems'. *Ecological Applications* vol. 9, no. 3, pp. 798–813.
Kimball, B.A., J.R. Mauney, F.S. Nakayama, et al. (1993). 'Effects of Increasing Atmospheric CO_2 on Vegetation'. *Plant Ecology* vol. 104-105, no. 1, pp. 65–75.
Köppen, W. (1931). *Die Klimate der Erde.* Walter de Gruyter, Berlin.
Le Houerou, H.N. (1996). 'Climate Change, Drought, and Desertification'. *Journal of Arid Environments* vol. 34, no. 2, pp. 133–185.
Le Houerou, H.N. (2002). 'Man-Made Deserts: Desertization Processes and Threats'. *Arid Land Research and Management* vol. 16, pp. 1–36.
Mainguet, Monique (1991). *Desertification: Natural Background and Human Mismanagement.* Springer, Berlin.
Mainguet, Monique (1999). *Aridity: Droughts and Human Development* T. Reimer, trans. Springer, Berlin.
Meigs, Peveril (1953). 'World Distribution of Arid and Semi-Arid Homoclimates'. UNESCO, Paris.
Mensching, H. (1990). *Desertifikation: ein weltweites Problem der økologischen Verwüstung in den Trockengebieten der Erde.* Winnenschaftliche Buchgesellschaft, Darmstadt.

Mortimore, Michael J. (1988). *Adapting to Drought: Farmers, Famines, and Desertification in West Africa*. Cambridge University Press, Cambridge.

Myneni, Ranga B., Sietse O. Los, and Compton J. Tucker (1996). 'Satellite-Based Identification of Linked Vegetation Index and Sea Surface Temperature Anomaly Areas from 1982 to 1990 for Africa, Australia, and South America'. *Geophysical Research Letters* vol. 23, pp. 729–732.

Nicholson, Sharon E., Compton J. Tucker, and M.B. Ba (1998). 'Desertification, Drought, and Surface Vegetation: An Example from the West African Sahel'. *Bulletin of the American Meteorological Society* vol. 79, no. 5, pp. 815–829.

Nicholson, Sharon E. (2001). 'Climatic and Environmental Change in Africa during the Last Two Centuries'. *Climate Research* vol. 17, pp. 123–144.

Niemeijer, David and Valentina Mazzucato (2002). 'Soil Degradation in the West African Sahel'. *Environment* vol. 44, no. 2, pp. 20–31.

Noy-Meir, Imanuel (1973). 'Desert Ecoysystems: Environment and Producers'. *Annual Review of Ecology and Systematics* vol. 4, pp. 51–58.

Otterman, Joseph (1974). 'Baring High-Albedo Soils by Overgrazing: A Hypothesized Desertification Mechanism'. *Science* vol. 186, no. 4163, pp. 531–533.

Parry, John T. (1996). 'Land Degradation in Tropical Drylands'. In M.J. Eden and J.T. Parry, eds., *Land Degradation in the Tropics: Environmental and Policy Issues*. Pinter, London.

Penman, H.L. (1948). 'Natural Evaporation from Open Water, Bare Soil, and Grass'. *Proceedings of the Royal Society of London, Series A* vol. 193, no. 1032, pp. 120–145.

Prince, Stephen D., E. Brown de Colstoun, and L.L. Kravitz (1998). 'Evidence from Rain-Use Efficiencies Does Not Indicate Extensive Sahelian Desertification'. *Global Change Biology* vol. 4, pp. 359–374.

Rosenfeld, Daniel, Yinon Rudich, and Ronen Lahav (2001). 'Desert Dust Suppressing Precipitation: A Possible Desertification Feedback Loop'. *Proceedings of the National Academy of Sciences of the United States of America* vol. 98, no. 11, pp. 5975–5980.

Schlesinger, William H., James F. Reynolds, Gary L. Cunningham, et al. (1990). 'Biological Feedbacks in Global Desertification'. *Science* vol. 247, no. 4946, pp. 1043–1048.

Schlesinger, William H. (2004). Communication at the American Association for the Advancement of Science. Seattle, 12–16 February.

Scoones, Ian (1994). *Living with Uncertainty: New Directions in Pastoral Development in Africa*. Intermediate Technology Publishing, London.

Thomas, David S.G. and Nick Middleton (1994). *Desertification: Exploding the Myth*. Wiley, Chichester.

Thornthwaite, C.W. (1948). 'An Approach toward a Rational Classification of Climate'. *Geographical Review* vol. 38, pp. 55–94.

Tiffen, Mary and Michael J. Mortimore (2002). 'Questioning Desertification in dryland sub-Saharan Africa'. *Natural Resources Forum* vol. 26, pp. 218–233.

Tucker, Compton J. and S.E. Nicholson (1999). 'Variation in the Size of the Sahara Desert from 1980 to 1997'. *Ambio* vol. 28, pp. 587–591.

United Nations Convention to Combat Desertification (1994). 'Text of the United Nations Convention to Combat Desertification'. <www.unccd.int/convention/text/convention.php> (December 2005).

United Nations Education, Science, and Cultural Organization. (1979). 'Map of the World Distribution of Arid Regions'. Man and the Biosphere Technical Notes 7. Paris.

United States Geological Survey (1997). 'Distribution of Non-polar Arid Land'. <pubs.usgs.gov/gip/deserts/what/world.html> (December 2005).

Walker, B.H., D. Ludwig, C.S. Holling, et al. (1981). 'Stability of Semi-Arid Savannah Grazing Systems'. *Journal of Ecology* vol. 69, pp. 473–498.

Westoby, M., B.H. Walker, and Imanuel Noy-Meir (1989). 'Opportunistic Management for Rangelands Not at Equilibrium'. *Journal of Range Management* vol. 42, pp. 266–274.

White, Gilbert F., ed. (1956). *The Future of Arid Lands: Papers and Recommendations from the International Arid Lands Meetings*. Washington DC.

White, Robin P. and Janet Nackoney (2003). 'Drylands, People, and Ecosystem Goods and Services: A Web-Based Geospatial Analysis'. World Resources Institute. <biodiv.wri.org/drylandsgis-pub-3813.html> (December 2005).

Whitford, Walter G. (2002). *Ecology of Desert Systems*. Academic Press, San Diego.

Wiens, John A. (1984). 'On Understanding a Nonequilibrium World: Myth and Reality in community Patterns and Processes'. In D.R. Strong, D. Simberloff, L.G. Abele et al., eds., *Ecological Communites: Conceptual Issues and the Evidence*, pp. 439–457. Princeton University Press, Princeton.

Wilhite, Donald A. and Michael H. Glantz (1985). 'Understanding the Drought Phenomenon: The Role of Definitions'. *Water International* vol. 10, no. 3, pp. 111–120.

Wilhite, Donald A., ed. (2000). *Drought: A Global Assessment*. Routledge, London.

Yin, Yan, Zev Levin, Tamir G. Reisin, et al. (2000). 'The Effects of Giant Cloud Condensation Nuclei on the Development of Precipitation in Convective Clouds'. *Atmospheric Research* vol. 53, pp. 91–116.

Chapter 3

Examining the Links between Poverty and Land Degradation: From Blaming the Poor toward Recognising the Rights of the Poor

Sally-Anne Way

This chapter examines the links between poverty and land degradation, with a special focus on marginal dryland areas, particularly in Africa. Traditional views on the links between poverty and the environment, or the 'poverty-environment nexus', locate the causes of desertification and land degradation in human activity, driven in particular by poverty that leads to the overexploitation of the land. Poor people are often blamed for unsustainable land use practices, leaving the land vulnerable to degradation and desertification. Poverty is seen as a key cause of land degradation. Poverty is also recognised as a consequence of land degradation. As land becomes increasingly degraded, the poor may be forced to extract even more from the land, leading to what has become know as a vicious cycle or downward spiral of poverty and land degradation. The spectre of this downward spiral of poverty and land degradation has long informed policy making.

More recently, however, a more complex picture of the poverty-environment nexus has emerged that suggests that such a downward spiral is not inevitable. Recent studies have shown that the poor are not always to blame for land degradation, but that they have an interest in protecting and investing in their land as far as possible. These studies highlight how many different factors mediate the links between poverty and land degradation, so that cause-and-effect linkages are more complex than frequently assumed. Assumptions of a downward spiral have sometimes rested on misunderstandings of traditional land use practices and processes of land degradation. Blaming the poor has also served to absolve other actors of their responsibilities, and has not taken account of inappropriate government policies, the lack of investment in dryland areas, as well as broader processes of exclusion and marginalisation that may be more to blame for land degradation than are poor people themselves. The current state of knowledge shows that it is possible to create a virtuous cycle that can reduce both poverty and land degradation.

Analyses of poverty and degradation in the drylands have increasingly come to revolve not only around agroclimatic processes and poverty, but also around issues of governance and poor people's rights over resources. This chapter suggests that there has been a marked shift away from blaming the poor for land degradation and desertification toward recognising the rights of the poor in marginal dryland areas. Research increasingly focuses on rights over resources — particularly rights to land, on the marginalisation of dryland areas, and on abuses of power by more powerful actors. Future research should in fact focus on exploring a rights-based approach to desertification and land degradation that recognises the human rights of people living in marginal dryland areas. Formulating a framework for a rights-based approach would mark a significant step forward in the fight against desertification.

Finally, this chapter suggests that implementation of the United Nations Convention to Combat Desertification (UNCCD) provides a key opportunity to address the complex socioeconomic causes that underlie both desertification and poverty. Measures to combat land degradation and desertification will be most effective when they seek to address poverty and land degradation simultaneously, but also recognise their context-specific underlying causes and respect the human rights of people living in marginal dryland areas. This may start with the unprecedented focus of the UN Millennium Project on the marginal lands and its call for the urgent need to increase investment in the drylands as part of the global effort to achieve the Millennium Development Goals (MDGs) on poverty and hunger (see, for example, UN Millennium Project 2004).

Reviewing the Extent of Hunger and Poverty in the Drylands

The UNCCD focusses specifically on dryland areas, as it defines 'desertification' in terms of land degradation in drylands.[1] As seen in Chapter 1, land degradation and desertification in dryland areas affect the livelihoods of 250 million people in more than a hundred countries. In the negotiations to draw up the convention, which is global in scope and intention, it was agreed that Africa required special attention, because of the extent of poverty in Africa, where more than 43 percent of the land area is classified as drylands and another 27 percent is desert (Ambler 1999).

Much of the international interest in desertification was originally stimulated by the tragic human, livestock, and environmental losses that occurred after repeated droughts in the Sudano-Sahelian regions of Africa during the early 1970s and over the last three decades. While more than a hundred countries in the Americas, Asia, Australia, and Africa lie entirely or partly within drylands, Africa is the continent with the greatest population living in dryland areas, with 325 million people, or 46 percent of its population living in drylands (Øygard, Vedeld, and Aune 1999; Ambler 1999).

As seen in Chapter 1, poverty and hunger are prevalent in dryland regions, where poor people live on lands that are highly prone to degradation and desertification. Few studies have examined in depth the extent of poverty in the drylands. However, one 1989 international study showed how poverty tends to be geographically concentrated in 'ecologically vulnerable areas' (Leonard 1989, cited in Barbier 1998). This study examined the distribution between 'low potential' and 'high potential' lands of the poorest 20 percent of the rural population in developing countries. It defined low-potential lands as resource-poor, marginal agricultural lands, with inadequate or unreliable rainfall, adverse soil conditions, limited fertility, and a topography that limits agricultural productivity and increases the risk of chronic land degradation (although this does not equate only to drylands). The study found that, more than 50 percent of Africa's poorest people are concentrated on low-potential lands that are prone to degradation, and the same was true in the rural areas of Latin America and Asia.

Other studies have shown also that poverty tends to be concentrated in drier areas where rainfall is low and uncertain (UN Development Programme's Office to Combat Desertification and Drought [UNSO] 1994). Poverty is particularly high in dryland countries that depend on agricultural economies. In Zimbabwe, for example, the overwhelming majority of the poor live in rural areas and poverty is most common and deepest in the low rainfall areas of Matabeleland South, Masvingo, and Matabeleland North provinces (International Fund for Agricultural Development [IFAD] 2001). In Chad, which has an agricultural economy reliant on the volatile rains, four fifths of the population is rural and an even higher proportion is poor. In Niger, where 100 percent of productive land is dryland, the population is predominantly rural, chronically poor, and subject to repeated food crises. Levels of undernourishment are particularly high across sub-Saharan Africa, with 34 percent of its people, or 186 million people, chronically undernourished (Food and Agriculture Organization [FAO] 2000a). Sub-Saharan Africa is also the only region in the world where food production per capita is not expected to be able to keep up with population growth and food insecurity is increasing (2000a; 2000b).

It is important to understand that hunger and poverty in the drylands are closely related to unpredictable and volatile rainfall and recurrent droughts, which can severely affect the resilience of the land and the resilience of people's livelihoods. The rains vary from season to season, and from year to year often bring risk and uncertainty, with years of plenty often followed by years of want (International Institute for Environment and Development 1995). Moving beyond hunger and poverty to look at concepts such as vulnerability and food insecurity helps to focus attention on the instability in people's capacity to feed themselves. People are food insecure not only because their food consumption level is low, but also because their access to food is variable and unpredictable over time, from one year or season to the next (Devereux 2000). Successive droughts and resulting degradation have contributed significantly to the erosion of people's assets and livelihood systems across the arid and semi-arid regions of sub-Saharan Africa (Ahmed 2000). The impact of wars and conflict can exaggerate the impacts of drought, as influxes of refugees can destabilise local coping mechanisms that allow people to respond to drought conditions and avoid destitution, as shown by one study in Western Darfur, Sudan (De Waal 1989).[2]

Critical Links between Poverty and Land Degradation

Poverty has long been understood to be a key cause of land degradation. This is based on the understanding that the poor may be forced to extract more from their lands than can be sustained in the long term. Faced with the imperative of short-term survival, poor people may have no other choice but to act against their long-term interests by degrading their land, as they strive to meet their short-term basic needs for food, shelter, and livelihood. In dryland areas, these processes can be accelerated by rain variability and uncertainty, as the poor may extract more in the good years, at the expense of land resilience in drought years (Devereux 2000).

Poverty can force people to engage in unsustainable land use practices, such as:

- overgrazing of pasture areas,
- expanding cultivation through extensification into unsuitable lands,
- pursuing land management and cultivation practices that deplete soils of their nutrient and organic matter content and promote acidification,
- reducing fallow periods so soils are inadequately rejuvenated,
- mining groundwater unsustainably,
- cutting but not replanting sufficient trees for fuelwood and other purposes, and
- unsustainable population growth (adapted from McCulloch, Babu, and Hazell 1999; Ahmed 2000).[3]

Poverty has also come to be understood as a key consequence of land degradation. As land becomes degraded, productivity may fall, resulting in lower incomes, producing greater poverty, and, therefore, accelerating pressure on lands. As land degradation usually has the greatest effects on the poorest, who depend on the land for their survival, this can lead to a vicious downward spiral of poverty and land degradation in which poverty is both cause and effect of land degradation, with the poor seen as both agents and victims of land degradation. These views have been expressed by the influential World Commission on Environment and Development (the Brundtland Commission) (1987) as follows:

> Poverty is a major cause and effect of global environmental problems. It is therefore futile to attempt to deal with environmental problems without a broader perspective that encompasses the factors underlying world poverty and international inequality ... Many parts of the world are caught in a vicious downwards spiral: poor people are forced to overuse environmental resources to survive from day to day, and their impoverishment of their environment further impoverishes them, making their survival ever more difficult and uncertain.

This view recognised that measures to combat environmental degradation would not be effective unless they simultaneously addressed both poverty and land degradation. In the past, policies to protect the environment had often been at the expense of the poor. Top-down, exclusionary policies tried to bar the access of populations to regions in order to improve the environment, by creating natural reserves. However, these were often implemented in draconian, policing style, excluding people from the resources on which they depended and therefore producing even greater poverty. Such policies often fail because they do not recognise the livelihoods and resource use of people living in these regions (Forsyth, Leach, and Scoones 1998).

At the same time, other policies that focus purely on poverty reduction without addressing the environment could have negative environmental effects and lead to even greater environmental degradation. For example, macroeconomic policies aimed at poverty reduction that fail to recognise the special problems of the dryland regions frequently create even greater environmental damage, contributing to the vicious downward spiral of poverty and land degradation (Forsyth, Leach, and Scoones 1998).

New Thinking on the Poverty-Environment Nexus

More recently, research on the poverty-environment nexus has shown that such a vicious downward spiral of poverty and land degradation is not always inevitable and does not always exist in practice. A number of studies have questioned assumptions that poor people are always engaged in land use practices that result in land degradation (see Leach and Mearns 1992; see also Forsyth, Leach, and Scoones 1998). Although poverty can certainly be both cause and effect of land degradation, these studies show that the relationship between poverty and land degradation is far more complex than is often assumed. These studies have questioned the direct link between poverty and land degradation, pointing to a whole range of different, context-specific factors at local, national, and global levels that mediate the poverty-environment nexus. This questioning has led to new awareness that it is vital to understand and identify the exact causes of land degradation in different localities, as policies based on an incorrect diagnosis of the problem may make the situation worse than before and can result in even greater poverty (Ambler 1999).

Research has shown that assumptions have been based on misunderstandings of traditional land use practices, particularly in vulnerable environments, such as drylands. In dryland regions, the dominant land use continues to be the grazing of livestock (88 percent of land) (United Nations Environmental Programme [UNEP] 1991, cited in Katyal and Vlek 2000); poor pastoralists have often been blamed for overgrazing by exceeding the 'carrying capacity' of the land and therefore contributing to desertification. However, recent studies have shown that overgrazing may be due more to the imposition of inappropriate models of land management (Scoones 1996; see also Øygard, Vedeld, and Aune 1999). These have shown that concepts such as carrying capacity have been conceived on the basis of Western models of management from stable, moderate, and equilibrium environments, and thus are wholly inappropriate for application to contexts of unstable, non-equilibrium environments that are very different and characterised by volatility and change (Scoones 1996); see also Chapter 2). As Ian Scoones (1996) has argued, the history of livestock development in Africa shows how Western models of fencing, ranch systems, and management concepts such as carrying capacity have been transferred from Western contexts, but have proven to be ill adapted to non-equilibrium environments. Enclosure imposed through ranching systems, fenced paddocks, and international borders has prevented pastoralists from following the rains and vegetation in a more sustainable way. This means that removing restrictions on the mobility of nomadic pastoralists and eliminating the restrictions imposed by fenced paddocks, ranches, and other restrictions on access to land can reduce the pressure on the land from overgrazing. Today, there is a new acceptance of nomadic pastoralism is a livelihood strategy that is actually very well adapted to utilising the marginal resources and managing the risks of the drylands (Scoones 1996; see also Øygard, Vedeld, and Aune 1999). Nomadic pastoralism is now understood as much better adapted to risky environments and far less destructive than previously thought.

Recent research questions whether traditional agricultural practices based on extensification have resulted in land degradation (see, for example, Barbier 1998). Poverty has been understood to drive extensification as poor households abandon degraded pasture and move on to new lands, on the assumption that it is less costly to bring additional land into production than to invest in maintaining or improving the long-term productivity of cultivated land. However, recent research suggests that traditional practices of extensification can in fact be a sustainable land use practice. For example, the bush-fallow crop rotation system practised traditionally across Africa has long been managed as a sustainable system, although it is also recognised that due to the pressure of rising rural population and less available land, that system is today less able to be conducted in a traditional sustainable manner (Barbier 1998).

Further research has also examined whether the poor should always be blamed for encouraging land degradation through the destruction of forests (Fairhead and Leach 1996).While the poor often do engage in deforestation, this is not always true, and there have been some misinterpretations of deforestation in some areas. For example, one study has shown the landscape of Kissidougou in the West African Republic of Guinea was usually interpreted as degraded and further degrading, but that a close examination of the history showed that this was not in fact the case. In this area of Guinea, patches of forest have long been understood to be the surviving remains of a once extensive virgin forest. It was assumed that this forest had been degraded to savannah by farmers forced into destructive, short-term practices by economic marginalisation and population pressure. However, using historical and photographical research, James Fairhead and Melissa Leach (1996) show that, in fact, in this region there was no such destroyed virgin forest, but rather that forest cover had increased on pre-existing savannah. They show how poor people in fact created the forest cover around their villages to meet their own needs, and did not destroy the forest as had been earlier assumed. This suggests that local ecology and patterns of perceived land degradation must be studied, rather than assumed. It also suggests that people can look after the environment, if the conditions are right. It has also been pointed that much deforestation is carried out not by poor people, but by corporations or governments engaged in profitable logging that does not respect environmental concerns.

Overpopulation among poor people has also long been seen as a key cause of land degradation, based on the assumption that the fragility of the land in arid regions means that there is a limited capacity to absorb increasing numbers of people (Campbell 1984). Kevin Cleaver and Götz Schreiber (1994) studied the interlinkages among population, agriculture, and the environment in sub-Saharan Africa and found that traditional methods — crop production, livestock husbandry, forest management practices, and land tenure systems — were well suited to Africa's fragile resource systems when population pressures were low, but have failed to evolve to keep up with increasing population pressure. However, this need not always be the case. Ester Boserup's (1965) classic study showed, for example, how increasing population density can create the conditions that induce innovation and intensification that increase agricultural productivity.

A more recent 1993 study on the semi-arid regions of Machakos district in Kenya showed that, despite a fast-growing population, the condition of the land is much better now than it was in the 1930s (Tiffen 1993). This study shows how, in the 1930s, the area was considered an environmental disaster, with a high prevalence of soil erosion, pasture degradation and deforestation, very low agricultural productivity and income, and a population considered to be well above its carrying capacity. However, by the 1990s, despite the population increasing more than fivefold over 60 years, the resource base had been rehabilitated and the value of agricultural output per head was three times greater (Tiffen 1993; Tiffen, Gichuki, and Mortimore 1994). Soil erosion had declined due to terraces put in place to protect arable land, and predictions of a fuelwood crisis were not fulfilled because a larger number of people farmed and protected trees. Many new agricultural technologies had been introduced, and the average income had increased with higher agricultural production per person and per hectare. This was due to improved education, local institutional development, good roads, and opportunities to grow high-value products for the Nairobi market, and access to capital for land-related investments (including terracing, tree growing, live fencing, and water harvesting) with capital coming from increased non-farm–related activities. The study emphasised that land resource management was closely tied to the overall development process in Kenya, but show that it is possible to create a virtuous cycle of reducing both poverty and degradation, if the conditions are right.

This implies ensuring that conditions are conducive to improving both poverty and land degradation. Yet recent research has shown how land degradation is frequently more the result of inappropriate government policies and weak institutions than the result of poverty itself. This is especially the case when government policy does not take into account its impact on poverty or land degradation. For example, one study argues that in Malawi government policies have encouraged monocropping of certain crops that has significantly contributed to land degradation (Barbier 1998). This study shows how economic policy and the shift in relative prices from traditional crops toward crops that have more erosive effects exacerbated land degradation.

Smallholders in Malawi traditionally relied on intercropping and relay cropping of maize, groundnuts, and pulses to meet their nutritional needs, which also helped to maintain soil fertility and conserve the soil. Crops such as pulses and groundnuts provide better ground cover, soil fertility maintenance, and soil structure cohesion than do cash crops such as maize, tobacco, and cotton. However, in response to government policy, and higher prices for these more erosive cash crops, farmers abandoned mixed-cropping systems for monocropping maize, tobacco, and cotton, contributing to degradation of the land (Barbier 1998; see also Chapter 12).

Further studies have identified numerous other factors that mediate the links between poverty and land degradation, such as land tenure, credit availability, access to markets, lack of rural infrastructure, transport, and water (Hoben 1996); they include wider national and international socioeconomic factors, such as the terms of trade, debt levels, conflict, or corruption (Leach and Mearns 1992); and they extend to include globalisation, the monetisation of local economies, urbanisation, sedentarization, and political marginalisation (Evers 1996). It has also been suggested that it is not the fault of the poor if they do not have the resources to invest in conservation. Indeed, the poor may be unable to invest in inputs such as fertilisers, especially when erratic rainfall in the drylands creates extreme variations in yield, so that investments in costly seed and fertiliser inputs are often considered too risky (Øygard, Vedeld, and Aune 1999).

Some suggest that land degradation is not only caused by the poor, because the rich usually have more livestock or more agricultural technology such as tractors and fertilisers, which can sometimes have negative effects on the environment (Reardon and Vosti 1996, see also Chapter 12). Others have even refuted altogether the hypothesis that poverty is the primary cause of degradation, arguing that the poor seldom initially or intentionally degrade the environment (Duraiappah 1998).

From Blaming the Poor toward Recognising the Rights of the Poor

As a result of this new thinking on the poverty-environment nexus, a number of international development agencies have begun to recognise that the cause-and-effect links between poverty and land degradation are complex and context specific. Assumptions of overgrazing, extensification, deforestation, and overpopulation and the links to land degradation must sometimes be questioned. At the same time, this new thinking suggests that there is no necessary vicious downward spiral of poverty and land degradation, but that a virtuous cycle can be created, if the conditions are right. The World Bank (2001, 140) in its 'Environmental Strategy' has argued for example that:

> Hypotheses abound, such as the theory that there is a vicious cycle of poverty, population growth, and environmental degradation. Some cases support that theory; others show quite the opposite. We have little empirical evidence that allows us to conclude with certainty that, in any particular circumstance, causality will go in one direction rather

than another. Several local factors — such as macroeconomic policies, the effectiveness of local institutions and property regimes, and gender relations — decisively influence the extent to which the poor have access to and control over natural resources and the potential to derive income from them.

This marks a vital shift away from blaming the poor for land degradation toward a focus on broader factors that mediate the link between poverty and the environment and that might be more to blame than poor people themselves. Examining factors such as the effectiveness of local institutions, property regimes, and gender relations marks the shift to growing awareness of the key role that issues of governance and rights over resources play in determining the relationship between poverty and land degradation.

Secure tenure and rights over land have been shown to be particularly important for preventing degradation. Research has shown that smallholders are more likely to degrade their land if they do not have clear rights to their land guaranteed by secure land tenure arrangements that allow them to invest in land for the future. Some studies have emphasised that it is not necessarily the absence of land tenure itself that can cause land degradation, but rather the constant fear of losing the land as it could be appropriated at any time by more powerful actors (Movik, Dejene, and Synnevåg 2003).

However, these studies also emphasise that secure land tenure arrangements do not necessarily require land privatisation, modelled after Western concepts of land ownership. In fact, even where land privatisation is introduced to support the poor, the result is often the reverse as privatised land eventually ends up concentrated in the hands of the non-poor as land is sold for survival (Jodha 1986). In some contexts, land privatisation has also wrested rights over land away from women, given that private land ownership has often been registered in the name of the husband (Griffin 2003). In the context of drylands, land privatisation may be inappropriate as it does not permit the mobility of livestock that is essential to managing the risks of unpredictable dryland environments (Scoones 1996). Other forms of land rights may therefore be more successful in guaranteeing the right to land of the poor. In the case of pastoralism, for example, open access regimes may not be the answer, but a more promising possibility may be common property ownership, where land rights are granted to a fixed group that can regulate the activities of its members, as has been suggested for the nomadic pastoralists that live in the arid steppes of Mongolia (Griffin 2003). Securing rights over land and regulating access can help to reduce land degradation.

Other studies show how without secure tenure people can be forcibly displaced from their lands by more powerful groups. This means that poor people are often living on marginal lands, including arid land, because they have been forced off more fertile lands (see Chapter 12). For example, William Adams (1991) argues that the displacement of poorer rural groups from their traditional farming and grazing lands is documented across Africa. He suggests that this problem is increasingly prevalent in river floodplain zones of the semi-arid regions of Africa, where different economic groups compete for land and water resources. Sometimes government investment schemes and policies also intersect with local power relations supporting the effective appropriation of water and land for one group of land users, at the expense of more vulnerable, and often poorer, land users who are then excluded from access to their land and resources. Robert Kates (1991) argues that throughout the developing world the poor are displaced from traditional entitlement to common resources, when development activities or appropriation of the resources by richer claimants take place. This suggests that there is a need for the protection of the existing assets and rights of the poor from more powerful groups (Ambler 1999).

A number of studies have further highlighted the power abuses of local or national elites or corporations against the poor that can create the conditions for land degradation (see Duraiappah 1998; Movik, Dejene, and Synnevåg 2003). Anantha Duraiappah has shown how, for example, large-scale commercial exploitation of forestland caused a fuelwood shortage in local communities. This forced farmers to switch from wood to manure for use as fuel, reducing the application of manure to the land, which eventually affected the soil nutrient balance and contributed to degradation of the land. This suggests how, in some cases, land degradation can be indirectly due to the actions of more powerful groups that effectively appropriate the resources of poorer farmers and indirectly cause land degradation. Frank Ellis (2000) has suggested that blaming the poor absolves other more powerful actors of responsibility:

> Making the poor the scapegoat for environmental deterioration merely lets off the hook the commercial and state behaviour responsible for the really big changes that result in switches in the dynamics of people in local environments.

Another study highlighted the marginalisation of dryland areas and the increasing marginalisation of poor people living in these areas who have no voice in government policy (Movik, Dejene, and Synnevåg 2003). Dryland regions are generally accorded a low level of priority, which is disproportionate to their size, population, and need. Despite large populations living in dryland areas, particularly in Africa, investment in these areas has been minimal and dryland areas are frequently left out of government planning, including poverty reduction strategies. This is evidenced in the generalised lack of investment in small-scale irrigation that would significantly reduce the vulnerability to drought of the poor in the drylands (UN Millennium Project 2004). Small-scale irrigation and water harvesting are often perceived as inefficient, and not perceived as a matter of rights over productive resources. As Synne Movik, Sileshi Dejene, and Gry Synnevåg (2003, 38) have argued:

> Correcting the symptoms (environmental degradation) without correcting what caused the symptoms in the first instance (e.g., institutional failures, insecure tenure, power abuse by political elites causing marginalisation of poor people) will not offer promising results.

Today, the primary direct causes of land degradation have therefore come to be increasingly understood as the result of limited rights over resources, inequitable processes of development, and unequal distribution of rights and power, as well as misguided policies. These new understandings mark an important a shift away from blaming the poor for land degradation toward recognising the rights of the poor in marginal dryland areas. In order to guarantee rights to water, food, and the livelihoods of people living in dryland areas, it is necessary to respect and protect existing rights to resources such as land and water, as well as accord greater priority to marginalised areas, including greater investment in water for irrigation and rural infrastructure. Preventing abuses of power by government and powerful groups, as well as issues of marginalisation and discrimination, are also important human rights issues. This shift toward recognition of power relations, good governance and rights, now needs to be integrated in order to formulate a rights-based approach to land degradation and desertification (see Chapter 8).

Some work on rights-based approaches to food security has already been done is closely linked to the issues related to land degradation and desertification. Approaches based on the right to food are currently emerging within international organisations, including the UN human rights bodies and the UN's Food and Agriculture Organization (FAO).[4] Under the definition of the Committee on Economic, Social, and Cultural Rights, a rights-based approach to food security emphasises the need to respect, protect, and fulfil the right to adequate food (UN Economic and Social Council 1999). This underscores the notion that governments should refrain from taking actions that negatively affect people's existing access to food and productive resources, and should protect people against the negative actions of other powerful actors that might deprive them of such. Finally, the right to food requires governments to take concrete action to improve the poor's access to food and other productive resources in order to reduce food insecurity and poverty progressively, especially for the most vulnerable groups.[5]

From a rights-based perspective, it therefore becomes an obligation to invest in ensuring food security for the poor, particularly the poorest who live in the most marginal lands, such as the drylands. Further work exploring the links among desertification, land degradation, and the right to food, the right to water, and other human rights is necessary to address issues of power relations and governance. Such issues should be a central theme of future research on the poverty-environment nexus. Formulating a framework for a rights-based approach to poverty and land degradation would mark a significant step forward in the fight against desertification.

The Convention to Combat Desertification and the Millennium Development Goals

Poverty and land degradation tend to persist in dryland regions, not only because of the relative low productivity of the lands and the threats of drought, but also because of the lack of adequate investment in these areas and the lack of recognition of the rights of the people who live in them. The UNCCD is important because it focusses on the specific and urgent challenges facing the dryland regions and desertification-affected communities around the world. The convention adds to the broader, international development debate on the eradication of rural poverty, by bringing this special focus on the need to invest in dryland regions, a focus that is now reflected in the debate on achieving the MDGs on hunger and poverty (see Chapter 1).

The UNCCD concentrates on the poverty-environment nexus, incorporating much of the new thinking by recognising that, in the past,

> most attempts to fight desertification have concentrated more on its symptoms than on its causes. They have contributed to mitigating its effects, and on reducing the human activities that seem to be immediately contributing to them. They have sought to tackle overcultivation, overgrazing, deforestation and faulty irrigation directly, but without addressing the underlying social and economic pressures that have produced them. This has often resulted, in effect, in blaming the victims of desertification for causing it, without making a serious attempt to understand the forces outside their control which are driving them to overexploit the land (UNCCD 1995, 22).

The convention therefore calls on governments to 'address the underlying causes of desertification and pay special attention to the socio-economic factors contributing to desertification processes'. It recognises that poverty and the environment must be addressed simultaneously and must address the broader factors and exclusionary processes that produce poverty and that can drive the poor to degrade their land.

The aims of the UNCCD fit well into current development policy goals for reducing hunger and poverty, especially government commitments toward the MDGs. The growing marginalisation of dryland areas is closely linked to the lack of investment in the drylands, which is coming to be acknowledged as a key reason for continued poverty. In the drive to implement the MDGs, there is growing international awareness of the importance of addressing the special problems of marginal areas in order to fight poverty.

The Millennium Project, a project set up to promote the implementation of the MDGs, has established the Task Force on Hunger, which is putting new stress on the urgent need to invest in marginal lands to reduce poverty and hunger among the very poorest. This recognises that poor people living in remote and marginal lands are not currently benefiting from development and calls for more investment in the poorest, marginal areas. It argues that there is a need to address the underlying factors that keep the poor in poverty in marginal areas, including the lack of markets, the lack of access to productive resources, including land and tenure insecurity, gender discrimination, the lack of investment in small-scale irrigation, and soil fertility. The Report of the Task Force on Hunger suggests that:

> The condition of land and people has much to do with governance and rights, and less to do with environment. As population densities increase, the value of scarce land increases as long as people have the necessary rights to use the land and have access to markets for their produce. Under these circumstances even drylands become valued.

While making development work in more marginal areas is clearly more challenging than in high potential areas, evidence from developed and developing countries shows that development of marginal areas is perfectly possible. To ignore marginal lands would consign millions to poverty and abandon vast areas with genuine potential (UN Millennium Project 2004, 167).

Clearly the table has now shifted from blaming the poor for environmental degradation to addressing the root causes of marginalisation. The UNCCD approach is in line with this new thinking.

Notes

1 The definition used in the convention is based on the definition agreed for Agenda 21, at the United Nations Conference on Environment and Development (UNCED) at Rio de Janeiro in 1992. Although, as seen in Chapter 2, this definition is still the subject of criticism and debate, it is the operational basis of the UNCCD regime.
2 For a detailed analysis of the migration-desertification nexus, see Chapter 4.
3 Most of these practices can also result from new pressures arising from agricultural trade liberalisation, as will be seen in Chapter 12.
4 See the FAO (2003). See also the reports of the Office of the UN Special Rapporteur of the Commission on Human Rights on the Right to Food at (2005).
5 These obligations apply to all governments that have ratified the International Covenant on Economic, Social, and Cultural Rights.

References

Adams, William M. (1991). *Green Development: Environment and Sustainability in the Third World*. Routledge, London.

Ahmed, Nessim (2000). 'Economic, Social, and Cultural Causes and Consequences of Drought and Desertification'. International Fund for Agriculture Technical Advisory Division. Rome. <www.unccd.int/knowledge/INCDinfoSeg/partiii.php#para1> (December 2005).

Ambler, John (1999). 'Attacking Poverty While Improving the Environment: Towards Win-Win Policy Options'. Poverty and Environment Initiative of the United Nations Development Programme and the European Commission. <www.undp.org/pei/pdfs/PEIPhase1SummaryPaper2.pdf> (December 2005).

Barbier, Edward B. (1998). 'The Economics of Land Degradation and Rural Poverty Linkages in Africa'. United Nations University Institute for Natural Resources in Africa Annual Lectures. <www.inra.unu.edu/documents/annual_lectures1998.htm> (December 2005).

Boserup, Ester (1965). *The Conditions of Agricultural Growth: The Economics of Agrarian Change under Population Pressure*. Aldine Publishing, Chicago.

Campbell, D.J. (1984). 'Response to Drought among Farmers and Herders in Southern Kajiado District, Kenya'. *Human Ecology* vol. 12, no. 1, pp. 35–64.

Cleaver, Kevin M. and Götz A. Schreiber (1994). *Reversing the Spiral: The Population, Agriculture, and Environment Nexus in Sub-Saharan Africa*. World Bank, Washington DC.

De Waal, Alexander (1989). *Famine that Kills: Darfur, Sudan, 1984–1985*. Clarendon Press, Oxford.

Devereux, Stephen (2000). 'Food Insecurity in Ethiopia'. Discussion paper prepared for the Department for International Development. <www.ids.ac.uk/ids/pvty/pdf-files/FoodSecEthiopia4.pdf> (December 2005).

Duraiappah, Anantha K. (1998). 'Poverty and Environmental Degradation: A Review and Analysis of the Nexus'. *World Development* vol. 26, no. 12, pp. 2169–2179.

Ellis, Frank (2000). *Rural Livelihoods and Diversity in Developing Countries*. Oxford University Press, Oxford.

Evers, Yvette D. (1996). *The Social Dimensions of Desertification: Annotated Bibliography and Literature Review*. United Nations Environment Programme, Nairobi.

Fairhead, James and Melissa Leach (1996). *Misreading the African Landscape: Society and Ecology in the Forest-Savanna Mosaic*. Cambridge University Press, Cambridge.

Food and Agriculture Organization (2000a). 'State of Food Insecurity in the World'. Rome. <www.fao.org/sof/sofi/index_en.htm> (December 2005).

Food and Agriculture Organization (2000b). 'Assessment of the World Food Security Situation'. 26th Session of the Committee on World Food Security, 18–21 September. Rome. <www.fao.org/docrep/meeting/x7921e.htm> (December 2005).

Food and Agriculture Organization (2003). 'Voluntary Guidelines to Support the Progressive Realization of the Right to Adequate Food in the Context of National Food Security'. 27–29 October. Rome. <www.fao.org/DOCREP/MEETING/007/J0492E.HTM> (December 2005).

Forsyth, Tim, Melissa Leach, and Ian Scoones (1998). 'Poverty and Environment: Priorities for Research and Policy'. Prepared for the United Nations Development Programme and European Commission, September. <www.eldis.org/fulltext/povenv.pdf> (December 2005).

Griffin, Keith, ed. (2003). *Poverty Reduction in Mongolia*. Asia Pacific Press, Canberra.

Hoben, A. (1996). 'The Cultural Construction of Environment Policy: Paradigms and Politics in Ethiopia'. In M. Leach and R. Mearns, eds., *The Lie of the Land: Challenging Received Wisdom on the African Environment*. International African Institute and Heinemann, Oxford.

International Fund for Agricultural Development (2001). 'Rural Poverty Report 2001: The Challenge of Ending Rural Poverty'. <www.ifad.org/poverty> (December 2005).

International Institute for Environment and Development (1995). 'The Desertification Convention: The Strategic Agenda for the EU'. EC Aid and Sustainable Development Briefing Paper No. 4. <europa.eu.int/comm/development/body/theme/environment/env_theme/land_resources_desertification/documents/04.htm> (December 2005).

Jodha, N.S. (1986). 'Common Property Resources and Rural Poor in Dry Regions of India'. *Economic and Political Weekly* vol. 21, no. 27, pp. 169–181.

Kates, Robert W. (1991). 'Hunger, Poverty, and the Human Environment'. DS-9. Center for Advanced Study of International Development.

Katyal, Jagdish C. and Paul L.G. Vlek (2000). 'Desertification: Concept, Causes, and Amelioration'. ZEF Discussion Papers on Development Policy No. 33. Center for Development Research (ZEF), Bonn.

Leach, Melissa and Robin Mearns (1992). 'Poverty and Environment in Developing Countries: An Overview Study'. Final Report to the Economic and Social Research Council and Overseas Development Administration. Institute of Development Studies, Sussex.

Leonard, H. Jeffrey (1989). *Environment and the Poor: Development Strategies for a Common Agenda*. Transaction Books, New Brunswick NJ.

McCulloch, Anna Knox, Suresh Babu, and Peter Hazell, eds. (1999). *Strategies for Poverty Alleviation and Sustainable Resource Management in the Fragile Lands of Sub-Saharan Africa*. International Food Policy Research Institute, Washington DC.

Movik, Synne, Sileshi Dejene, and Gry Synnevåg (2003). 'Poverty and Environmental Degradation in the Drylands: An Overview of Problems'. Noragric Working Paper No. 29. Agricultural University of Norway. <www.eldis.org/static/DOC13294.htm> (December 2005).

Office of the United Nations High Commissioner for Human Rights (2005). 'Special Rapporteur of the Commission on Human Rights on the Right to Food'. <www.ohchr.org/english/issues/food/index.htm> (December 2005).

Øygard, Ragnar, Trond Vedeld, and Jens Aune (1999). 'Good Practices in Drylands Management'. Paper prepared for the World Bank Dryland Program. Noragric Agricultural University of Norway. <www.eldis.org/static/DOC8322.htm> (December 2005).

Reardon, Thomas and Stephen A. Vosti (1996). 'Link between Rural Poverty and the Environment in Developing Countries'. Centro Latinoamericano para el Desarrolo Rural. <www.rimisp.org/webpage.php?webid=117> (Deecember 2005).

Scoones, Ian (1996). 'Politics, Polemics, and Pastures: Range Management Science and Policy in Southern Africa'. In M. Leach and R. Mearns, eds., *The Lie of the Land: Challenging Received Wisdom on the African Environment*. International African Institute and Heinemann, Oxford.

Tiffen, Mary (1993). 'Productivity and Environmental Conservation under Rapid Population Growth: A Case Study of Machakos District'. *Journal of International Development* vol. 5, no. 2, pp. 207–223.

Tiffen, Mary, Francis Gichuki, and Michael J. Mortimore (1994). *More People, Less Erosion: Environmental Recovery in Kenya*. J. Wiley, Chichester.

United Nations Convention to Combat Desertification (1995). 'Down to Earth: A Simplified Guide to the Convention to Combat Desertification, Why It Is Necessary, and What Is Important and Different About It'. <www.unccd.int/publicinfo/downtoearth/downtoearth-eng.pdf> (December 2005).

United Nations Development Programme's Office to Combat Desertification and Drought (UNSO) (1994). 'Poverty Alleviation and Land Degradation in the Drylands: Issues and Action Areas for the International Convention on Desertification'. Paper produced by the United Nations Sudano-Sahelian Office, Food Studies Group, University of Oxford. <www.undp.org/seed/unso/text/public/pov-eng.htm> (December 2005).

United Nations Economic and Social Council (1999). 'The Right to Adequate Food (Article 11)'. E/C.12/1999/5, Committee on Economic, Social, and Cultural Rights, 5 December. Geneva.

United Nations Environmental Programme (1991). 'Status of Desertification and Implementation of the United Nations Plan of Action to Combat Desertification'. Nairobi. <www.na.unep.net/des/uncedtoc.php3> (December 2005).

United Nations Millennium Project (2004). 'Halving Hunger by 2015: A Framework for Action'. Task Force 2 on Hunger, New York. <www.unmillenniumproject.org/documents/tftwointerim.pdf> (December 2005).

World Bank (2001). 'Making Sustainable Commitments: An Environment Strategy for the World Bank'. Washington.

World Commission on Environment and Development (1987). *Our Common Future* (Brundtland Report). Oxford University Press, Oxford.

Chapter 4

Desertification and Migration

Michelle Leighton

There are currently 175 million migrants spanning the globe (International Organization for Migration [IOM] 2005). The International Federation of Red Cross and Red Crescent Societies (2003, ch. 5) estimated in 2003 that 25 million were environmental migrants fleeing natural disasters, including flood, drought, and desertification, a figure exceeding all other categories of global refugees. One researcher, using Essam El-Hinnawi's classification of such migrants as environmental refugees, estimated that the numbers could double by 2010 (Myers 1997).[1]

While some theorists find the characterisation of environmental refugees legally unhelpful, few doubt that the disruption of the natural environment contributes to migration (see, for example, Keane 2004; Castles 2001; Leighton 1997, esp. annex 1; Wood 2001; Bilsborrow 1992). Research within the past 20 years suggests that the onset of desertification and drought can play a significant role in both temporary and permanent migration (Hugo 1995, 105, 109).

This chapter explores the desertification and drought-related migration phenomena considering various studies geographically. The findings relate to the effects of land degradation and water scarcity on population movements and the reverse — how migration itself has contributed to desertification. They illustrate the importance of remittances from migration as a means of survival for affected communities. They also provide some insight into the link between desertification, migration, and conflict.

The research presented is illustrative but in no way exhaustive of the scientific efforts to explain this phenomenon. Human migration describes a dynamic process of population movement, including by persons who are uprooted, displaced, or refugees, as well as by so-called economic migrants.[2] It is an ancient phenomenon, as Tom Farer (1995, 258) says:

> Indeed, one could structure an entire history of the human race in terms of great migrations — of Vandals and Visogoths, Arabs, Mongolians, and Turks, Jews and Huguenots, Spanish and English, Zulus and Dutch, Irish and Italians, Russians and Armenians, impelled at a certain moment (by duress, ambition, dreams) to move from the known place to one that could at best be imagined.

As international law has developed, a number of distinctions and definitions have emerged. An international migrant is a person who voluntarily leaves his or her country to reside elsewhere (UN Economic and Social Council 1998), while an internal migrant is one who crosses a political boundary, district, or area to reside in another part of the same country (United Nations Education, Science, and Cultural Organization [UNESCO] 2005). A different type of legal protection is afforded a refugee who has left his or her country because of a well-founded fear of persecution.[3]

The complex dynamic of migration, suggesting not one causal factor but many, has led to new characterisations of migrants. An economic migrant, for example, may refer to a person leaving his or her country in search of better employment, whereas an environmental migrant may be forced to leave due to serious environmental change or natural disaster affecting that individual's survival (UNESCO 2005).

Migration related to desertification is concerned with both economic and environmental change. It can take root as a coping strategy for households in drought-prone or desertified areas, as where migrants send home remittances and periodically return home, or as a permanent survival strategy (UNESCO 2005; see also Suhrke 1993). The cases reflect the differing nature of opportunity for migrants to relocate to cities within their country of origin or internationally, whether crossing the ocean from West Africa to Europe or crossing the border between Mexico and the United States. The precise quantification of these population movements on a global scale is uncertain.[4] This chapter instead examines a series of acute local and regional problems in Africa, Latin America, and Asia.

Finally, this chapter considers how migration issues were incorporated into the negotiation process for the United Nations Convention to Combat Desertification (UNCCD) and in the convention's text itself. It concludes with an observation about the convention's potential role in addressing this phenomenon.

The Role of Desertification in Migration

The degradation of agricultural lands can contribute directly to migration through its impact on poverty. Low agricultural yields resulting from poor soil or lack of water due to drought can reduce the household income for families dependent on agriculture (Leighton 1997; see also Leighton 1998). Poverty is a major determinant of migration (see chapters 1, 3, and 12). As such, families may be forced or may choose to migrate to supplement family income. While Africa is the most acutely affected region, the problem is becoming better understood in Latin America and Asia. Some believe that if this problem is left unattended it will grow and conflict may be inherent (Hugo 1995, 120).

Africa

Three quarters of all agricultural drylands are now degraded to some degree in Africa, the region of greatest concern to the UNCCD (2005b; 2005a). The relationship between drylands degradation and human migration has been studied in several African countries but, as no continent-wide study has yet been undertaken, the quantification of this migration is speculative. In considering the research, one author believes that migration related to desertification must be greater than that of other regions: 'Of all the continents, Africa, a land where poor soils and variable rainfall pose a harsh climate for agriculture, has spawned the most environmental refugees. Most came from the Sahel' (Jacobson 1988, 11, cited in Hugo 1995, 115).[5]

Whether this is correct, the role of desertification and drought, particularly in Sahel countries, remains important as a factor — as does conflict — in newer migration trends. In 2000, the IOM (2000) proclaimed that the East African subregion 'is experiencing movements of refugees and internally displaced persons (IDPs), as a result of environmental disasters (drought and desertification), as well as civil and guerrilla wars in the Horn of Africa, in Somalia, Eritrea, Ethiopia and southern Sudan' (see also IOM 2005).

Historically, periodic droughts in the Sahel lasting years or even decades have played a durable role in the region's migration patterns (Sahel Club 1984, 1, 86, cited in Tamondong-Helin and Helin 1991). The droughts of 1968–73 and 1982–84 contributed to poverty from the Atlantic to the Horn of Africa, including Senegal, Mauritania, Mali, Burkina Faso, Niger, Nigeria, Chad, Sudan, Ethiopia, Eritrea, Djibouti, and Somalia. These droughts were coupled with high rates of population growth, intensification of agriculture, and lack of sustainable land and water management policies. Millions of people migrated during these periods, with one million estimated to have left Burkina Faso alone during the 1968–73 drought (Sahel Club 1984, cited in Tamondong-Helin and Helin 1991). Research suggests that the migration of men from rural Burkina Faso to other areas in the country is more likely to be permanent now in response to repeated drought.[6]

Research in Ethiopia has found that individuals from poor households in ecologically fragile communities have a higher propensity to migrate compared to those from wealthier households in less ecologically vulnerable communities (Ezra 2001).[7] Sample data from selected drought-prone areas in Amhara and Tigray, the regional states with nearly one third of Ethiopia's total population, demonstrate that rural out-migration is largely a consequence of environmental degradation and poverty, that are structural and institutional in origin, rather than from individual choice.[8]

Case studies of communities in Mali and Senegal are also instructive. Migration in Mali has arisen as both an immediate and a long-term response to threat of recurrent droughts (Findley 1994, 539). During the 1969–74 drought, populations moved from the more arid zones bordering the Sahara to cities in the Sahel reflecting temporary migration patterns (542).[9]

Later, during the 1983–85 drought in Mali, short-term migration more than doubled. Rainfall was 30 percent below the 50-year average and most families, unable to produce enough cereals and livestock, were forced to buy up to 60 percent of their food requirements (Findley 1994, 543). Some 63 percent of families during that period depended for their survival on remittances from those who had already migrated, mostly to France (544–546).[10]

Several studies in Senegal have also found that persistent droughts and land degradation contributed to both seasonal and permanent migration. In the southeastern Tambacounda region, an area comprising 30 percent of Senegal's territory, the economy depends entirely on agriculture, particularly peanuts, cotton, and subsistence crops.[11] Since the 1960s, crop production has declined because of soil erosion. The agricultural declines and lack of other job opportunities led to a large emigration to Dakar, other urban centres, and, later, other countries in Africa and Europe, particularly France.

By the early 1990s, rural migration had become a structural phenomenon in Tambacounda: 90 percent of the region's men between 30 and 60 years old had migrated at least once in their lifetime. This loss of men posed challenges for efforts to rehabilitate degraded lands and increased the economic burden on the remaining women and children. Remittances became critically important to the community for survival, contributing 75 percent of family incomes in 1993 and helping to finance schools, post offices, and social service centres (Seck 1996).

The pattern of relying on internal and international migration as a coping strategy has also been documented among the Kaskas, the Soninke, the Sere, and the Haal Puular (Dia 1992, 57–64, cited in Knerr 2004). With an average of 1.5 migrants per household, the Kaskas use remittances from seasonal migration to support agricultural activities that generate negative income. Of the Haal Puular, more than 90 percent of men between 30 and 60 years of age have migrated at least once in their life, with 58 percent of migrants moving to towns within Senegal, 35 percent moving to Mauritania, and 6 percent moving further away. Since the 1980s, the Sere community has begun to use migration as a coping strategy for drought, with 48 percent moving predominantly to Dakar, while the Soninke migrated to France.

Migration can also play a role in exacerbating desertification processes (African Union [AU] 2005, 30). A study in Morocco provides one example. The migration of men away from traditional villages, along with the decline of nomadism, has caused a neglect of lands and hydraulic systems, leading to sand accumulation in once active irrigation furrows (de Haas 1998, 12). Desertification is now becoming more widespread. Some families have begun to address this problem by using remittances from migrant family members to invest in water-harvesting technologies that can help revitalise farming in the area (de Haas 2003, 260).

The Americas

As both an environmental factor and an economic factor contributing to migration, desertification has become a rising concern in the Western hemisphere, particularly in the countries of Latin America where 1.4 million square kilometres of land are undergoing some form of degradation and one third is severely degraded (Natural Heritage Institute 1996).[12]

In North America, the relationship between desertification and migration may be most poignant for rural Mexicans. Most of Mexico's drylands, representing 70 percent of its land territory, are undergoing desertification (Leighton 1997, 16). For the farm communities in these areas, migration is often a critical economic choice. As agricultural productivity lessens because of desertification processes, internal and cross-border migration to the United States correspondingly become more economically attractive for rural households (24).

In 1997, the Mexican government and other researchers estimated that between 600 000 and 700 000 people were migrating from these areas annually (Leighton 1997, 27). Rural agricultural migrants are developing more sophisticated social networks in Mexico's major cities and the U.S., reducing the costs and risks associated with future migration. Importantly, the states undergoing rapid deforestation and desertification are also those with rapidly rising rates of migration, particularly Oaxaca and Tamaulipas.

Studies suggest that in South America internal migration is more predominant. The periodic droughts and desertification plaguing northeast Brazil contributed to factors causing 3.4 million people to emigrate between 1960 and 1980 (Sanders 1990-91): the region comprises 63 percent of the rural poor and 32 percent of all poor people in Brazil.[13] Desertification has been exacerbated by people clearing the caatinga (semi-arid brush forests), savannah, and forests so that crops could be planted and more animals could graze, without allowing the traditional fallowing period for the lands. The forests, caatinga, and savannah that had occupied 65 percent of northeast Brazil occupied only 35 percent in 1990 ('Desertificao do Nordeste e Tema de Debate' 1991). Large-scale landowners halted agricultural production, eliminating subsistence sharecroppers and increasing the necessity for migration among unemployed farm workers.[14]

Research into Argentina's rural arid regions suggests similar characteristics. Its drylands cover 70 percent of its territory. One study found that there is a growing demographic suffering the effects of desertification, poverty, and migration (Adamo 2003). In the Jáchal area of the San Juan province, households that depend more on farming and livestock raising than do other sectors of society are more vulnerable to desertification (or land degradation) and to resorting to migration as a strategy to cope with desertification.[15]

While many rural dryland families depend on remittances for their survival, in some areas this migration safety net has not materialised, creating despair for those remaining behind. The clearing of forests and burning of fields in the Valle Grande region of Bolivia has left only 10 percent of land that is arable (Müller 1993, cited in Knerr 2004). Many people have migrated to escape the resulting steep economic and social decline but have not sent remittances back to the community. This suggests a more permanent resettlement: 76 percent of those remaining have no access to non-agricultural income.

Asia and the Middle East

Although Asia and the Middle East are less documented than other regions, several studies suggest that desertification in Asian countries has also played a part in international migration and internal displacement (Goria 1998). Research studies in Uzbekistan, Kazakhstan, Syria, Iran, Bangladesh, and India are informative. Large-scale water diversions, monocultural cropping, and excessive use of fertilisers have left the Aral Sea and surrounding regions vulnerable to toxic windstorms and intense salinisation. Straddling Uzbekistan and Kazakhstan, the area now suffers a declining fish industry and high infant mortality rates; approximately one million migrants annually have fled the area to other parts of the north and west of Kazakhstan (Glazovsky and Shestakov 1994).

In Kazakhstan, 'the main push factor for migration is … a deterioration of the living standards and living conditions following environmental deterioration which leads to the degradation of pasture and arable lands and severely polluted drinking water' (Shestakov and Streletsky 1998, 68).[16] Migrants move internally but there is a high incidence of international migration among non-native Russian speakers and people of German extraction (69).

Desertification in Syria has contributed to more historical patterns of international migration. Early migration was forced on the Syrian Druze by drought, poor agricultural production, and population growth (Escher 1994). Desertification continues in the Gabal al-Arab settlements, which were built upon ancient Roman ruins and lava rock. Water is scarce, arising from a few shallow springs that dry up entirely in the summer. Droughts are common. Communities rely now on remittances from international migration to survive and to reinvest in agricultural development (7). The destination of migrants has broadened, too, reaching as far as Argentina, Venezuela, Nigeria, Senegal, and Lebanon (3–5).

In south Asia, Bangladesh has experienced precipitous declines in arable land, a determinant of migration to the eastern borders of India such as Assam (Hazarika 1993, 45). Population growth, land exploitation, and both flooding and unsustainable water management are noted causes (Homer-Dixon, Boutwell, and Rathjens 1993, 40, cited in Hazarika 1993, 49).

Desertification in India's drylands has also led to migration, although mostly domestic in nature. Nearly half of its territory is prone to drought (Maloney 1991). In Rajasthan state, with a third of its people living under the poverty line, millions have survived the recurring droughts and land degradation by seasonally migrating.

The Role of Desertification and Migration in Conflict

Desertification and migration can intensify the competition for arable soil and water resources. This competition, coupled with ethnic and border tensions, has led to violent conflict and further population displacement. The IOM (2005, 37) has determined that 'initially concentrated in East Africa, where environmental disasters, such as drought and desertification, and ethnic, border and guerrilla wars in the Horn of Africa (Somalia, Eritrea, Ethiopia and southern Sudan) gave rise to large movements of refugees and IDPs in the 1980's and 1990's, in recent years [the problem] has shifted to Central and West Africa'. The cases of Ethiopia, Sudan, Mali, and Nigeria provide some insight into the role of desertification and migration in conflict.

Ethiopia's resettlement policies in response to severe desertification and drought in the early 1980s, for example, led to further land degradation, migration, and social conflict in the areas of resettlement (Westing 1994). Officials relocated 70 000 people from the highlands to Gambella, where they outnumbered the local Nilotic Anuak tribes (Stiles 1997).[17] Officials forcibly conscripted the Anuak to clear large forests and build houses. This intense exploitation of the area's natural resources led to deforestation, soil erosion, and the siltation of rivers, in turn reducing the community's fish yields and food availability. Maize crops failed after being cultivated on unproductive land. Together, the political repression of the Anuak and the consequences of desertification engendered violent conflict among the Anuak tribes and relocated settlers. Eventually, the Anuak fled to Sudan (Stiles 1997).

Sudan has suffered its own cycle of conflict, migration, and land degradation. The Darfur region is a continuing example with its progressively worsening droughts and desertification. The United Nations Office of Humanitarian Assistance (2003) has concluded that this situation has affected the population's livelihood and has forced people to migrate. Competition between pastoral communities and settled communities for scarce resources has also led to conflict and inter-tribal disputes.

In West Africa, following the 1984–85 Malian drought, armed conflict erupted between the Tuareg and other ethnic groups such as the Songhai and Fulani in part due to the control of scarce lands (Stiles 1997).[18] During this drought, the Tuareg were forced to move their herds south or sell them. Some Tuareg migrated, but those who remained fell into poverty. They were disenfranchised from the other ethnic groups who bought up and controlled the lands and livestock trade. The Songhai and Fulani became the predominant herd managers, intensifying livestock production and grazing to maximise profits. These practices, while enriching the Songhai and Fulani, were not sustainable. Tensions mounted over land use among the impoverished Tuareg and the more prosperous ethnic groups, resulting in armed conflict. Some Tuareg were arrested or killed, while others fled to neighbouring countries.

Competition for scarce land resources is now leading to migration and conflict in Nigeria as well. Dozens of people have died and hundreds have been displaced in fighting between local farming communities and nomadic Fulani herders in the Mambilla plateau in northeastern Nigeria. Many Fulani herdsman fled across the border into Cameroon for fear of reprisals. In reporting on this conflict, the UN Office for Coordination of Humanitarian Affairs (2002) has observed that this may be part of a broader phenomenon: 'Clashes between pastoral and farming communities linked to disputes over grazing land, have become frequent in parts of central and northern Nigeria in recent years. Some analysts have blamed the trend on increasing desertification, which is pushing herders southwards in their search for pasture, often putting them in conflict with farmers.'

The UNCCD and Its Relationship to Migration

The UNCCD is the first United Nations environmental treaty to recognise the link between environmental degradation and migration. It is also the first to incorporate provisions for addressing the adverse impacts of this phenomenon.

Prior to 1994, international agencies, nongovernmental organisations (NGOs), and scientists laid the foundation for the negotiating parties' examination of this issue. The United Nations Environment Programme (UNEP), in a 1985 report, discussed the global consequences of a growing number of 'environmental refugees' that included people forced to move because of desertification (El-Hinnawi 1985). The Worldwatch Institute issued a report with similar global warnings in 1988 (Jacobson 1988). In 1991, the United Nations Population Fund warned that there could be major shifts of population distribution due to environmental causes leading to global instability (Myers 1991, 63). Just before the UN's Rio Conference on Environment and Development (UNCED), the IOM and the Refugee Policy Group issued a report evaluating the problem, finding that there were continuing impacts of desertification on internal displacement and international migration. Subsequent reports from the Norwegian Refugee Council and the World Foundation for Environment and Development supported these findings (Trolldalen et al. 1992). In 1992, these concerns were incorporated into Agenda 21, the UNCED's comprehensive summit report.[19]

These efforts paved the way for a discussion within the context of the UNCCD negotiations. In February 1994, the government of Spain and the secretariat for the UNCCD's Intergovernmental Negotiating Committee on Desertification (INCD) sponsored an international symposium in Almeria, Spain, to explore the issue and provide recommendations to the negotiating parties. The gathered experts agreed on the critical importance of desertification as a cause and consequence of both internal and international migration. Their findings provided governments with direct evidence that desertification, having transboundary migration impacts, should not be addressed only as a domestic issue but should be taken up by the international community in co-operation. The experts determined that 35 percent of the estimated 100 million migrants in 1994 were found in sub-Saharan Africa, the most important region for the UNCCD, and that half of those affected by desertification lived in the Sahel.[20]

The Almeria Declaration on Migration and Desertification, adopted by the International Symposium on Desertification and Migrations (1991), states in part:

> The concern of the Symposium is that forced migrations resulting from arid-land degradation, and their consequences, often exhibit undesirable dimensions, both at the poles of origin and at the sites of relocation ... These increases are largely of rural origin and related to land degradation. It is estimated that over 135 million people may be at risk of being displaced as a consequence of severe desertification.

In recommending new policy reforms, the declaration focussed on preventive action. It suggested that the key is to promote sustainable agriculture and range management in arid lands through the provision of financing, technology, capacity building, and economic incentives. It stated that 'there is a need to review the legal status and regulatory regime governing the ever-growing number of environmental migrants and displaced persons and to identify options for their protection and relief, helping returnees and assisting others with their integration into host communities'.[21]

In addition, early warning systems can allow governments to intercept the problem before it expands to crisis proportions or leads to acute conflict.[22]

The declaration proved to be a successful organising tool for governments and nongovernmental groups advocating for specific provisions within the convention. Armed with its recommendations, NGOs advocated strongly for treaty provisions to recognise the issue more fully within the convention mandate. In a statement to the fourth session of the INCD, delivered both orally and in writing to delegates, NGOs identified their support for the Almeria recommendations stating: 'We urge governments and other parties involved in the negotiation of the Convention to Combat Desertification to give serious consideration to the relationship between desertification and migration, and to the action priorities identified in the Almeria Declaration, in developing policies and programs to address severe land degradation and drought.'[23]

Although governments did not adopt all of the recommendations made at the Almeria Symposium, specific provisions were included in the preamble to the UNCCD and in the sections pertaining to national action planning and to scientific research.[24] Recognising that governments tend to address environmental issues separately from emigration and immigration problems, articles 10 and 17 encourage governments to understand these links better and to collaborate across borders.

In article 10.3 the convention states that national action programmes (NAPs) may include 'establishment and/or strengthening, as appropriate, of early warning systems, including local and national facilities and joint systems at the subregional and regional levels, and mechanisms for assisting environmentally displaced persons' (UNCCD 1994). This provision follows the Almeria recommendation that, in developing NAPs, officials need to look beyond the physical environmental problems associated with desertification toward the social and economic impacts related to cross-border migration and the internal displacement of those affected by desertification. Where the problem is known, governments should focus policies on communities that have a significant desertification and migration problem, and co-operate across borders to seek solutions.

Acknowledging the Almeria recommendation for further research and monitoring of desertification-related migration, article 17.1(e) urges governments to 'take into account, where relevant, the relationship between poverty, migration caused by environmental factors, and desertification' (UNCCD 1994). This provision seeks to encourage governments to undertake and support further investigation into these relationships with a view to determining the best policies and programmes needed within the context of programs to address desertification. Based on the discussions in Almeria, the provision also encourages collaboration with international agencies of expertise, such as the IOM and the United Nations High Commissioner for Refugees (UNHCR), that recognise the problem and offer assistance and technical expertise (Lohrmann 1994).[25] Indeed, two years after the conclusion of the treaty, on 22–24 April, 1996 in Geneva, the IOM and UNHCR, along with the Swiss government, held the International Symposium on Environmentally Induced Population Displacement and Environmental Impacts Resulting from Mass Migration, a joint meeting of experts to explore ways and means of addressing these issues (see Lohrmann 1996).

Conclusion

Ten years after adopting the UNCCD, the secretariat's 2004 Desertification Day focussed on the issue of poverty and migration. Yet it remains unclear whether many of the 191 parties affected by desertification-related migration are adequately addressing this problem or are seeking to do so under the convention's framework. In one project related to desertification and its socioeconomic consequences being undertaken in Mexico, Chile, and Brazil (all parties to the convention), many officials clearly seek to address the environmental issues associated with desertification but are less willing or able to promote solutions within an inter-agency context, for example, among and between social welfare and agricultural ministries, impeding potential development of durable policy solutions (see UNEP 2003).

A number of opportunities for international and inter-agency collaboration may be emerging. In 2003, the UN Secretary General organised the Global Commission on International Migration, a body of experts working with governments, intergovernmental organisations, and NGOs to garner attention and co-operation in addressing migration issues within an interdisciplinary context.[26] The African Union (2005) is now establishing an initiative to co-ordinate policies and programmes to address both international migration and internal displacement, including the causes and consequences of migration related to soil degradation and other environmental factors. The IOM (2005) is collaborating broadly with governments and other international organisations in the establishment of the processes of interdisciplinary dialogue and policy development throughout the African continent and globally.

Newer bilateral agreements, such as Senegal's collaborative programmes with Italy, Spain, and Germany and with the European Commission, may provide reforms worth further evaluation. In 1998, an agreement between Senegal and France initiated a new programme of co-operation for those Senegalese seeking repatriation (see Diatta and Mbow 1999). This programme seeks to address concerns in both the North and South about migratory flows and to promote 'migrants as agents of development through the skills that they can transfer to their countries of origin' (246–249).[27]

It could be expected that the institutions of the UNCCD could play a role in these various initiatives, in particular to provide specialised expertise on the desertification-related issues of migration. Such participation in co-operative activities could increase the attention paid to the development and funding of early warning and prevention systems explicitly to identify and intercept desertification-related migration patterns and to the impact of migratory flows on desertification in host countries. It is possible that this type of work would be considered highly relevant to the operational programme on desertification of the Global Environment Facility (GEF). The UNCCD's Committee on Science and Technology has analysed and encouraged early warning systems more generally for drought, and its expertise would be instructive.

Whether governments will embrace the UNCCD's institutional structures to play a more significant role in this regard is uncertain. It is clear that the textual provisions of the convention provide ample support for a heightened institutional presence in this area.

Notes

1 The term 'environmental refugees' has been attributed to Essam El-Hinnawi (1985), who described categories of refugees forced to leave their homelands because of different environmental phenomena.
2 See the report of the Working Group of Intergovernmental Experts on the Human Rights of Migrants submitted in accordance with the Commission on Human Rights resolution, which adopts the IOM's definition of migrants and migration (United Nations Economic and Social Council 1998).
3 Convention Relating to the Status of Refugees, entered into force 22 April 1954, 189 UNTS 150, and its Protocol, entered into force 4 October 1967, 267 UNTS 267.
4 Suhrke (1993, 12) writes: 'Although no one has tried to estimate how much of the contemporary rural-to-urban migration in the developing world is specifically due to land degradation in sending areas, there is rich case study material that documents the dynamic.'
5 The IOM (2005, 36) estimates that 42 percent of the 16 million migrants within Africa live in West Africa and those along border regions live in unstable conditions due in part to war, ethnic fighting, or drought.
6 Sabine, Henry, Bruno Schoumaker, and Cris Beauchemin (2003), 'Do Environmental Conditions Influence Migration? An Event-History Anaysis of the Determinants of Village Departure in Burkino Faso', on file with author. The study uses data from surveys and a national representative sample survey conducted of 8644 individuals aged 15 to 64 undertaken in 2000 by University of Ougadougou, the Demography Department of Université de Montréal and the Centre d'Études et de Recherche sur la Population pour le Développement of the Sahel Institute in Bamako, Mali. See also Sabine, Schoumaker, and Beauchemin (2004).
7 The study by Markos Ezra (2001) examined patterns in 2000 of households from 40 villages with migration surveys in 1994–95. Ezra considers the challenges of distinguishing migration from war, such as the 1998–2000 border war with Eritrea, and the four major famines in the past several decades, have contributed to land degradation.
8 In areas where ecological degradation is greatest, scarcity of arable land combined with population growth led to surfeit labourers on small landholdings (Ezra 2001).
9 Sally Findley (1994) evaluates data on migration patterns during the Malian drought between 1983 and 1985. Her study area is the Upper Senegal River Valley from Matam, Senegal to Diamou, Mali.
10 While very few of the household heads explicitly reported drought or famine as the reason for migration, Findley (1994) found this to be expected because households were interviewed after the drought and subject to reporting and reinterpretation biases.
11 This information is taken from findings from an investigation in Senegal presented by Masse Lo at the Almeria Symposium in 1994, on file with author. See also Seck (1996).
12 Central American countries have also suffered desertification and migration. See, for example, the study of the hill region of Las Ayuamas in the Dominican Republic by Mark Zweifler, Michael Gold, and Robert Thomas (1994).
13 In the 1960s net emigration was 4.1 million, rising to 4.6 million a decade later (Projeto Áridas 1995, 47).
14 This was found by the Natural Heritage Institute in collaboration with the Esquel Group Foundation of Brazil, under a project sponsored by the Global Environment Facility (GEF) on drylands biodiversity indicators (materials on file with the author). See also Sanders (1990–91).

15 The findings of the study by Susana Adamo (2003) suggest that desertification can play a more significant role in migration for those families highly dependent on agriculture for their livelihood rather than those households with off-farm income, where desertification may play a limited or nonexistent role.
16 The ecological 'push' factors were recognised by the Kazakh government in its Outline of a State Programme for Population Migration in 1995, which identified the need to reduce the environmental degradation that caused displacement.
17 This resettlement may have been politically motivated.
18 Historically, deeper social conflicts have existed between ethnic groups of this region, exacerbated by drought and lack of arable land.
19 In 1993, the issue was raised at preparatory meetings for the UN International Conference on Population, with one expert stating: 'Land degradation drives people from their homes to find fertile soil that can sustain their basic needs and it is considered the primary reason for migration in the southern hemisphere' (Fornos 1993).
20 The author participated in both the symposium and the drafting of the final declaration. At the symposium, Reinhard Lohrmann (1994) cited an alarming study by Dieter Weiss that indicated that because of decline in cereal production and the stagnation of yields per hectare of land in parts of Africa, as well as a decline in per capita arable land, out-migration could be 60 million within three decades.
21 The International Symposium on Desertification and Migrations (1991) also recommended that 'the negotiation process of the Convention to Combat Desertification give greater attention to the phenomenon of desertification induced migration, at the local, regional and global levels'.
22 The United Nations Institute for Training and Research (UNITAR) suggested such a system could be used to explore different economic and land-use planning options for mitigating or preventing the effects of desertification on migration. The discussion was based on the model's application to Crete in considering rural to coastal migration, which led to a decline in terrace maintenance and an increase in soil erosion.
23 The statement by the NGOs dated 23 March, 1994 was signed by 48 organisations from 30 countries. It was an early mobilising tool to bring NGOs together during the negotiations. In 1995, many of these same NGOs would serve as the core that organised into a global network of several hundred known as the Réseau international des ONG pour la désertification (RIOD, or the International Network on Desertification). It also served to catalyse U.S. NGOs into a coalition co-led by the author, which successfully promoted the U.S. ratification of the UNCCD.
24 The preamble states: 'Mindful that desertification and drought affect sustainable development through their interrelationships with important social problems such as poverty, poor health and nutrition, lack of food security, and those arising from migration, displacement of persons and demographic dynamics' (UNCCD 1994).
25 Lohrmann's (1994) paper included policy recommendations to address desertification and migration, including recommending the sharing of information among government agencies and the promotion of multilateral co-operation to support scientific data collection; the analysis of meteorological, agricultural, and demographic data to consider international population movements; regularised scientific policy dialogue; and the intensification of research to track the pressures on environment that do or could lead to migration and other early warning systems to avert humanitarian crisis.
26 See the Global Commission on International Migration materials presented at <www.gcim.org/en> (December 2005).

27 Some of the early training programmes were not as effective for reasons presented in the report (Diatta and Mbow 1999).

References

Adamo, Susana B. (2003). 'Vulnerable People in Fragile Lands: Migration and Desertification in the Drylands of Argentina — The Case of the Department of Jáchal'. Ph.D. thesis. University of Texas at Austin.

African Union (2005). 'Draft Strategic Framework for Policy on Migration in Africa'. Third Ordinary Session of the African Union Labour and Social Affairs Commission, 18–23 April. Johannesburg.

Bilsborrow, R.E. (1992). 'Rural Poverty, Migration, and Environment in Developing Countries: Three Case Studies'. Country Economics Department Paper No. 1017. World Bank, Washington DC.

Castles, Stephen (2001). 'Environmental Change and Forced Migration'. Paper prepared for the Westmorland General Meeting 'Preparing for Peace' Initiative, 6 December. <www.preparingforpeace.org/castles_environmental_change_and_forced_migration.htm> (December 2005).

de Haas, Hein (1998). 'Socio-economic Transformations and Oasis Agriculture in Southern Morocco'. In P.M. Blaikie and L. de Haan, eds., *Looking at Maps in the Dark: Directions for Geographical Research in Land Management and Sustainable Development in Rural and Urban Environments of the Third World*. Royal Dutch Geographical Society and Faculty of Environmental Sciences, University of Amsterdam, Utrecht and Amsterdam.

de Haas, Hein (2003). 'Migration and Development in Southern Morocco: The Disparate Socio-economic Impacts of Out-Miration on the Todgha Oasis Valley'. Ph.D. thesis thesis. University of Amsterdam.

'Desertificao do Nordeste e Tema de Debate'. (1991). *Jornal de Brasil*, 1 April.

Dia, I. (1992). 'Les migrations comme stratégie des unités de production rurale: Une étude de cas au Sénégal'. In A. Blokland and F. van der Staay, eds., *Sustainable Development in Semi-Arid Sub-Saharan Africa*. Ministry of Foreign Affairs, Netherlands, The Hague.

Diatta, Marie Angelique and Ndiaga Mbow (1999). 'Releasing the Development Potential of Return Migrants: The Case of Senegal'. *International Migration* vol. 37, no. 1, pp. 243–266.

El-Hinnawi, Essam (1985). 'Environmental Refugees'. United Nations Environmental Programme, Nairobi.

Escher, Anton (1994). 'Migrant Network: An Answer to Contain Desertification; A Case Study of Southern Syria (Gabal al-Arab)'. Paper presented at the International Symposium on Desertification and Migrations, 9–11 February. Almeria, Spain.

Ezra, Markos (2001). 'Ecological Degradation, Rural Poverty, and Migration in Ethiopia: A Contextual Analysis'. Policy Research Division Working Paper No. 149. Population Council, New York. <www.popcouncil.org/publications/wp/prd/149.html> (December 2005).

Farer, Tom (1995). 'How the International System copes with Involuntary Migration: Norms, Institutions, and State Practice'. In M.S. Teitelbaum and M. Weiner, eds., *Threatened Peoples, Threatened Borders: World Migration and U.S. Policy*. W.W. Norton, New York.

Findley, Sally (1994). 'Does Drought Increase Migration? A Study of Migration from Rural Mali during the 1983–1985 Drought'. *International Migration Review* vol. 28, no. 3, pp. 539–553.

Fornos, Werner (1993). 'Population Distribution and Migration'. Proceedings for United

Nations Expert Group Meeting on Population Distribution and Migration, 18–22 January 1993, Santa Cruz, Bolivia.

Glazovsky, Nikita and Alexander Shestakov (1994). 'Environmental Migration Caused by Desertification in Central Asia and Russia'. Paper presented at the International Symposium on Desertification and Migrations, 9–11 February. Almeria, Spain. <geographytsu.freehomepage.com/CIS%20env%20migration.htm> (December 2005).

Goria, Alessandra (1998). 'Desertification and Migration in the Mediterranean: An Analytical Framework'. Fondazione Eni Enrico Mattei, Milan.

Hazarika, Sanjoy (1993). 'Bangladesh and Assam: Land Pressures, Migration, and Ethnic Conflict'. Occasional Paper No. 3. Project on Environmental Change and Acute Conflict.

Homer-Dixon, Thomas, Jeffrey H. Boutwell, and George Rathjens (1993). 'Environmental Change and Violent Conflict'. *Scientific American* vol. 268, no. 2, pp. 4, 38–45.

Hugo, Graeme (1995). 'Environmental Concerns and International Migration'. *International Migration Review* vol. 30, no. 1, pp. 105–142.

International Federation of Red Cross and Red Crescent Societies (2003). 'World Disasters Report: Focus on Ethics in Aid'. Geneva. <www.ifrc.org/publicat/wdr2003> (December 2005).

International Organization for Migration and Refugee Policy Group (1992). 'Migration and the Environment'. Geneva.

International Organization for Migration (2000). 'IOM Migration Policy Framework for Sub-Saharan Africa'. MC/INF/244, 17 November. Geneva. <www.iom.int/en/PDF_Files/other/policysubsaharan.pdf> (December 2005).

International Organization for Migration (2005). 'World Migration 2005: Costs and Benefits of International Migration'. Geneva.

International Symposium on Desertification and Migrations (1991). 'The Almeria Statement on Desertification and Migration'. 9–11 February. Almeria, Spain. <www.unccd.int/regional/northmed/meetings/others/1994AlmeriaSpain.pdf> (December 2005).

Jacobson, Jodi (1988). 'Environmental Refugees: A Yardstick of Habitability'. World Watch Paper No. 86. World Watch Institute, Washington DC.

Keane, David (2004). 'The Environmental Causes and Consequences Migration: A Search for Meaning of "Environmental Refugees"'. *Georgetown International Environmental Law Review* vol. 16, no. 2, pp. 209–214.

Knerr, Béatrice (2004). 'Desertification and Human Migration'. In D. Werner, ed., *Biological Resources and Migration*, pp. 317–338. Springer, Berlin.

Leighton, Michelle (1997). 'Environmental Degradation and Migration: The U.S./Mexico Case Study'. December. Natural Heritage Institute.

Leighton, Michelle (1998). 'Environmental Degradation and Migration: The U.S.-Mexico Case Study'. *Environmental Change and Security Project Report* no. 4, pp. 61–67.

Lohrmann, Reinhard (1994). 'The Need for Enhanced International Cooperation in Addressing Environmental Migration Issues'. Paper presented at the International Symposium on Desertification and Migrations, 9–11 February. Almeria, Spain.

Lohrmann, Reinhard (1996). 'Environmentally Induced Population Displacements and Environmental Impacts from Mass Migrations: Conference Report'. *International Migration* vol. 34, no. 2, pp. 335–339.

Maloney, Clarence (1991). 'Environmental Displacement of Population in India'. Field Staff Reports No. 14. UFSI Inc. and National Heritage Institute.

Müller, Peter M. (1993). 'Tragfähigkeitsveränderung durch Bevölkerungsverlust'. *Geographische Rundschau* vol. 45, no. 3, pp. 173–179.

Myers, Norman (1991). *Population, Resources, and the Environment: The Critical Challenges*. United Nations Population Fund, New York.

Myers, Norman (1997). 'Environmental Refugees'. *Population and Environment* vol. 19, no. 2, pp. 167–182.

Natural Heritage Institute (1996). 'North-South NGO Forum on Desertification in the American Hemsiphere'. 10–12 November. San Francisco.

Projeto Áridas (1995). 'A Strategy for Sustainable Development in Brazil's Northeast'. Ministry of Planning and the Budget, Brasilia.

Sabine, Henry, Bruno Schoumaker, and Cris Beauchemin (2004). 'The Impact of Rainfall on the First Out-Migration: A Multi-level Event-History Analysis in Burkino Faso'. *Population and Environment* vol. 25, no. 5, pp. 423–460.

Sahel Club (1984). *Environmental Change in the West African Sahel (Transformation de l'environnement dans le Sahel ouest africain)*. Club de Sahel, Paris.

Sanders, Thomas G. (1990-91). 'Northeast Brazilian Environmental Refugees: Part I — Why They Leave'. Field Staff Reports No. 20. National Heritage Institute.

Seck, Emmanuel S. (1996). 'Désertification: effets, lutte et convention'. ENDA TM, Dakar.

Shestakov, Alexander and Vladimir Streletsky (1998). *Mapping of Risk Areas of Environmentally Induced Migration in the Commonwealth of Independent States*. United Nations High Commissioner for Refugees, International Organization for Migration, and Refugee Policy Group, Geneva.

Stiles, Daniel (1997). 'Linkages between Dryland Degradation and Migration: A Methodology'. *Desertification Control Bulletin* no. 30, pp. 9–18.

Suhrke, Astri (1993). 'Pressure Points: Environmental Degradation, Migration, and Conflict'. American Academy of Art and Science. Cambridge. <www.cmi.no/publications/1993%5Cpressure_points.pdf> (December 2005).

Tamondong-Helin, Susan and William Helin (1991). 'Migration and the Environment: Interrelationships in Sub-Saharan Africa'. Field Staff Reports No. 22. UFSI Inc. and National Heritage Institute.

Trolldalen, John Martin, Nina Birkeland, Jan Borgen, et al. (1992). 'Environmental Refugees: A Discussion Paper'. Paper prepared for the United Nations Conference on Environment and Development, 3–14 June. World Foundation for Enviornment and Development and Norwegian Refugee Council.

United Nations Convention to Combat Desertification (1994). 'Text of the United Nations Convention to Combat Desertification'. <www.unccd.int/convention/text/convention.php> (December 2005).

United Nations Convention to Combat Desertification (2005a). 'Reports Submitted by Africa'. <www.unccd.int/cop/reports/africa/africa.php> (December 2005).

United Nations Convention to Combat Desertification (2005b). 'Combating Desertification in Africa'. <www.unccd.int/publicinfo/factsheets/showFS.php?number=11> (December 2005).

United Nations Economic and Social Council (1998). 'Measures to Improve the Situation and Ensure the Human Rights and Dignity of All Migrant Workers'. Report of the Working Group of Intergovernmental Experts on Human Rights of Migrants E/CN.4/1998/76, 10 March. <www.unhchr.ch/Huridocda/Huridoca.nsf/TestFrame/e696166bf66373f3c12566180046b9c6> (December 2005).

United Nations Education, Science, and Cultural Organization. (2005). 'Migrant/Migration'. <www.unesco.org/shs/migration/glossary> (December 2005).

United Nations Environmental Programme (2003). 'An Indicator Model for Dryland

Ecosystems in Latin America'. Final report submitted to the Global Environment Facility. GF/1040-00-10.

United Nations Office for the Coordination of Humanitarian Affairs (2002). 'Nigeria: Dozens Reported Dead in Clashes between Farmers, Herders'. IRINews, 8 January. <www.irinnews.org/print.asp?ReportID=18545> (December 2005).

United Nations Office for the Coordination of Humanitarian Affairs (2003). 'Agreement Reached Allowing Humanitarian Access to Darfur Region of Sudan'. Press release AFR/701 IHA/795, 17 September. <www.un.org/News/Press/docs/2003/afr701.doc.htm> (December 2005).

Westing, Arthur (1994). 'Findings regarding the Ethiopian-Somali War of 1977–78'. Presented at the International Symposium on Desertification and Migrations, 9–11 February. Almeria, Spain.

Wood, William B. (2001). 'Ecomigration: Linkages between Environmental Change and Migration'. In A.R. Zolberg and P. Benda, eds., *Global Migrants, Global Refugees: Problems and Solutions*, vol. 42–61. Berghahn Books, New York.

Zweifler, Mark O., Michael A. Gold, and Robert N. Thomas (1994). 'Land Use Evolution in Hill Regions of the Domincan Republic'. *Professional Geographer* vol. 46, pp. 39–53.

Chapter 5

Negotiating Desertification

Adil Najam[1]

The United Nations Convention to Combat Desertification (UNCCD) tends to evoke a strong reaction among those who study it. On the one hand there are those who applaud it as the very embodiment of the aspirations and promise of the 1992 Earth Summit at Rio and as the first 'sustainable development treaty' (Chasek 1997). Others are less celebratory about what they describe as 'Rio's stepchild' (Sharma 1999), and dub it a 'second class convention' and 'poor relation' to other Rio treaties (Agarwal, Narain, and Sharma 1999). Yet even the sceptics acknowledge the UNCCD's promise as perhaps the first treaty that 'provides the world with a framework for global and national action that is both just and bold' (Agarwal, Narain, and Sharma 1999, 186). And even its supporters acknowledge that at times (particularly on North-South financial tensions), 'negotiations seemed to be nothing more than a debate among fluent tongues and deaf ears; the spirit of the Earth Summit was apparently an ephemeral spurt' (Kassas 1995, 177).

Given the strong — and sometimes divergent — reactions that this subject attracts, the goal of this chapter is to provide a brief negotiation history (from 1989 to 1996) of the UNCCD and to derive lessons from it for other North-South environmental negotiations. It is useful to revisit this history a decade after the entry into force of the convention, because there is the danger of losing the historical perspective on why it is structured the way it is, and also because this history can highlight lessons that could be useful to future negotiations. This chapter will focus on the 'collective South', represented in international negotiations through the Group of 77 and China (G77),[2] which is the negotiating caucus of more than 130 developing countries in the United Nations system. It is important to use such a conceptual lens to understand the UNCCD negotiations, because they were manifestly driven by the South, because they saw uncommonly intense South-South bargaining within the context of a global environmental negotiation, and because they also revealed intense North-South differences.

The following sections will first provide an abbreviated history of desertification as an international policy issue, including the UNCCD negotiations. The chapter will then analyse in greater detail the dynamics of the negotiations during the 1989–96 period. It will conclude with lessons that can be derived from this analysis.

Desertification as an International Policy Issue

The history of desertification as an international policy issue predates the history of the desertification convention (see Table 5-1). Although the Great Sahelian Drought (1968–73) was devastating large parts of Africa at that very time, desertification as an issue was not on the international environmental policy radar at the 1972 United Nations Conference on the Human Environment at Stockholm. However, given that the United Nations Environment Programme (UNEP) soon became the first — and remains the only — UN agency to be housed in Africa (or for that matter anywhere in the developing world), it was not surprising that the new organisation soon turned its attention toward the environmental havoc that had been wrought by the drought. The United Nations Conference on Desertification (UNCOD), held in Nairobi in September 1977, adopted the Plan of Action to Combat Desertification (PACD) with

Table 5-1   UNCCD Timeline

Pre-negotiation Phase				Negotiation Phase		Interim Phase		
1989	1990	1991	1992	1993	1994	1995	1996	1997
UNGA (Resolution 44/228) New York, Dec/89 Meeting of African Environmental Ministers Abidjan, Nov/91 UNCED PrepCom-IV New York, Mar/92 UNCED Rio de Janeiro, Jun/92 UNGA New York, Dec/92				Organisational Session New York, Jan/93 INCD-1 Nairobi, Jun/93 INCD-2 Geneva, Sep/93 INCD-3 New York, Jan/94 INCD-4 Geneva, Mar/94 INCD-5 Paris, Jun/94		INCD-6 New York, Jan/95 INCD-7 Nairobi, Aug/95 INCD-8 Geneva, Feb/96 INCD-9 New York, Sep/96 INCD-10 New York, Jan/97 UNGASS New York, Jun/97 INCD-10 (resumed) Geneva, Aug/97		
UNGA decision to negotiate a convention on desertification 22 December 1992				UNCCD adopted 17 June 1994		UNCCD comes into force 26 December 1996		

Notes:
INCD: Intergovernmental Negotiating Committee on Desertification
PrepCom: Preparatory Committee
UNCCD: United Nations Convention to Combat Desertification
UNCED: United Nations Convention on Environment and Development
UNGA: United Nations General Assembly
UNGASS: United Nations General Assembly Special Session

the objectives of preventing and arresting the advance of desertification, reclaiming desertified land for productive use, and sustaining and promoting, within ecological limits, the productivity of areas vulnerable to desertification in order to improve the quality of life of their inhabitants (Burns 1995).

The nonbinding PACD not only served as the precursor to the legally binding UNCCD but also laid down the foundation of the latter's substance and architecture. Although a number of attempts were made to create functional institutions to implement the spirit of the PACD — and although there was a brief second spurt of global interest in the issue when disaster struck Africa yet again in 1983–84 — chronic disinterest in the rest of the world, institutional turf battles, and a failure to raise funds for required action doomed the PACD, which has been described as an 'abject failure' (Burns 1995, 852). The lessons from its demise were clear: if anything meaningful were ever to be done about desertification, the issue would have to attain higher international profile as a truly global issue, it would have to be dealt with within a legally binding framework, and a dependable and sufficient supply of funds would have to be procured.

The African countries launched a move to actualise such an agreement in the run-up to the 1992 United Nations Conference on Environment and Development (UNCED), the Earth Summit at Rio de Janeiro. After intense deliberations at the conference about whether there should be a global treaty on desertification, the Earth Summit agreed that there should indeed be one, and the UN General Assembly (UNGA) established the Intergovernmental Negotiating Committee on Desertification (INCD) in late 1992. After a first organisational session, held in January 1993 in New York, the INCD met five times between June 1993 and June 1994 to prepare the text of the convention, officially titled the 'United Nations Convention to Combat Desertification in Those Countries Experiencing Serious Drought and/or Desertification, Particularly in Africa'.

The convention (and its four regional annexes) was adopted at the fifth session of the INCD at Paris in June 1994 and opened for signature later that year.[3] It was decided that the committee would continue to meet until the requisite number of ratifications were received and until the first meeting of the Conference of the Parties (COP). A total of six additional meetings were held in this interim period, with the last meeting held in August 1997, right before the first COP. The convention legally came into force when the 50th instrument of ratification was formally received (from Chad) on 26 December 1996. Table 5-1 depicts the UNCCD negotiation process as organised according to key negotiation phases. Post-negotiation COPs were held annually from 1997 to 2001 (at Rome, Dakar, Recife, Bonn, and Geneva). It was decided that subsequent COPs would be held biennially, interchanging with sessions of the Committee for the Review of the Implementation of the Convention (CRIC). The first meeting of CRIC was held in Rome in 2002, and the sixth meeting of the COP was held in Havana in 2003, which also served as the second CRIC meeting. The third CRIC meeting was held in Bonn in 2005 and the seventh Conference of the Parties was held in Nairobi in October 2005.

UNCCD: The Story of Some Tortured Negotiations

Even more than most other global environmental negotiations, North-South differences have remained a key fault line in UNCCD deliberations. It is equally significant that the UNCCD negotiations were different from other recent environmental negotiations because of the intensity of South-South differences. Given that desertification was an eminently Southern issue, it should not be surprising that the developing countries were much more engaged and animated in these negotiations. This resulted in more intense South-South negotiation. What is surprising, however, is that by the end of the negotiations the major South-South differences had been meaningfully resolved but the most important North-South tensions remained unresolved.[4]

UNCED: How Africa Got 'Its' Convention

Even though the main proponents of UNCED were from the North, UN resolution 44/228, which set the mandate for the conference, was a creature of Southern interests (Najam 1995). More significantly, on the insistence of African members of the G77, desertification was afforded a special place of prominence in this resolution, calling upon UNCED to 'accord high priority to drought and desertification control'. UN resolution 44/228 gave much more importance to desertification than to, for example, climate change, biodiversity, or deforestation. Despite this mandate, the UNCED preparatory process paid little attention to desertification; certainly, the first three of the four UNCED Preparatory Committee meetings (PrepComs) all but ignored it (Corell 1999; Long 2000).

Unhappy at the treatment that 'their' issue was receiving, a special meeting of 40 African environment ministers held in Abidjan in November 1991 (right before PrepCom IV) adopted the African Common Position on Environment and Development and the Abidjan Declaration. Both documents called for an international convention to combat desertification. However, the proposal got lukewarm reception from the rest of the G77 when the entire group met in Kuala Lumpur early in 1992 and failed to endorse the African call in its communiqué (Porter, Brown, and Chasek 2000, 133). Although the issue had not yet become an open dispute within the G77, the seeds of dissent had been sown.

The issue came to a head at the fourth UNCED PrepCom at New York in March 1992. Consensus on most elements of what eventually became Chapter 12 of Agenda 21 (Managing Fragile Ecosystems) was achieved relatively easily, but the African demand for a desertification convention was fiercely resisted by the North. The United States, in particular, was fearful that a desertification convention would trigger calls for additional funding under the Global Environment Facility (GEF) and doubted, along with the European Community, the validity of desertification as anything more than a cluster of local environmental issues. However, what began as a straightforward North-South tussle soon took a different twist, when the Europeans decided to dangle 'the desertification convention as a carrot to induce the South to agree to a forest convention' by openly offering to accept the desertification convention in exchange for the forest convention (Agarwal, Narain, and Sharma 1999, 169; see also Porter, Brown, and Chasek 2000, 133; Corell 1999).

The European Union's move not only drove a wedge within the G77 but also split Africa, since a number of African countries with large tracts of tropical forests, such as Cameroon and Gabon, were also opposed to a forest convention that would focus principally, if not solely, on tropical forests (Agarwal, Narain, and Sharma 1999, 229). The G77 leadership saw this as a sabotage attempt and took steps to regroup. Hectic internal negotiations were held among various G77 groups and, ultimately, the entire G77 ended up standing behind the African demand for a convention and Africa decided not to give in on deforestation. As Elisabeth Corell (1999, 75) writes,

> [the] African countries, supported by the rest of the developing world, forcefully pushed the desertification issue against resistance from [the North]. The turning point came when the United States agreed to support the proposal and the pressure became too strong for the European Community to resist.

Six months later, UNGA adopted resolution 47/188 establishing the INCD. In reflecting the sense of urgency demanded by Africa, the resolution stated that the negotiations should be finalised by June 1994.

Various reasons have been suggested for why the U.S., and later the European Community, changed its position even though they clearly had not been convinced of the need for such an instrument. These include the following: Singapore's Tommy Koh, chair of the UNCED PrepCom, had taken personal responsibility for the issue and used his office to press for an agreement; the U.S. decision was meant to deflect the negative publicity it was getting at home for its other positions at UNCED; the North gave in because it wanted to keep the African and other developing countries engaged in the Rio process; and, finally and importantly, the African group had indicated throughout PrepCom IV and UNCED that it would insist on a desertification convention before approving any other UNCED document — an African defection, could therefore, have jeopardised all other Rio agreements (Corell 1999, 75; Agarwal, Narain, and Sharma 1999, 171).

Whatever the reasons for the North's turnaround, the G77 emerged galvanised by the challenge. Ironically, the blatancy of the Northern attempt to 'break' the G77 by using the desertification-forests card spurred Southern unity. Northern intransigence turned desertification, once again, into a developing country issue rather than just an African issue. By the time the dust settled, developing countries were clearly happy at having resisted Northern 'browbeating' and forcing the North to 'backtrack' (Agarwal, Narain, and Sharma 1999, 171); the G77 was demonstrating a higher level of internal negotiation and co-ordination; and international action on desertification, although still an 'African priority', had assumed broader support.

INCD-1: The Nightmare in Nairobi

When delegates met for the first substantive session of the INCD at Nairobi in May 1993, most of them expected rather staid and straightforward information sharing. However, neither the G77 nor the INCD secretariat anticipated the reaction that awaited a chance suggestion that the INCD chairman, Bo Kjellén of Sweden, had thrown out toward the very end of the organisational session:

> The Chairman proposed that the special concern for Africa be expressed in the form of an annex or protocol to the convention and that other parts of the world could have protocols or annexes in the future. However, this caused a split in the G77 ... Latin American and Asian delegates in particular resisted this proposal, insisting that instruments for other regions should be negotiated simultaneously. They argued that the problem was worldwide and that their efforts to combat desertification deserved equal attention to those of Africa. This disagreement was so strong in the last hours of INCD-1 that the entire negotiation seemed to be in jeopardy (Corell 1999).

The issue of granting priority to Africa became the defining issue at INCD-1. It caused a greater fissure within the G77 than it otherwise might have because Colombia, which had recently assumed the G77 chair, had not shown up for the negotiation and, in its absence, Brazil was serving as the acting chair. Brazil, with significant drylands of its own, was one of the most vocal proponents of all regional annexes being negotiated simultaneously. The Africans, already incensed at what they considered to be an attempt to highjack 'their' convention, saw this as a blatant conflict of interest and refused to meet under Brazilian leadership. The G77 was essentially immobilised as a vehicle of Southern co-ordination for the first time ever in any international environmental negotiation. The *Earth Negotiation Bulletin* team reporting on the meeting ended its summary assessment of the negotiations with this uncharacteristically emotional commentary: 'the first session of the INCD ended on a discordant note, exactly one year after the Earth Summit in Rio. Whatever was embodied in the "Spirit of Rio" — compromise, consensus and the desire to avoid failure — was lost in Nairobi' (International Institute for Sustainable Development [IISD] 1993a).

Although the epicentre of the feud was very much within the G77, posturing from the North further fuelled the tensions. The U.S. and European delegates actively prodded the African group to negotiate independently, in what some saw as an attempt to scuttle Southern unity (Agarwal, Narain, and Sharma 1999, 173). Indeed, from a Southern perspective, the North continued trying to divide the South along Africa/non-Africa lines throughout the negotiations, including on issues related to financing and treaty administration (Agarwal, Narain, and Sharma 1999, 171; see also IISD 1993b).

INCD-2: Patching Up the G77

The tensions from Nairobi spilled over into INCD-2 at Geneva. Colombia, still the G77 chair, had originally intended not to attend the meeting but was forced to rush a delegate to the meeting once it became apparent that the G77 would not be able to meet under Brazil's leadership. Sensing crisis, a number of larger G77 countries — especially Malaysia — intervened to maintain group solidarity. The highlight of INCD-2 was not simply that the G77 was able to bounce back, but that intense South-South negotiation resulted in a formula that was able to break the looming impasse. The G77 members agreed that the convention and the regional instrument for Africa would be finalised by June 1994 and the other regional instruments would be negotiated in the interim period pending the entry into force of the convention. This innovative solution allowed Africa to retain its sense of priority while ensuring for the other regions that the convention would remain global in character and would enter into force with multiple regional instruments in place (IISD 1993c).

The key item on INCD-2's agenda was the single compilation text prepared by the secretariat. The secretariat had sifted through nearly 700 pages of comments to produce a synthesis document of 86 pages, and this became the basis of the negotiations. The developing countries, particularly from Africa, contributed significantly to this process. The Organization for African Unity (OAU) submitted a complete draft convention text; this was complemented by texts from Burkina Faso, Egypt, Mali, Senegal, and Tunisia (IISD 1993c). As Marybeth Long (2000) writes,

> [North-South] divergence occurred in discussions regarding the 'global' nature of desertification. Southern countries supported use of the word global because it emphasized the importance of desertification throughout the world and not only in Africa. Northern states opposed the term global. Northern delegates believed that the word global connoted global warming and suggested Northern responsibility for desertification's solution. Furthermore, industrialized countries feared that emphasis on desertification, as a global problem linked to climate change would enable anti-desertification projects to access new funding via GEF.

The discussion on financing also began at INCD-2 and it became clear that the North had no inclination to agree to any additional and systematic sources of funds. In short, the issues that came to the forefront of the agenda at INCD-2 only served to remind the Southern countries of the broader goals that had brought the G77 together in the first place and nudged them to reassert their unity. The G77 left Geneva bruised, but recovering.

INCD-3, INCD-4, and INCD-5: Money Matters

The rationality of why the developing countries were sticking together under the G77 umbrella came into sharp relief at the third, fourth, and fifth substantive sessions of the INCD held, respectively, in New York, Geneva, and Paris in January, March, and June of 1994. During these meetings, the G77 — now with an African country, Algeria, as its chair — progressively reaffirmed its role as the principal voice of the South. Although a whole host of issues was discussed and decided over these three sessions, the most compelling debates were on state obligations and financial provisions. The convention defines three types of obligations — general, affected country, and developed country obligations. Under general obligations, the U.S. and the EU succeeded in thwarting Southern attempts to link desertification directly to concerns about international trade, debt, and poverty eradication. Discussions on the obligations of affected parties were relatively straightforward except for some rhetorical sparring on the part of the U.S., which insisted that the focus not be limited to affected developing countries.

Much more entrenched debate ensued in crafting the obligations of developed countries. As one observer wrote, 'the trouble began at INCD-3, when Article 6 in the first draft of the convention was titled "Obligations of country Parties in a position to provide assistance." The outspoken Ambassador Wen Lian Ting from Malaysia immediately ... argued that this was a blatant attempt to expand the traditional donor community to include nondeveloped countries. Most of the Asian and Latin American countries concurred' (Chasek 1997, 160). However, 'for many African countries, the money was important, not its source ... [and] many Francophone African countries, including Burkina Faso, Cameroon, Mali, Mauritania and Senegal supported the Norway-UK-US proposal of expanding the source of assistance' (Agarwal, Narain, and Sharma 1999, 175). For other G77 members, the real issue nonetheless related to precedence and maintaining the 'common but differentiated responsibility' principle (IISD 1994a). The difference within the G77 at this point was between those most focussed on the specific issue of desertification and those more concerned about broader G77 goals beyond this particular one. The G77 was ultimately able to arrive at compromise language, which deleted the precedent-setting tone of the draft and retained an encouragement to all parties to provide 'voluntary' assistance (Chasek 1997).

North-South financial tensions proved more difficult to resolve. The lesson from the 1977 UNCOD was that no international effort could ever be implemented without a stable and significant source of independent funding. For developing countries, this was the ultimate test for the UNCCD. Yet every attempt they made toward this goal was shot down: seeking a new GEF funding window for desertification, seeking a new fund for desertification, and seeking the actualisation of the North's Rio commitment of 0.7 percent of gross national product (GNP) for official development assistance (ODA).

At INCD-4, the two articles on financial issues contained approximately 65 sets of brackets; positions were clearly polarised along North-South lines. Algeria was proclaiming that the regional annexes (and, by default, the UNCCD) would be 'pointless' without new and additional financial resources and without a financial mechanism that was independent, identifiable, and capable of mobilising such resources (IISD 1994b). On the other hand, 'the industrialized countries were a united veto coalition in rejecting any provision for new and additional financing' (Porter, Brown, and Chasek 2000, 134).

With the clock set by UNGA ticking fast, INCD-5 began shrouded in thick North-South tension. The talks were deadlocked with the U.S.-Europe blocking coalition unwilling to budge and the South unable to give in. The chair's contact group (eight members each from North and South) struggled with the issue, meeting around the clock for three days. With nothing else to hold on to, the group began to concoct the concept of something called a 'global mechanism' that would be charged with promoting actions leading to the mobilisation of funds (see Chapter 9). Although there was no clear sense of how it would do so, or any enthusiasm for whether it would be able to, this was the one idea the contact group had to offer when time ran out (IISD 1994c).

Technically, the deadline for concluding the negotiations had already passed at midnight on 17 June, but frantic negotiation continued on the elements of the financial compromise. It was in the early hours of 18 June that such a compromise seemed in hand:

> The regional groups quickly gathered to hear the results. Although the OECD [Organisation for Economic Co-operation and Development] group of countries said they were, reluctantly, going to accept the contact group's text, the African Group and the G77 had more difficulty. At one point, it looked as though the G77 was going to reject the compromise and the negotiations would end in failure, but at 3:45 a.m., the G77 finally emerged [from its meeting] and said that they would accept the compromise, although a number of its members had reservations. When the final plenary meeting began at 4:00 a.m., Uganda, Ethiopia, Sudan, India, and Djibouti all stated their reservations for the record (Chasek 1997, 162).

As the session finally wrapped up at 8:00 a.m. on the morning of 18 June 1994, after having negotiated and adopted the convention, the *Earth Negotiation Bulletin* reported that many delegates 'were too exhausted to celebrate' (IISD 1994c); others, of course, may not have wanted to celebrate what was a compromise-laden convention that left just about all parties seriously unsatisfied with at least some of its elements.

The Interim Phase: Going Around in Circles

The UNCCD was negotiated in record time and at breakneck speed over a period of just 13 months. Things slowed down significantly immediately after INCD-5. It was decided at Paris that the INCD would continue to meet until the requisite 50 ratifications were received and the first COP was held. Six such meetings were held between January 1995 and August 1997.

The stated purpose of continuing INCD meetings during the interim phase was to design and decide the modalities of the Global Mechanism (GM), to initiate preparations for the first meetings of the COP and the Committee on Science and Technology (CST), to decide on a location and institutional arrangements for a permanent secretariat for the convention, and to facilitate the implementation of the Resolution on Urgent Action for Africa. The issue of particularity for Africa versus the universality of the convention as an international instrument continued to raise its head periodically. Tensions around the provision of meaningful financial mechanisms were as confrontational as ever, with the G77 pushing for an independent and well-endowed Global Mechanism but the U.S. and EU insisting on limiting its scale and scope.[5]

INCD-6 was a relatively low-key event that focussed primarily on organising for the interim phase (IISD 1994d). INCD-7 was even less eventful, resulting mostly in procedural decisions (IISD 1995). At INCD-8, negotiations became earnest. Two competing visions of the GM were in place. Most Southern delegates supported a central fund with its own resources, while most Northern delegations sought something that 'would provide motivation and be an information source that would leave funds in existing bilateral and multilateral funds' (IISD 1996a). To the North, the developing countries of the G77 were again trying to expand the scope of the convention. To the South, the industrialised countries were showing that they were less serious about desertification than about issues such as biodiversity or climate change. When the delegates reassembled in September 1996 for INCD-9 in New York, they replayed the same debate, again with no resolution in sight. Prophetically, one delegate was reported as commenting that the 'Mechanism would forever haunt the Convention' (IISD 1996b).

INCD-10 also ended in deadlock and decided to pass on the most contentious issues — those related to the GM, the permanent secretariat, and the composition of the COP bureau — to the first COP. The session also decided to 'resume' its meeting in August, two months prior to the first COP, to try to resolve other pending issues (IISD 1997a). Desperate, the G77 tried one more card. In June 1997, the so-called 'Rio+Five' was convened in New York as a UN General Assembly Special Session (UNGASS) to review the progress on decisions made during the Earth Summit. The South hoped that this high-level meeting would 'remind the Northern countries of their Rio commitments' (Agarwal, Narain, and Sharma 1999, 175). However, attempts by the G77 to try to get the North to commit to fund the UNCCD Global Mechanism went 'unrewarded' (Corell 1998, 223). The last, resumed, meeting of the INCD only served to show that INCD-5 at Paris would not have ended differently even if there had been more time. The blocking coalition of the U.S. and the EU adamantly vetoed any proposal except its own. Not surprisingly, the G77's search for a well-defined and well-endowed GM proved futile (IISD 1997b).

Arguably, the North-South polarisation at the end of the interim phase was more acute than at the end of INCD-5. The failure to raise the profile of either the convention as a whole or of the GM had raised serious doubts about just how seriously the world takes the UNCCD. This sense of being treated like a second-class convention has become a continuing theme since then (Agarwal, Narain, and Sharma 1999, 175).

Lessons and Conclusions

There are three types of conclusions that can be derived from this retelling of the negotiation history that are potentially relevant to future negotiations on desertification or other global environmental issues: about managing South-South frictions, about managing North-South frictions, and about the particularities of the UNCCD negotiations.

Managing South-South Frictions

On at least two important occasions (first at PrepCom IV/UNCED and then at INCD-1/INCD-2), the Southern collective was faced with an important threat to its group solidarity. In each case, the G77 was able to act quickly to avert the threat. At least two important lessons for the South can be derived from this.

First, although the G77 may have been able to rise above the lowest common denominator in these instances, it did so under conditions of threat. The good news is that, when pushed, the South is capable of meaningful internal negotiations that do rise above the lowest common denominator. The bad news is that this case does not suggest a latent disposition within the G77 for investing in more structured internal negotiations. The lesson to highlight is about the need for systematic and routine South-South negotiations prior to the larger North-South dialogue, particularly on issues such as desertification, where interests within the South are highly differentiated.

The second lesson is that arriving at internal agreements within the group requires hard work and constant vigilance. Given the chronic resource constraints faced by the G77 (and the sheer size of the collective), it is not surprising that the group is unable to invest in pre-negotiation preparation. For the G77 the lesson from the UNCCD is that the challenge is quite clearly the challenge of internal G77 organisation and group management.

Managing North-South Frictions

In relation to North-South tensions, the debates were mostly familiar, largely on issues related to financing and regime governance, and ultimately less conclusive. Unlike most other global environmental negotiations, this was a case where it was the South calling for a treaty and the North resisting. This odd and unusual situation had at least two important impacts on the general nature of North-South relations. First, since the North has never fully accepted the rationale for a global convention, it has continued to fight for restricting the scale and scope of the UNCCD. The North has been far more successful in this blocking coalition role because the balance (or imbalance) of power in terms of resources and expertise is so heavily tilted in its favour. Second, in its unfamiliar role as regime leader, the South was no longer able to retreat to the comfortable position of simply adopting a reactive strategy. The entire debate on financing was indicative of the G77's inability to convince the U.S. and the European Community to adopt the South's preferred options. The GM is much more a vindication of the North's position that no new institutions or funding are needed, than a response to the G77's deeply held belief that without a stable and significant source of financing the fate of this convention will be no different from that of the PACD before it.

Two noteworthy lessons do emerge from this experience. The most important lesson, once again, is that of preparation, including systematic South-South negotiations prior to any North-South negotiation. Better pre-negotiation preparation is the one strategic step that would not only strengthen the G77's internal unity but would also prepare it for more proactive negotiation with the North.

UNCCD negotiations demonstrated that the G77 is, in fact, a fairly stable collective. Situations that served to place a wedge within the South, failed to break Southern solidarity even when the payoffs were relatively high. From the North's perspective, a South that is negotiating in coalition mode may well be a tougher negotiator because it is better prepared and better organised; it is also likely to be a better partner in the search for innovative options for tricky problems, because of the same reasons.

UNCCD: A Convention Full of Surprises

Finally, let it not be forgotten that the very fact that a convention on desertification exists is somewhat of a surprise and something worth celebrating. In fact, the entire story of the UNCCD is full of surprises. Let us highlight just a few of the big ones:

- Given the contents of UN Resolution 44/228, it was a surprise that the UNCED preparatory process did not include discussions on a desertification convention.
- Given the then prevailing attitude of the North as well as the African South, it was a surprise that UNCED did eventually call for a convention on desertification.
- Given the high level of unity and co-ordination that the G77 had shown at the 1992 UNGA discussions on desertification and subsequent organisational meetings, it was indeed a surprise — in fact, a shock — that the G77 broke up along regional lines at INCD-1 and was totally paralysed well into INCD-2.
- Given the intensity of the South-South disputes at INCD-1 and INCD-2, it was a surprise that the G77 was able to emerge again as a united, fairly well co-ordinated, and even effective caucus of all developing countries by INCD-3 and beyond.
- Given the entrenched North-South differences on issues of finance up to the very closing moments of INCD-5, it was a major surprise that a convention was actually adopted in June 1994, after only 13 months of active negotiation.
- Given all of the above, it is a surprise that the UNCCD has not just survived but, as some of the later chapters in this volume discuss, has even thrived in certain dimensions.
- Not surprisingly, however, unresolved tensions, including on the issue of financing, continued to affect the implementation of the UNCCD and discussions among parties in the following COPs (see chapters 9 and 13).

For the developing countries, the desertification negotiations were a roller-coaster ride that they had not expected, for which they were not prepared, but which they did survive. The UNCCD negotiations demonstrated the G77's considerable survival skills and its ability to respond to internal crises. In short, UNCCD negotiations seem to have demonstrated that the South, as represented by the G77, is far more adept at maintaining its internal unity than it is capable of changing the terms of the debate in its discourse with the North.

Notes

1   This chapter is based and builds upon Najam (2004).
2   Although it is normally referred to as 'The Group of 77' of G77, the official designation of this developing country caucus is 'Group of 77 plus China' (G77+China). For ease of readership, this chapter will use G77 to refer to this collective.
3   The four regional annexes are for Africa, Asia, Latin America and the Caribbean, and Northern Mediterranean. A fifth annex for Central and Eastern Europe was added at the fourth Conference of the Parties (COP).
4   The factual description of events in the next three sections relies heavily on the narrative of the UNCCD negotiation process available in Burns (1995), Danish (1995), Chasek (1997), Corell (1999), Agarwal (1999), and Long (2000). In addition, reports from the Earth Negotiation Bulletin have also been used <www.iisd.ca/vol04> (December 2005).
5   This tension is reflected in the mandate of the GM as well as in COP decisions that have often failed to provide clear guidance on its role (see Chapter 9).

References

Agarwal, Anil, Sunita Narain, and Anju Sharma, eds. (1999). *Green Politics: Global Environmental Negotiations I*. Centre for Science and Environment, New Delhi.
Burns, W.C. (1995). 'The International Convention to Combat Desertification: Drawing a Line in the Sand?' *Michigan Journal of International Law* vol. 16, no. 3, pp. 831–882.
Chasek, Pamela S. (1997). 'The Convention to Combat Desertification: Lessons Learned for Sustainable Development'. *Journal of Environment and Development* vol. 6, no. 2, pp. 147–169.
Corell, Elisabeth (1998). 'North-South Financial Tensions: Desertification after UNGASS'. *Environmental Politics* vol. 7, no. 1, pp. 222–226.
Corell, Elisabeth (1999). 'The Negotiable Desert: Expert Knowledge in the Negotiations of the Convention to Combat Desertification'. Ph.D. thesis. Linköping University, Linköping, Sweden.
Danish, K.W. (1995). 'International Environmental Law and the "Bottom-Up" Approach: A Review of the Desertification Convention'. *Indiana Journal of Global Legal Studies* vol. 3, no. 1, pp. 133–176.
International Institute for Sustainable Development (1993a). 'Summary of the First Session of the INCD: 24 May–3 June 1993'. *Earth Negotiations Bulletin* vol. 4, no. 11. <www.iisd.ca/vol04/0411000e.html> (December 2005).
International Institute for Sustainable Development (1993b). 'A Brief History of the INCD'. *Earth Negotiations Bulletin* vol. 4, no. 45. <www.iisd.ca/vol04/0445000e.html> (December 2005).

International Institute for Sustainable Development (1993c). 'Summary of the Second Session of the INC for the International Convention to Combat Desertification: 13–24 September 1993'. *Earth Negotiations Bulletin* vol. 4, no. 22. <www.iisd.ca/vol04/0422000e.html> (December 2005).

International Institute for Sustainable Development (1994a). 'Summary of the Third Session of the INC for the Elaboration of an International Convention to Combat Desertification: 17–28 January 1994'. *Earth Negotiations Bulletin* vol. 4, no. 34. <www.iisd.ca/vol04/0434000e.html> (December 2005).

International Institute for Sustainable Development (1994b). 'Summary of the Fourth Session of the INC for the Elaboration of an International Convention to Combat Desertification: 21–31 March 1994'. *Earth Negotiations Bulletin* vol. 4, no. 44. <www.iisd.ca/vol04/0444000e.html> (December 2005).

International Institute for Sustainable Development (1994c). 'Summary of the Fifth Session of the INC for the Elaboration of an International Convention to Combat Desertification: 6–17 June 1994'. *Earth Negotiations Bulletin* vol. 4, no. 55. <www.iisd.ca/vol04/0455000e.html> (December 2005).

International Institute for Sustainable Development (1994d). 'Summary of the Sixth Session of the INC for the Elaboration of an International Convention to Combat Desertification: 9–18 January 1995'. *Earth Negotiations Bulletin* vol. 4, no. 65. <www.iisd.ca/vol04/0465000e.html> (December 2005).

International Institute for Sustainable Development (1995). 'Summary of the Seventh Session of the INC for the Convention to Combat Desertification: 7–17 August 1995'. *Earth Negotiations Bulletin* vol. 4, no. 75. <www.iisd.ca/vol04/0475000e.html> (December 2005).

International Institute for Sustainable Development (1996a). 'Summary of the Eighth Session of the INC for the Convention to Combat Desertification: 5–15 February'. *Earth Negotiations Bulletin* vol. 4, no. 86. <www.iisd.ca/vol04/0486000e.html> (December 2005).

International Institute for Sustainable Development (1996b). 'Summary of the Ninth Session of the INC for the Convention to Combat Desertification: 3–13 September 1996'. *Earth Negotiations Bulletin* vol. 4, no. 95. <www.iisd.ca/vol04/0495000e.html> (December 2005).

International Institute for Sustainable Development (1997a). 'Tenth Session of the INC for the Convention to Combat Desertification: 6–17 January 1997'. *Earth Negotiations Bulletin* vol. 4, no. 105. <www.iisd.ca/vol04/04105000e.html> (December 2005).

International Institute for Sustainable Development (1997b). 'Summary of the Resumed Tenth Session of the INC for the Convention to Combat Desertification: 18–22 August 1997'. *Earth Negotiations Bulletin* vol. 4, no. 106. <www.iisd.ca/vol04/enb04106e.html> (December 2005).

Kassas, Mohammed (1995). 'Negotiations for the International Convention on Desertification'. *International Environmental Affairs* vol. 7, no. 2, pp. 176–186.

Long, Marybeth (2000). 'Grains of Truth: Science and the Evolution of International Desertification Policymaking'. Ph.D. thesis. Massachussetts Institute of Technology, Cambridge MA.

Najam, Adil (1995). 'An Environmental Negotiation Strategy for the South'. *International Environmental Affairs* vol. 7, no. 3, pp. 249–287.

Najam, Adil (2004). 'Dynamics of the Southern Collective: Developing Countries in desertification Negotiations'. *Global Environmental Politics* vol. 4, no. 3, pp. 128–154.

Porter, Gareth, Janet Welsh Brown, and Pamela S. Chasek, eds. (2000). *Global Environmental Politics*. Westview Press, Boulder CO.

Sharma, Anju (1999). 'Rio's Stepchild'. *Down to Earth* vol. 7, no. 17, pp. 24–25.

Chapter 6

The United Nations and the Fight against Desertification: What Role for the UNCCD Secretariat?

Steffen Bauer[1]

The dramatic imagery of wandering sand dunes that expand to irreversibly devour ever more fertile land represents a gross distortion of the actual phenomenon of desertification. This notwithstanding, it provides for a valid illustration of the transboundary nature of ecological deterioration. Indeed, in our world of 'global transformations', environmental problems are arguably the most blatantly ignorant of national jurisdictions and state borders.[2] It is hardly surprising then, that the challenges humankind faces in the field of environmental policy have been driving the research of complex interdependencies in world politics and transnational policy making. Hence, the analysis of international environmental co-operation has advanced our understanding of the institutionalisation of international politics tremendously (see Zürn 1998; Mitchell 2002). This holds in particular for the prolific branch of scholars that has dedicated itself to the analysis, first, of the emergence and, then, the effectiveness of international institutions. Many of these scholars derive their theoretical understanding of international institutions from the empirical analysis of numerous case studies on environmental regimes (see Haas, Keohane, and Levy 1993; Young, Levy, and Osherenko 1999; Miles 2002).

Virtually all of these regimes, generally built around a multilateral environmental agreement (MEA), include some kind of a secretariat that is supposed to co-ordinate the efforts being made to address the policy challenge in question. Such environmental treaty secretariats generally take shape as small intergovernmental bureaucracies that are run by international civil servants under the formal control of predefined multilateral governmental mechanisms. Often perceived as a minor feature of a wider institutional setting, few scholars have bothered to take a closer look as to their specific functions and role within that setting.[3] This is odd, not least in a view of the vibrant debates about how global environmental governance is in need of reform.[4] Whether perceived as an environmental convention or a sustainable development convention, the complex governance structure of the United Nations Convention to Combat Desertification (UNCCD) is necessarily included in these debates. In an attempt to rectify the neglect of treaty secretariats in the analysis of global environmental governance, this chapter will explore what role the convention's permanent secretariat is playing in the institutionalisation and implementation of the treaty it was created to serve.

'Stage Hands or Actors?' Some Thoughts on Environmental Treaty Secretariats[5]

Following a definition by Steinar Andresen and Jon Birger Skjaerseth (1999, 2), a typical convention secretariat can be understood as a specific type of international organisation 'established by the relevant parties to assist them in fulfilling the goals of the treaty'. However, such organisations are not passive tools in the hands of the relevant parties — normally state governments — but actively interact with their environment as the bureaucratic personalities that they tend to become.[6]

Such actor quality in an international organisation is generally concentrated in its top executives, such as Secretary General Kofi Annan, who lends his face to the bureaucratic personality that is universally recognised as the United Nations secretariat. In other words, international civil servants are the people who act on behalf of the international organisation and thus interact with the organisational environment. It is fundamental in this respect to recognise that international civil servants are commonly found to pursue the objectives of their organisation rather than being a puppet of the national government of their origin.[7] This is not to deny that the freedom to act of an international organisation is considerably circumscribed by national governments. However, just as the bureaucrats of a national ministry act *vis-à-vis*, for instance, parliaments at the domestic level, international treaty secretariats can be trusted to use, at least potentially, the leeway they have to interact within their respective policy domains.

In the realm of international environmental politics, Rosemary Sandford (1994, 19) has traced a three-step evolution of environmental treaty secretariats, beginning with the administrative embodiments of the early treaties spawned by the UN Conference on the Human Environment in Stockholm in 1972. She distinguishes those secretariats that were typically helped into being by the United Nations Environment Programme (UNEP) during the 1980s as a second generation of treaty secretariats and, then, the distinct bureaucracies that emerged from the so-called 'Rio conventions' in the aftermath of the 1992 UN Conference on Environment and Development (UNCED).

Moreover, when analysing secretariat performance, Sandford (1992; 1994) distinguished between some secretariats as being passive and others as being active. While this distinction may be useful analytically, there is in fact no such thing as a passive secretariat. Clearly, some secretariats are more active than others, but even those secretariats that would describe themselves as passive are not passive to the extent where they would totally abstain from political interaction within their respective regime (Bauer forthcoming-a, forthcoming-b; Busch forthcoming). Secretariat staff often have to rely on activities behind the scenes to feed their own views into a regime process, but this does not preclude them from shaping regime outcomes. In short, secretariats may be cautious in their actions, but they are hardly passive. Indeed, the need for cautiousness and sophistication reflects the 'delicate balance' between the activism that is required for secretariats to influence regime outcomes and the risk to be chastised as an impertinent agent messing with the manifest interests of principals (Andresen and Skjaerseth 1999, 7).

The UN and the Fight against Desertification 75

As discussed below, the UNCCD secretariat has turned out to be a particularly proactive player in the arena of global environmental governance. First, it is necessary to give a brief summary of how it is historically and institutionally situated within the international regime to curb land degradation and to outline the mandate and functions it has been formally assigned.

The Institutional History and Environment of the UNCCD Secretariat

The UNCCD is still comparatively new, having been formally adopted in Paris on 17 June 1994 and entering into force by the end of 1996. However, the specific discourse and subsequent agenda setting about how to frame the environmental problem addressed by the convention began in the early 1970s. It is important to recall the quintessential prerequisites for the global convention that eventually emerged in the wake of the Rio conference. In particular, the assumed causes of the problem identified under the subject heading 'desertification' and their interlinkages to socioeconomic development, as well as the prospective solutions, are salient issues regarding the political relations between developing and developed countries.[8] Not least, with a view to the role of the UNCCD secretariat, there is considerable continuity regarding the people who facilitated the pre-convention negotiations and are now central in sustaining the convention process. Notably, before becoming the head of the UNCCD secretariat, the incumbent executive secretary, Hama Arba Diallom, served as the secretary to the intergovernmental negotiations that eventually brought about the convention.

With a membership of now 191 signatories, the convention represents the closest approximation to universal membership any multilateral environmental agreement has achieved to date (UNCCD 2005). The complex institutional setting of the desertification regime, which involves numerous UN agencies such as the Food and Agriculture Organization (FAO) and UNEP as well as regional institutions and a range of banks and funding agencies, has been described elsewhere in more detail (Chasek and Corell 2002). However, as no study to date has focussed systematically on the UNCCD secretariat, the following will elaborate on how the provisions of the convention relate to the specific role of the convention secretariat.

The UNCCD as a Rio Convention

Following Sandford's (1994) typology, the UNCCD belongs to the third generation of international environmental treaties that has emerged from the 1992 UNCED in Rio de Janeiro. What is more, the status of a UN convention has been attributed to the UNCCD — like the UN Framework Convention on Climate Change (UNFCCC). Hence, the UNCCD secretariat enjoys a somewhat elevated status compared to most other environmental treaty secretariats, including the secretariat of the Convention on Biological Diversity (CBD), which also serves a Rio Convention but is formally a part of the UNEP (see Siebenhüner forthcoming). For instance, the UNCCD executive secretary is ex-officio assistant secretary general to the United Nations. Another notable implication of the UNCCD's post-Rio emergence is that it has been framed as a sustainable development treaty rather than as an environmental treaty in the narrow sense, a point that is emphasised time and again by UNCCD officials and parties from the developing world. Indeed, poverty eradication — a policy objective central to developing countries — is prominently anchored in the convention as an essential precondition for the 'combat' against desertification to be effective (UNCCD 1994; see also UNCCD 1995).

At the same time, the UNCCD does not provide for a distinct financial mechanism.[9] Of course there is the Global Mechanism (GM), which is meant to facilitate money flows between the donor countries and affected country parties. However, the absence of a substantive fund to combat desertification is an issue that has generated considerable controversy among parties and a great deal of frustration in the secretariat. While this has been somewhat ameliorated by the incorporation of sustainable land management into the portfolio of the Global Environment Facility (GEF), many still feel that the UNCCD is the 'poor sister' of the three Rio conventions.[10]

As a notable distinction from old established international environmental agreements the UNCCD places a strong and explicit emphasis on the involvement of nongovernmental stakeholders, particularly community-based organisations. This, too, can be traced to the pre-convention negotiation process, in which nongovernmental organisations (NGOs) had been involved to an unprecedented extent (see Chapter 7; see also Corell and Betsill 2001). The maintenance of links between the local and global levels of the convention process has thus become one of the additional tasks of the secretariat.

Form and Functions of the UNCCD Secretariat

Following Decision 5 of the first Conference of Parties in Rome in 1997, the interim secretariat of the 'United Nations Convention to Combat Desertification in Those Countries Experiencing Serious Drought and/or Desertification, Particularly in Africa' was transformed into a permanent one (UNCCD 1997, 31). In January 1999 it moved from its preliminary Geneva site to the United Nations' premises in the German city of Bonn, next door to the so-called 'Climate Secretariat' that serves the UNFCCC.

It is the secretariat's official mandate to provide the necessary services to the parties as determined in article 23.2 of the convention (UNCCD 1994). Therein it has been assigned:

a) to make arrangements for sessions of the Conference of the Parties and its subsidiary bodies established under the Convention and to provide them with services as required;
b) to compile and transmit reports submitted to it;

c) to facilitate assistance to affected developing country Parties, on request, particularly those in Africa, in the compilation and communication of information required under the Convention;

d) to coordinate its activities with the secretariats of other relevant international bodies and conventions;

e) to enter, under the guidance of the Conference of the Parties, into such administrative and contractual arrangements as may be required for the effective discharge of its functions;

f) to prepare reports on the execution of its functions under this Convention and present them to the Conference of the Parties; and

g) to perform such other secretariat functions as may be determined by the Conference of the Parties.

From this follows that like all other international treaty secretariats, the UNCCD secretariat was set up first and foremost to administer the continuous negotiation process of the parties and to help them implement the agreements they have committed to by ratifying the convention. As such the secretariat is at the service of the Conference of Parties (COP) as the convention's principal governing body, the Committee on Science and Technology (CST), a multidisciplinary expert body open to government representatives of all parties, and the Committee for the Review of the Implementation of the Convention (CRIC). Moreover, the secretariat maintains a roster of independent experts on whom the convention parties may draw as needed and has the responsibility to assist developing country parties in meeting the requirements of the convention, notably by facilitating the compilation and communication of information that needs to be reported to the COP.[11]

In 2003 the UNCCD secretariat was run by a staff of 67 directed by the executive secretary of the convention (Schram Stokke and Thommessen 2003). Since the post was established in 1999, it has been occupied by Hama Arba Diallo of Burkina Faso, who had hitherto served as the head of the secretariat of the Intergovernmental Negotiating Committee on Desertification (INCD) from 1993 to 1994 and the Geneva-based interim secretariat that served the first COPs (from 1994 to 1998).[12]

With regard to budgetary requirements, the first COP determined in its Decision 3 that the secretariat must 'enjoy the administrative and financial autonomy necessary to ensure efficient servicing of the Convention and of its implementation' (UNCCD 1997, para. 4). The regular budget for the secretariat to administer the convention amounts to US$17 million for the 2004–05 biennium (United Nations Convention to Combat Desertification 2003a, 8).[13] Additionally, the Special Trust Fund for Participation, the Trust Fund for Supplementary Activities, and the Trust Fund for the Supplementary Contribution to the Convention Activities by the Host Government (the so-called 'Bonn Fund') have been created mainly to cover for travel costs of developing country delegates and affected-country NGOs.[14]

The organisational structure of the UNCDD secretariat reflects very well the importance attributed to the convention's regional annexes insofar as it provides for distinct administrative divisions to serve affected country regions — a notable difference from other environmental treaty secretariats. Namely, the secretariat includes distinct Action Facilitation Co-ordinators for Africa, Asia, Latin America and the Caribbean, and the Northern Mediterranean. It is their specific task to maintain close links to the national focal points of the parties and to the Regional Co-ordination Units (RCUs), which they themselves have helped to set up.[15] These regional co-ordinators are supervised by the office of the Principal Co-ordinator for the Facilitation of Implementation of the Convention.

Moreover, there are three functional units: one is at the service of the COPs and its subsidiary bodies, one in charge of external relations and public information, and one for general matters pertaining to administration and finance. Not least, of course, is the office of the executive secretary and its deputy at the head of the secretariat to guide and supervise the work of the secretariat as well as to represent it externally. Finally, the secretariat dispatches so-called liaison officers to some of the major UN locations, notably Rome — the seat of the FAO and the International Fund for Agricultural Development (IFAD) — and UN headquarters in New York. An additional liaison officer in Brussels acknowledges the importance of the European Union as one of the main supporters of the UNCCD process.

Out of Africa? Activities and Impact of the UNCCD Secretariat

It may be seen as one of the more peculiar ironies of the UN system that the secretariat of the only global convention that explicitly deals with drylands and deserts is situated in the rain-prone Rhineland region of Germany. This notwithstanding, the UNCCD secretariat takes great care to advance the cause of the convention and its manifest emphasis on the prioritisation of Africa.[16] The following assessment of the various activities of the UNCCD secretariat is structured along four broad conceptual themes that would be replicable for most other international treaty secretariats: the shaping of the global discourse on desertification, the advancement of the institutionalisation of the convention process, the facilitation of capacity building at national and regional levels, and, finally, the political conduct of the secretariat within the overall desertification regime.[17]

Desertification Discourse

The role of UNCCD officials in the global discourse on desertification is an interesting object of study. It started long before the secretariat was formally established, but, again, there is considerable continuity as the negotiation process evolved into the institutionalisation and implementation of the convention. The framing of desertification — as opposed to land degradation — bears strong implications for the interpretation of the convention and how its objectives are to be translated into on-the-ground policy implementation. In particular, the use of the terminology affects how the non-expert stakeholder perceives the problem being addressed (Corell 1999, 53). It is precisely because of this perception that the secretariat actively promotes the desertification label, sanctioned not least by the title of the convention. Its motivation is as simple as it is straightforward: 'Desertification has a political appeal that land degradation does not have' (Diallo, as cited in Corell 1999, 65).

That this distinction is at odds with the broad scientific consensus that desertification is itself a somewhat misleading term for the environmental phenomenon at stake — dryland degradation — is deliberately accepted, at least implicitly.[18] In contrast, most bilateral donor organisations and intergovernmental implementing agencies such as the FAO, the United Nations Development Programme (UNDP), or the World Bank show increasingly consistent avoidance of the desertification label, but speak of land degradation instead. Even the parlance of the UNEP, which once pushed the issue of desertification onto the international agenda, has adjusted to the terminology of land and soil degradation (see UNEP 2004). The UNCCD secretariat, however, is effectively conserving the desertification trademark.

This protection is predominantly achieved by general awareness raising and public outreach activities. Efforts include the secretariat's 'Down to Earth' newsletter and other flagship publications, annual celebrations of the World Day to Combat Desertification, or extraordinary schemes such as the International Year of Deserts and Desertification, which has been called for 2006.[19] Using its discursive capacity, the secretariat has also played a lead role in shifting the global understanding of desertification from being a regional problem into a global commons problem. This discursive transformation is a striking example for the 'power of discourse' and bears tangible material implications.

Only because of the incremental reframing of desertification as a global problem, projects that pertain to the implementation of the UNCCD have become eligible for funding through the resourceful GEF. The establishment of its Operational Programme 15 (Sustainable Land Management) reflects nothing less than a concession of the donor community to the developing world (see Chapter 9). Affected country parties, notably from Africa, have pushed for the GEF to address desertification ever since it was established in 1994. While it is not feasible to measure the specific impact of the UNCCD secretariat in bringing about this concession, it has played its part to keep the issue on the agenda of the GEF council and backed developing countries' efforts in this respect. Indeed, the secretariat is happy to accept credit for having contributed to this end and UNCCD officers have expressed the view that conceding accessibility to the GEF has been an overdue step to make up for lack of a genuine financing mechanism of the UNCCD (see Bauer forthcoming-a).

Progressive Institutionalisation

In line with its mandate and the general objectives of the convention, the secretariat's efforts to further and solidify the institutionalisation of the convention process are characterised by a strong regional focus. Indeed, the establishment and maintenance of strong links with 'affected regions' — co-ordinated by the secretariat's respective Regional Action Facilitators for Africa, Asia, and Latin America — have helped the UNCCD secretariat to advance the institutionalisation of the convention. Pushing ahead, the secretariat seeks to expand and strengthen the RCUs that have been developed to improve intra-regional co-operation with a view to the implementation of regional action plans. While the secretariat has been expectedly successful in mustering support for this endeavour among affected country regions, donor country parties are wary of institutional duplication and question the necessity and added value of RCUs. Consideration of whether and how the role of existing units should be expanded and whether additional ones are desirable will likely be prominent issues in the near future (International Institute for Sustainable Development [IISD] 2003).

The creation of the CRIC is even better suited to illustrate the active role of the secretariat in furthering the institutionalisation of the convention. As with the concept of the RCUs, the idea to set up a distinct subsidiary body that would assist parties to review progress in implementing the convention was first suggested from within the UNCCD secretariat. It thus triggered the creation of an additional institution that was initially perceived to be at odds with the interests of major donor countries and the EU but is now — despite continuous criticism of a number of details — widely accepted as a potentially useful complementation to the overall regime.

After the first CRIC had convened in Rome in November 2002, even its critics emphasised the positive and co-operative spirit of the meeting, although a certain measure of ambiguity as to its mandate prevailed (IISD 2002). Since then, two more meetings of the new subsidiary body have been held with a mixed record. CRIC-2, which was held back to back with COP-6 in Havana, unsurprisingly suffered from the highly politicised atmosphere of the overall meeting. CRIC-3, held in Bonn in May 2005, saw little substantive progress with regard to the implementation of the convention, but was nonetheless felt to be a helpful exercise to reinvigorate momentum after the dubious outcome of the Havana meeting.

While COP-7 has been mandated to decide whether the CRIC should be maintained as an additional subsidiary body, its work so far has, on balance, advanced the institutionalisation of the UNCCD process. At the very least it can be judged to have helped governments to move from institutionalising the convention process to implementing the convention's objectives.[20]

Capacity Development

Convention secretariats are not designed to act as capacity builders in the sense that implementing agencies such as the World Bank or the UNDP are. The UNCCD secretariat is no exception. It is neither a funding agency nor does it have the infrastructure or resources to provide for on-the-ground activities. However, the UNCCD secretariat is expected to play an enabling role for capacity development at local and national levels — at least indirectly. Basic information materials on the convention and its sociopolitical implications are provided in all five UN languages and in recent years have been expanded to include a teachers' kit specifically developed to target elementary schools. Likewise the secretariat provides comprehensive background documentation to the UNCCD process, thus fulfilling a major service as requested by the parties. Notably, all COP decisions as well as documents pertaining to the CRIC and CST meetings can be downloaded from the secretariat's multilingual homepage.[21]

Beyond these rather basic functions, there is an additional role for the secretariat to contribute to capacity development. This is inherent in its mandate to facilitate the development of National Action Programmes (NAPs) in affected country parties. In line with the convention

> National institutions and legal frameworks are to be built up and new ones created when needed. National training and research capacities and strategic planning and management are also to be strengthened and members of rural organisations are to be trained in participatory approaches. Strengthening the capacity of local people to draw up programmes to manage their resources is perhaps the single most important step in constructing a healthier relationship between people, power and the environment (UNCCD 1995, 26).

To this end, the secretariat is concerned with the empowerment of national focal points in affected country parties, which are often ill prepared to maintain eye-to-eye relationships with partners at the international level.[22] One important activity of the secretariat in this regard is to link donors with suitable addressees in affected countries. For instance, the UNCCD secretariat, in collaboration with national focal points, helps bilateral agencies and private partners identify the very people who are eventually invited to participate in specific training programmes and so on. In a more direct manner, the officers of the secretariat's regional action divisions themselves assist national focal points in meeting their reporting requirements to the COPs and the CRIC by means of providing tailor-made help guides and individual advice upon immediate requests from the countries.

Occasionally, a more direct involvement of secretariat staff 'on the ground' does occur, too, although only under extra-budgetary schemes and in co-operation with other agencies. Commonly these would be UN agencies such as the UNDP or pertinent international NGOs, such as the World Agroforestry Centre. These joint projects are typically intended to sensitise national focal points for funding opportunities, such as the innovative GEF's Operation Programme 15, or to diffuse best practice experiences through sub-regional workshops. Such activities, however, are the exception to the rule, as the secretariat has neither the human resources nor the mandate to deploy missions for capacity building at the local level.

Political Conduct

Once international secretariats are analytically treated as actors within a respective governance structure, it becomes particularly interesting to analyse how they behave politically. Other actors' judgements on their political conduct will necessarily have strong implications on whether the actions of international secretariats are perceived as desirable and legitimate. Again, the UNCCD secretariat provides an illuminating example.

A striking point in case is the secretariat's deliberate endeavour to raise the political profile of the convention process. For instance, when the secretariat arranged for a so-called High-Level Segment to be included at the sixth COP in Havana, this was meant to increase public attention and to generate additional momentum for successful implementation of the convention around the world. Instead, the special segment triggered considerable diplomatic turmoil. The nine heads of state and government that constituted the High-Level Segment included Fidel Castro of host country Cuba, Hugo Chávez Frías of Venezuela, and Robert Mugabe of Zimbabwe but no representative of the developed world. Unsurprisingly, the resulting 'Havana Declaration of Heads of States and Governments' was hardly representative of the UNCCD's overall constituency and perceived as an affront by developed country parties (UNCCD 2003b, 13–16).

In the wake of the Havana meeting, the secretariat drew heavy criticism for the High-Level Segment, even from affected country parties who are naturally anxious not to offend donor countries. The UNCCD secretariat, however, has been unwilling to accept the blame for the failure of developed country parties to be represented adequately at the COPs. While this is a legitimate point to make, the experienced executives of the secretariat might have known better than to bring on the High-Level Segment in the first place. With hindsight, at least, it appears an unnecessary political risk. It not only failed to pay off but it also diminished an already modest eagerness of developed country parties to prioritise the UNCCD process.[23]

Similarly critical to the reputation of the UNCCD secretariat have been recurrent allegations of a lack of transparency in its operations. Issues that were raised at the Havana COP related to allegations of formal irregularities regarding the election of CRIC officials as well as to the controversial financial support granted by the secretariat to selected NGOs. While executive officers of the secretariat have offered plausible explanations for the emergence of these controversies and reject the allegation of lack of transparency, some damage has been done. Consequently, the reputation of the secretariat has suffered and found its manifest expressions. Among other things, the secretariat has been subject to a formal *démarche* of the EU, and Canada declared to resort to bilateral efforts in addressing land degradation unless the ways in which the convention is being administered would be rectified. Moreover, the COP decided for the secretariat to undergo a comprehensive review by the UN's Joint Inspection Unit in time for its next meeting (IISD 2003).

Conclusion

This chapter has shone a spotlight on the specific contribution made by the secretariat of the UNCCD as a distinct player in the international regime to halt the degradation of drylands around the world. This analysis does not to imply that the success of the convention relies on the capabilities and fortunes of international technocrats. Even in the context of global environmental governance, thus assuming a multiplicity of actors in world politics, it is intergovernmental politics — the deliberate interaction of nation-states — that predominantly determines the outcomes of international treaties. It is for this reason precisely that most of the international relations literature attributes little attention or significance to the activities of international secretariats.

However, convention secretariats can be more active players in global environmental governance than governments and most scholars of international relations or international law would have it. Not only do such secretariats fulfil basic technocratic functions pertaining to the administration of intergovernmental negotiation and policy making, but they also use their roles to become influential stakeholders in their own right. Whether this is good or bad, or effective and legitimate with regard to the objectives of a multilateral environmental agreement that has been negotiated and agreed upon by nation-states, is an altogether different question.

As illustrated above, the UNCCD secretariat has had (and continues to have) its effects on the evolution of the desertification regime evolves and on the implementation of the convention to combat desertification. It has influenced the global discourse effectively on the very issue of desertification; it has been influential in advancing the institutionalisation of the convention as well as, crucially, how it is being institutionalised; and — although to a minor extent — it actively contributes to the building of convention-related capacities in affected regions. Moreover, it has been shown to exert influence deliberately at the political level of the convention process. Indeed, initiatives such as the staging of the High-Level Segment at a COP are highly political. While the latter has proven to be a boomerang that has arguably damaged the secretariat's reputation as a neutral administrator of the UNCCD, it strongly substantiates the assumption of international secretariats being actors of international governance.

In sum, the UNCCD secretariat emerged as a small but central player in the complex web of informal institutions, formal organisations, and parties to the convention that has come to represent the global regime to combat desertification around the world. The case of the UNCCD thus reveals that treaty secretariats are a relevant factor with regard to how MEAs are put into practice in international politics.

At the same time, one must keep in mind that treaty secretariats are but one of many factors in the system of global environmental governance. They can hardly be held directly responsible for either the success or the ineffectiveness of an environmental regime. Notwithstanding, they can potentially trigger outcomes that have not been anticipated by the parties when crafting a treaty and setting in place a bureaucracy such as the UNCCD secretariat. However, these outcomes can make a meaningful difference to the ways in which an MEA is implemented.

While states remain the formal masters of any given international regime, they will often find themselves reacting to incentives by bureaucratic personalities as represented, for instance, by treaty secretariats and the international civil service that runs them. The latter will be the more influential, the more they are accepted as authoritative and legitimate in their capacities and actions. In order to make a meaningful difference to global environmental governance, treaty secretariats must thus be careful of tiping the delicate balance of proactively using their latitude and the authority and legitimacy upon which such governance is built.

Notes

1. The author thanks Per-Olof Busch, Karel Mayrand, Adil Najam, and the anonymous reviewers for their comments and suggestions. He is particularly grateful to Hama Arba Diallo, Executive Secretary of the United Nations Convention to Combat Desertification, and the senior officers whom he was privileged to interview at the secretariat in Bonn in November-December 2003. Funding by the Volkswagen Foundation is gratefully acknowledged.
2. On global transformations in the environmental sphere, see Held (1999, ch. 8).
3. An exception to the rule, Jorgen Wettestad (2001) included treaty secretariats as one of six key factors in his analytical framework to explain regime effectiveness.
4. For a comprehensive survey on the debate surrounding organisational reform in global environmental governance see Biermann and Bauer (2005b).
5. Although the 'stage hands or actors' metaphor is borrowed from Rosemary Sandford (1994), the following conceptualisation of international treaty secretariats draws from Steffen Bauer (2006, in press).
6. While this statement will likely be acknowledged with a snigger among practitioners, it is anything but trivial in a context of international relations research, where the notion that there may be relevant actors in world politics other than nation-states is a relatively new phenomenon. For an elaborate conceptualisation of the 'power and pathologies' of international organisations as actors in global politics see Barnett and Finnemore (Barnett and Finnemore 1999; 2004).
7. See, for instance, the comparative study of five international treaty secretariats by Sandford (1994, 25). But see Thomas G. Weiss (1982) for a more critical stance, albeit against a background of Cold War politicking.
8. For a point in case see Adil Najam's (1995) study on the exemplary character of the intergovernmental negotiations on desertification in terms of North-South bargaining (see also Chapter 5).
9. A notable deviation from the third-generation criteria listed by Sandford (1994, 18), namely 'the use of the framework convention as a treaty prototype; the *incorporation of a financial mechanism in treaties*; and the increasing emphasis on the value of regional approaches to the implementation' (emphasis added).
10. For an elaborate survey on the UNCCD financing structure, see chapter 9.
11. For details on institutional and procedural arrangements of the convention see UNCCD (1995, in particular 30–31) and Chasek and Corell (2002).
12. Diallo's term as Executive Secretary extends to the 2005–07 period.
13. The budget was US$14 million for 2000–01 and US$15.3 million for 2002–03 (Schram Stokke and Thommessen 2003, 215).
14. These funds combined amounted to roughly US$6 million for the biennium 2002–03 Schram Stokke and Thommessen (Schram Stokke and Thommessen 2003, 215).

15 A fifth regional annex for Central and Eastern Europe (Annex V) was adopted at COP-4 in Bonn (UNCCD 2001, 22–27). However, this is not yet reflected in the secretariat structure.
16 On the political causes for the special status of Africa in the UNCCD process, see Corell (2003, 13–15); see also Chapter 5.
17 For further-reaching discussions on how to assess the functions and effects of international environmental organisations systematically and an elaborate analytical framework, see Biermann and Bauer (2004) and Biermann and Bauer (2005a) respectively.
18 For elaborate discussions of how to define desertification and land degradation, see *inter alia* Chapter 2, as well as Middleton and Thomas (1992), Corell (1999), and Reynolds and Stafford Smith (2002).
19 The World Day to Combat Desertification is celebrated on 17 June each year to commemorate the adoption of the convention on 17 June 1994.
20 For detailed assessments of the three CRIC meetings see IISD (2002; 2003; 2005).
21 See the UNCCD at <www.unccd.int> (December 2005).
22 The generalised weakness of existing national focal points is discussed Chapter 10.
23 For a more detailed discussion of the Havana High-Level Segment and its political implications see Bauer (forthcoming-b).

References

Andresen, Steinar and Jon Birger Skjaerseth (1999). 'Can International Environmental Secretariats Promote Effective Co-operation?' Paper presented at the United Nations University's International Conference on Synergies and Co-ordination between Multilateral Environmental Agreements, 14–16 July. Tokyo. <www.geic.or.jp/interlinkages/docs/Andresen.PDF> (December 2005).
Barnett, Michael N. and Martha Finnemore (1999). 'The Politics, Power, and Pathologies of International Organizations'. *International Organization* vol. 53, no. 4, pp. 699–732.
Barnett, Michael N. and Martha Finnemore (2004). *Rules for the World: International Organizations in Global Politics*. Cornell University Press, Ithaca.
Bauer, Steffen (2006, in press). 'Does Bureaucracy Really Matter? The Authority of Intergovernmental Treaty Secretariats in Global Environmental Politics'. *Global Environmental Politics* vol. 6, no. 1.
Bauer, Steffen (forthcoming-a). 'The Secretariat of the United Nations Convention to Combat Desertification'. In F. Biermann and B. Siebenhüner, eds., *Managers of Global Change: Effectiveness and Learning of International Environmental Organizations*,
Bauer, Steffen (forthcoming-b). 'The Ozone Secretariat: Administering the Vienna Convention and the Montreal Protocol on Substances That Deplete the Ozone Layer'. In F. Biermann and B. Siebenhüner, eds., *Managers of Global Change: Effectiveness and Learning of International Environmental Organizations*.
Biermann, Frank and Steffen Bauer (2004). 'Assessing the Effectiveness of Intergovernmental Organisations in International Environmental Politics'. *Global Environmental Change* vol. 14, no. 2, pp. 189–193.
Biermann, Frank and Steffen Bauer (2005a). 'Managers of Global Governance: Assessing and Explaining the Influence of International Bureaucracies'. Global Governance Working Paper No. 15. Amsterdam, Berlin, Oldenburg, Potsdam. <glogov.org> (December 2005).

Biermann, Frank and Steffen Bauer, eds. (2005b). *A World Environmental Organization: Solution or Threat to Effective International Environmental Governance?* Ashgate, Aldershot.

Busch, Per-Olof (forthcoming). 'Making a Living in a Strait-Jacket: The Secretariat to the United Nations Framework Convention on Climate Change'. In F. Biermann and B. Siebenhüner, eds., *Managers of Global Change: Effectiveness and Learning of International Environmental Organizations*,

Chasek, Pamela S. and Elisabeth Corell (2002). 'Addressing Desertification at the International Level: The Institutional System'. In J.F. Reynolds and D.M. Stafford Smith, eds., *Global Desertification: Do Humans Cause Deserts*. Dahlem University Press, Berlin.

Corell, Elisabeth (1999). 'The Negotiable Desert: Expert Knowledge in the Negotiations of the Convention to Combat Desertification'. Ph.D. thesis. Linköping University, Linköping, Sweden.

Corell, Elisabeth and Michele M. Betsill (2001). 'A Comparative Look at NGO Influence in Intenrational Environmental Negotiations: Desertification and Climate Change'. *Global Environmental Politics* vol. 1, no. 4, pp. 86–107.

Corell, Elisabeth (2003). 'Dryland Degradation — Africa's Main Environmental Challenge: International Activities from the 1970s to the 1990s and the Future of the United Nations Convention to Combat Desertification'. In B. Chaytor and K. Gray, eds., *International Environmental Law and Policy in Africa*. Kluwer Academic Publishers, Dordrecht.

Haas, Peter M., Robert O. Keohane, and Marc A. Levy, eds. (1993). *Institutions for the Earth: Sources of Effective International Environmental Protection*. MIT Press, Cambridge MA.

Held, David, Anthony G. McGrew, David Goldblatt, et al. (1999). *Global Transformations: Politics, Economics, and Culture*. Polity Press, Oxford.

International Institute for Sustainable Development (2002). 'Summary of the First Session of the Committee for the Review of the Implementation of the Convention to Combat Desertification: 11–22 November 2002'. *Earth Negotiations Bulletin* vol. 4, no. 162. <www.iisd.ca/vol04/enb04162e.html> (December 2005).

International Institute for Sustainable Development (2003). 'Summary of the Sixth Conference of the Parties to the Convention to Combat Desertification: 25 August–6 September 2003'. *Earth Negotiations Bulletin* vol. 4, no. 173. <www.iisd.ca/vol04/enb04173e.html> (December 2005).

International Institute for Sustainable Development (2005). 'Summary of the Third Session of the Committee for the Review of hte Implementation of the Convention to Combat Desertification: 2–11 May 2005'. *Earth Negotiations Bulletin* vol. 4, no. 175. <www.iisd.ca/vol04/enb04175e.html> (December 2005).

Middleton, Nick and David S.G. Thomas (1992). *World Atlas of Desertification*. Edward Arnold, London.

Miles, Edward L. (2002). *Environmental Regime Effectiveness: Confronting Theory with Evidence*. MIT Press, Cambridge MA.

Mitchell, Ronald B. (2002). 'International Environment'. In W. Carlsnaes, T. Risse-Kappen and B.A. Simmons, eds., *Handbook of International Relations*. Sage, London.

Najam, Adil (1995). 'An Environmental Negotiation Strategy for the South'. *International Environmental Affairs* vol. 7, no. 3, pp. 249–287.

Reynolds, J.F. and D.M. Stafford Smith (2002). 'Do Humans Cause Deserts?' In J.F. Reynolds and D.M. Stafford Smith, eds., *Global Desertification: Do Humans Cause Deserts?* Dahlem University Press, Berlin.

Sandford, Rosemary (1992). 'Secretariats and International Environmental Negotiations: Two New Models'. In L. Susskind, E.J. Dolin and J.W. Breslin, eds., *International Environmental Treaty Making*. Harvard Law School, Cambridge MA.

Sandford, Rosemary (1994). 'International Environmental Treaty Secretariats: Stage-Hands or Actors?' In H.O. Bergesen and G. Parmann, eds., *Green Globe Yearbook of International Co-operation on Environment and Development 1994*. Oxford University Press, Oxford.

Schram Stokke, Olav and Oystein B. Thommessen, eds. (2003). *Yearbook of International Co-operation on Environment and Development 2003/2004*. Earthscan, London.

Siebenhüner, Bernd (forthcoming). 'Lean Shark in a Regulatory Jungle? The Case of the CBD Secretariat'. In F. Biermann and B. Siebenhüner, eds., *Managers of Global Change: Effectiveness and Learning of International Environmental Organizations*,

United Nations Convention to Combat Desertification (1994). 'Text of the United Nations Convention to Combat Desertification'. <www.unccd.int/convention/text/convention.php> (December 2005).

United Nations Convention to Combat Desertification (1995). 'Down to Earth: A Simplified Guide to the Convention to Combat Desertification, Why It Is Necessary, and What Is Important and Different About It'. <www.unccd.int/publicinfo/downtoearth/downtoearth-eng.pdf> (December 2005).

United Nations Convention to Combat Desertification (1997). 'Report of the Conference of the Parties on Its First Session, Held in Rome from 29 September to 10 October 1997: Part Two — Actions Taken by the Conference of the Parties at Its First Session'. ICCD/COP(1)/11, 29 December. <www.unccd.int/cop/officialdocs/cop1/pdf/11add1eng.pdf> (2005 December).

United Nations Convention to Combat Desertification (2001). 'Report of the Conference of the Parties on Its Fourth Session, Held in Bonn from 11 to 22 December 2000: Part Two — Action Taken by the Conference of the Parties at Its Fourth Session'. ICCD/COP(4)/11/Add.1, 25 January 2001. <www.unccd.int/cop/officialdocs/cop4/pdf/11add1eng.pdf> (December 2005).

United Nations Convention to Combat Desertification (2003a). 'Programme and Budget'. ICCD/COP(g)/2/Add.1, 23 May 2003. <www.unccd.int/cop/officialdocs/cop6/pdf/2add1eng.pdf> (December 2005).

United Nations Convention to Combat Desertification (2003b). 'Report of the Conference of the Parties on Its Sixth Session, Held in Havana from 25 August to 5 September 2003: Part One — Proceedings'. ICCD/COP(6)/11, 3 November 2003. <www.unccd.int/cop/officialdocs/cop6/pdf/11eng.pdf> (December 2005).

United Nations Convention to Combat Desertification (2005). 'Status of Ratification and Entry into Force'. <www.unccd.int/convention/ratif/doeif.php> (December 2005).

United Nations Environmental Programme (2004). 'UNEP's Strategy on Land Use Management and Soil Conservation: A Strengthened Functional Approach'. UNEP/GC.22/INF/25, 4 December. <www.unep.org/GC/GC22/Document/k0263473.pdf> (December 2005).

Weiss, Thomas G. (1982). 'International Bureaucracy: The Myth and Reality of the International Civil Service'. *International Affairs* vol. 58, no. 2, pp. 287–306.

Wettestad, Jorgen (2001). 'Designing Effective Environmental Regimes: The Conditional Keys'. *Global Governance* vol. 7, no. 317–341.

Young, Oran R., Mark A. Levy, and Gail Osherenko, eds. (1999). *Effectiveness of International Environmental Regimes: Causal Connections and Behavioral Mechanisms*. MIT Press, Cambridge MA.

Zürn, Michael (1998). 'The Rise of International Environmental Poliitcs'. *World Politics* vol. 50, no. 4, pp. 617–649.

Chapter 7

Civil Society's Role in Negotiating and Implementing the UNCCD

Friederike Knabe

One of the great legacies of the 1992 United Nations Conference on Environment and Development (UNCED) is the recognition that nongovernmental organisations (NGOs) and other civil society groups had played an essential role in global policy development. This recognition would prove increasingly significant in the years that followed, as it resulted in new alliances between the United Nations and civil society. The UN had entered into a 'second generation' relationship with civil society, broadening the range of NGOs with access to UN consultations from a select group of international NGOs to a myriad of regional and local organisations (Hill 2004). The United Nations Convention to Combat Desertification (UNCCD) benefited considerably from this favourable situation. When the convention's Intergovernmental Negotiating Committee on Desertification (INCD) began its work in May 1993, the long-term field experience garnered by NGOs and other community-based organisations justified their place at the negotiating table.

The UNCCD is notable among the multilateral environmental agreements (MEAs) in its inclusion of civil society. This chapter analyzes the role played by NGOs in both the INCD and the UNCCD processes. This analysis covers understanding how the UNCCD distinguishes itself with regard to NGO participation, how NGO involvement has influenced intergovernmental decision making, and how this enhanced participation translates into actions to implement the convention.

Definitions of what comprises 'civil society' have varied considerably depending on the context and interpretation.[1] While the term is not used in the text of the UNCCD, the convention's secretariat has since commended the partnership with civil society.[2] In the UNCCD context 'nongovernmental organisation' covers a spectrum of non-state sectors, even including the private sector.[3] In this chapter, the term 'NGO' refers to any non-profit organisation accredited by the UNCCD; the term 'civil society' applies to other non-state groupings, in particular the affected populations.

Three phases of NGO participation are distinguished in this chapter. These are as follows: involvement in the initial processes of the INCD, participation in negotiations subsequent to the adoption of the UNCCD in June 1994, and activities in the reporting and monitoring processes of the UNCCD along with an increasing focus on implementation of the convention at the national level through National

Action Programmes (NAPs). This chapter also examines the internal NGO networking operations within their national and international contexts.

Participation in UNCCD Negotiations

Several factors led to the 'Rio spirit' of NGO participation in the INCD process. The UN resolution that established the INCD sought constructive contributions from all NGOs, especially those from developing countries (United Nations General Assembly [UNGA] 1992, para. 19). From the beginning, NGOs were supported by government delegations, by INCD chairman Bo Kjellén, and by the INCD secretariat.

The first meeting of the INCD opened with a one-week information-sharing session with 31 presentations by UN technical advisors, research institutions, and governments.[4] Topics included local experience with drought and desertification, physical and socioeconomic consequences, and international co-operation. Relevant NGO programmes were included in the presentations. The dialogue revealed the global extent of the problem, while emphasising Africa's special requirements. The need for stressing local participation and the full integration of women in combating desertification was made apparent. The chair, in summarising, called for the 'dream' of Rio to be turned into human-centred reality. This was, he said, the first chance since UNCED to reflect the central role of people's contribution (International Institute for Sustainable Development [IISD] 1993).

For that first INCD session, 84 NGOs and other civil society organisations had been accredited: 40 African, 20 international or North-based, 18 from Asia, and 7 from Latin America. An NGO joint statement accentuated the role of communities and the need for targeting women, the group most vulnerable to the impacts of desertification. The NGOs further contended that their role in community mobilisation ought to rank them as equal partners in UNCCD implementation. To foster strategic NGO input into the negotiations, the INCD secretariat organised an NGO consultation prior its next meeting, held in Bamako, Mali, in 1993. At that consultation, the topics of local implementation, global strategies, and financial concerns were discussed. The latter dealt with NGO response to the reluctance of donor governments to provide 'new money' for the convention. The final statement addressed six main issues: land tenure, awareness and information, education, community and NGO empowerment, science and technology, and international relationships (Gadgil undated).

Elaboration of the convention's text rested on government submissions compiled by the INCD secretariat (see UNCCD 1993). NGO and civil society's role was emphasised from the preamble onward. A 'bottom-up' approach to desertification assigned priority to affected local populations, and special attention was given to the socioeconomic aspects of desertification. Proposed NAPs added strength by emphasising participation by NGOs in supporting local organisations and initiatives. The final text of the convention includes 21 references to NGOs, plus others in the regional annexes.[5] The multi-level role of NGOs from local to national to international participation was recognised in full.

Influencing Decision Making

With the presence of NGOs approved from the outset, the question of their influence deserves further scrutiny. NGO delegates were officially observers. At working group sessions they could make interventions by invitation from the chair. Written submissions were channelled through selected delegates from different regional groups. One feature of NGO participation was the dominance by Southern members, particularly from Africa. A core group of NGOs attended most, if not all, the INCD sessions, offering a consistent perspective based on their own analysis and consensus building. They tracked each of the major INCD topics, building negotiation skills and confidence progressively (Corell 1999, 139, 140). Delegates were contacted and relationships nurtured through ongoing dialogue in formal and informal sessions.

Comments by government delegates on the influence of the NGOs have ranged from their participation being useful to it being invaluable. Some contended the convention text was rendered more meaningful by active NGO participation. Nonetheless, some governments looked on this participation less favourably. Questions were raised about NGO legitimacy in representing local populations, as well as about their professional qualifications and competencies.

Michelle Betsill and Elisabeth Corell (2001b) present a systematic study of NGO influence on the both the desertification convention and the United Nations Framework Convention on Climate Change (UNFCCC). They contend that information about civil society's role in both these issues is extensive and requires disciplined methodologies to assess it, and suggest applying specific measurement criteria to appraising NGO influence on negotiations. 'Influence' must be assessed within the specific scope of an NGO's role in the negotiation process (Betsill and Corell 2001a, 65).

With treaty negotiations being the responsibility of governments, NGO influence can be measured according to the level of interaction with parties and the adoption of changes to drafted texts reflecting stated positions of civil society. As observers, NGOs depend on the space they are allotted by governments. Participation does not automatically translate into influence. A major tool for NGOs to exert influence is information: competent and detailed information on the effects of desertification on communities, on the abilities of those communities to adapt and on their know-how in controlling or reversing land degradation. By providing concrete information on local conditions and actions, NGOs can bridge the gaps between local communities, governments, and international policy makers. As stakeholders in the process reaching from negotiation to implementation, their effectiveness must be assessed fully, including through responses from all participating groups (Betsill and Corell 2001a, 75).

During the INCD negotiations, NGOs from all regions developed effective and co-ordinated strategies based on their common objectives and priorities. Joint statements reflected the consolidated positions and led to the perception among delegates that NGOs spoke with a single voice. This solidarity was recognised as an important aspect of their heightened influence in negotiations. Although direct interventions were limited, NGOs lobbied government delegations in various ways, successfully introducing several modifications into the text of the convention. For example, NGO lobbying led to the inclusion of the National Desertification Fund (NDF) in UNCCD funding mechanisms (Scherl 1996; see also the United Nations Development Programme's Office to Combat Desertification and Drought (UNSO) 1999).

With many process-related matters still unresolved at the closing INCD session, operational details — including financial mechanisms — became pre-eminent.[6] The final NGO statement praised the growing co-operation between themselves and governments.[7] Parties were called on to continue in that spirit, while NGOs pledged full co-operation in projects combating desertification and improving the quality of life in drylands.

Effectiveness studies have noted that NGO membership cohesion in the INCD is a signal characteristic for their influence (Betsill and Corell 2001a). It was reflected in the consistency in presenting their three main objectives: the use of a participatory bottom-up approach; inclusions of traditional and local knowledge, including that of women; and provision of new and additional financial resources for dryland management. Partly, this cohesion coincided with the shift in representative majorities to countries of the South. Being closely connected to the challenges faced by the local communities, NGOs understood their responsibility to act as conduits to the international negotiations and advocates for the UNCCD. The means for their influence was expert information and competencies in local know-how (see Corell 1999, 158–161).

Comparing NGO Participation in the Three Multilateral Environmental Agreements

While the legal basis for NGO participation was identical for the MEAs emanating from the Rio process, practical arrangements for the INCD resulted in a different composition of civil society representation.[8] Between 1993 and 1997 (from the first INCD meeting to the first Conference of the Parties [COP]), 187 NGOs remained active in the UNCCD process. With a majority of these representing grassroots interests, in particular many from Africa (around 50 percent), some 40 organisations participated in the meetings. Representatives from larger international environmental or development NGOs, however, were noticeably absent (Betsill and Corell 2001b, 93).

NGO participation in the two other MEAs is a sharp contrast. Nearly 240 NGOs had been participating in the UNFCCC; by 1997, they were dominated by NGOs from developed countries (91 percent) (Paoletto and Schroeder 1997). Similarly, at Conferences of the Parties of the Convention on Biological Diversity (CBD), civil society participation consisted of North-based NGOs, academic institutions, and research centres or networks.[9]

One reason for the different weighting in the UNCCD processes in favour of Southern NGOs, particularly African ones, was a special funding provision (Oberthür et al. 2002, 111). Voluntary contributions by donor parties financed the participation of some NGOs from affected developing countries. This facilitated their active and ongoing involvement. However, this process has not been without difficulties. Variations in funding levels and selection criteria have at times impaired both participation in both INCD meetings and COPs, as well as affected inter-NGO relationships.

Establishing the NGO Network

In order to improve co-ordination among participating NGOs and other groups, the Réseau international des ONG pour la désertification (RIOD) was established in 1994.[10] With the goal of ensuring full involvement in all phases of UNCCD implementation, RIOD defined information sharing and program coordination on a global scale as its objectives (Environment and Development Action in the Third World [ENDA] 1999). Using 'focal points' as its core structure, RIOD linked participating NGOs and encouraged community-based organisations and smaller groups to join in (Environmental Liaison Centre International [ELCI] 1994a).[11]

At the international level, RIOD's role was to support NGOs while endorsing UNCCD provisions at the regional and national levels. Promoting emerging NAP processes, RIOD encouraged information and communication activities locally. Women's involvement, better interaction among groups, and UNCCD implementation monitoring were major thrusts. National NGO Co-ordinating Committees on Desertification (NCCD) were established to facilitate interaction between governments and groups. RIOD saw itself as an information channel drawing on local input and forwarding it to international negotiations. Monitoring and evaluation activities might also draw on these resources. Flexibility was the key in maintaining two-way communication flows (see Figure 7-1) (ELCI 1994b).

By 1996, RIOD faced its first organisational crisis. Rapid growth had produced the need for a structural revision, with the original loose network possibly being replaced by a more formal organisation, and had resulted in imbalanced representation, flawed financial mechanisms, and interrupted communications. While most NGO delegates' energy and attention were focussed on UNCCD substantive and procedural issues, RIOD's organisational issues nevertheless intruded on INCD sessions, leading NGOs to organise special meetings to deal with them. The issue of the RIOD's international structure simmered during the first phase of its development. On substantive issues, differences of opinion among NGOs emerged on major negotiation topics, such as on the role of the Global Mechanism (GM). Here, compromise positions were worked out thanks to experienced NGO leadership.[12]

Fostering Regional Contributions

In 1996, during the preparations for the first COP, several regional meetings took place to discuss convention implementation. NGOs were involved in a number of these meetings and also organised their own with funding support from a range of donors and the interim UNCCD secretariat. Two of the underlying themes were improved co-operation in North-South and multi-stakeholder relations (Ledgar 1997).

The 1996 Afro-Asian Conference in Hyderabad, India, provides a good illustration of the level of consultation among NGOs and other UNCCD stakeholders at that time. More than 100 delegates from 62 countries attended. At the plenary several country initiatives were discussed for designing and formulating options for NAP implementation based on several reports by participants on the status of the process in their countries. In these examples, reflecting the spirit of the convention, central co-ordinating bodies for NAPs had been created, underlining a multi-sectoral, inter-ministerial approach.

According to participants, the highlight was the round tables on key aspects of implementing the UNCCD. These round tables covered topics such as the role of different stakeholders in the implementation, the specific role and status of women in the struggle against land degradation and in sustainable dryland development, networking and partnership building, and the important role of media and advocacy.[13]

Highlighting the Role of Women

The vital role played by women in combating desertification is stressed throughout the UNCCD (see Chapter 8). Women are part of the local community groups identified

Figure 7-1   RIOD's Focal Point Structure 1997/98

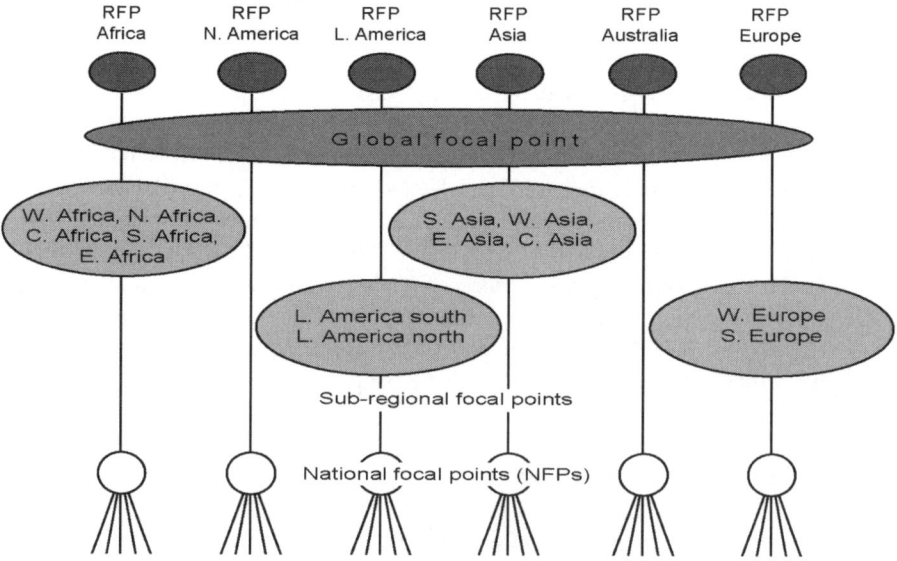

Source: Holmén (1997).

in the convention, such as farmers and pastoralists, but they are also recognised as a special group particularly vulnerable to desertification and poverty (see Chapter 3). However, this chapter cannot address in depth the the equal participation of women in the implementation of the UNCCD, or the related issues.[14]

The significance of the Women and Desertification Caucus, which the NGOs established at INCD-7 in 1995, warrants separate mention. The caucus contributed significantly to raising awareness of the need for the equal participation of women at the INCD/COP sessions and the vital importance women carry at the community level and for the NAP process. NGO women participants in particular reminded delegates of the leadership role expected of them in furthering the application of the respective UNCCD provisions. In the final NGO statement at the closing plenary of the last INCD session in 1997, Friends of the Earth stated that the role foreseen for women in the convention's text was not sufficiently reflected in the number and quality of their representation in its structures and mechanisms (IISD 1997).

In summary, on the topic of NGO involvement in the INCD process, the question remains about the influence that civil society representatives brought to the elaboration of the convention. NGO influence rested primarily on three aspects: first, the confirmation of the essential role of NGO participation in all UNCCD-related sessions; second, the reinforcement of the bottom-up approach throughout the convention text and implementation mechanisms, in particular at the national level; and third, the recognition of NGOs as conduits between local populations and effective anti-desertification programming with a specific role in the National Desertification Fund. In all these aspects the NGO objectives were achieved. NGOs had established themselves as experts in their field and competent partners for the implementation phase.

Participating in the Conferences of the Parties

The first COP expanded the dialogue between governments and NGOs by way of a special one-day forum. This, and the subsequent half-day open dialogue session resulted in the COP's decision to build two half-day sessions into the official structure of future COPs (UNCCD 1997, 95–96). Observers of international environmental negotiations have pointed out that the concept of open dialogue sessions is a unique feature among MEAs (Oberthür et al. 2002, 69).

The NGO forum provided stakeholders with a platform for discussing issues of common interest and concern. The different workshops addressed topics such as local experiences and traditional knowledge as well as building partnership mechanisms. Improving communications and gender issues in desertification, drought, and poverty were also discussed.[15] NGOs also voiced their demands during the session: full participation in all COP deliberations, inclusion of NGO experts in the new roster of experts, and civil society involvement in the NAP processes.

Encouraged by the COP-1 experience, NGOs have organised open dialogue sessions at all subsequent COPs in co-operation with the UNCCD secretariat.

Whenever possible, the selected themes have followed the official session agenda, aiming at providing valuable insights from grassroots perspectives and resulting recommendations.

Raising Issues with the Committee on Science and Technology

From the outset, NGOs participated in the negotiations on the composition and work program of the Committee on Science and Technology (CST). It was evident to NGOs that scientific research, while playing an important role in implementing the UNCCD, had to be combined with local know-how and expertise to be beneficial for affected populations.

The composition of the CST was the result of intense intergovernmental negotiations. The NGO position was to insist on full participation. The final CST structure, consisting of government delegates, complemented by a roster of independent experts and optional ad-hoc panels drawn from the roster, suggested that NGOs would be included (UNGA 1996). Optimism was dampened, however, when NGOs realised that their participation was curtailed due to the selection criteria established for nominating experts. These criteria required accreditation by internationally recognised research institutions, thus excluding NGO or community-based expertise.

In its leading role of setting standards for UNCCD implementation and reporting, NGOs regarded the CST as an important force. NGO proposals for the work plans emphasised the integration of local traditions and knowledge, more attention to women-focussed programming, and the establishment of explicit benchmarks and indicators to monitor civil society involvement in the NAPs and other UNCCD provisions. To have specific benchmarks for civil society participation incorporated required forceful and repeated campaigning. Drawing on the UNCCD principles, NGOs continued pressuring delegations to avoid top-down approaches that ignored local expertise garnered over long duration.

Traditional knowledge was a major topic in the negotiations for the CST work programme. A draft secretariat text for COP-2 called for synergy between local knowledge and academic or industrial science, and for shared knowledge and improved communications (Canadian International Development Agency [CIDA] 1998). Women's role in preserving traditional knowledge was acknowledged, and the need to protect that role was stressed (see also IISD 1998).

Throughout the early COPs, NGOs have consistently argued for meaningful participation of local populations and community-based organisations. A report on the consideration of local knowledge in national and regional action programmes published by ENDA states: 'The present state of affairs reveals that the ways in which the issue of local knowledge is considered when designing national action programmes varies greatly, and is tackled using contrasting methodologies' (Sokona and Seck 2000). Four scenarios — from the issue not being taken into account at all to the valorisation of indigenous knowledge in the development of action guidelines and projects — are summarised.

During COP-5, NGOs again reminded delegates that 'most of TK [traditional knowledge] has developed and is held by communities and their populations ... Therefore, to access this knowledge and to investigate its loss, information has to be gathered in close collaboration with the local communities and their CBOs ... A feedback loop that allows the data collected to be shared for the benefit of the local populations has to be built into any information network on TK'.[16]

Several decisions taken by COP-6 have provided new opportunities to expand the scope of NGO participation. In particular, in part as a result of the NGO delegates' lobbying the CST, Decision 13 regarding the roster of experts recalls, among other issues, the important role played by NGOs in implementing the UNCCD (UNCCD 2003a, 30–31). Noting their under-representation in various disciplines, the conference requested the secretariat 'to revise the format for the curricula vitae contained in annex II of ICCD/COP(5)/5/Add.1, taking into account non-governmental organizations' (30). It also encouraged parties to propose new candidates for the roster and to achieve better representation 'of women and a more balanced geographical distribution of non-governmental organizations' (30).

Also highly relevant in this context is Decision 16 on traditional knowledge (UNCCD 2003a, 36). It invites parties 'to involve relevant governmental and non-governmental organizations, research institutions and local and indigenous communities in formulating views on how traditional knowledge can contribute to fulfilling the objectives of the Convention, and especially on the elements proposed for establishing a network on traditional knowledge to combat desertification' (36). It further invites parties to involve NGOs, among others, in compiling case studies and lessons learned on the management and protection of traditional knowledge. NGOs, for their part, are called upon to contribute to the implementation of this decision through case studies and analysis in the fields of their expertise.

If one looks back at the involvement and influence of NGOs from COP-1 to COP-6, it appears that the relationship between parties and civil society has changed over time. As the COPs increasingly dealt with policy issues related to implementation, the NGOs' contributions also reflected these developments. The initial goal of the NGOs to become full and equal participants, while not reached completely, has been met to an ever increasing degree. There were considerable advances in the priority areas addressed by the CST. The bottom-up approach was reaffirmed, respect for local expertise enhanced, and NGOs inclusion in the roster of experts enabled.

Still, the NGO participation has not been without challenges and difficulties. This has been noticeable in particular since the COPs moved to a two-year cycle in 2001. The preparation and follow-up for the sessions have been neither consistent nor fully satisfactory. Funding difficulties as well as local programming priorities have prevented ongoing exchanges among accredited NGOs and with the UNCCD secretariat. As a result, the quality of contributions at the COPs has been uneven and cohesion among NGO positions has loosened.

Enhancing the Co-ordination of Nongovernmental Organisations

At the time of COP-1, 452 NGOs were accredited, sending about 100 delegates to the meeting. COP-6 saw more than 150 NGO attendees from 50 countries (of about 700 accredited NGOs).[17] This increase brought opportunities and challenges for RIOD as well as for any other form of NGO co-ordination. While the contributions to the COP agenda took precedence, RIOD members attempted to address the increasing structural weaknesses of the network that have been a permanent feature since its inception. Many delegates and NGO participants noted that the NGO network was in need of serious revision and renewal.

As early as 1999, NGOs had identified three major organisational challenges: ensuring representational participation of NGOs involved in the day-to-day struggle against desertification, adapting the network to the implementation phase of the convention, and addressing organisational change resulting from the network's fast growth.[18]

To assist RIOD in defining its future direction, UNSO initiated a study on NGO networking for sustainable dryland development with emphasis on the RIOD experience (Holmén and Jirström 1997). Taking into account the concerns and tensions that had arisen within the network, it surveyed a substantial number of NGOs active in RIOD or involved in the UNCCD. The authors concluded that there was no consensus from within and outside the network on its strengths and on whether it added value to NGO participation. The study also revealed that most RIOD NGOs were not primarily working at the grassroots level but instead were more concerned with advocacy. In fact, with RIOD the notion of splitting the organism into several networks was considered.[19] At the RIOD Global Meeting in 1999, the majority agreed that the it should remain a loose, decentralised network with national and regional focal points as its core elements.[20] Consequently, the Action Plan emphasised the focal points' networking and communication responsibilities. Regional focal points would elect a RIOD Facilitation Unit for global level co-ordination.[21]

The growing number of NGO delegates participating in the COPs has also meant a diversification into groupings and networks beyond RIOD, reflecting similar, parallel, regional, or specific subject focus. With a considerable number of first-time representatives at each session, new perspectives emerged, as did new needs. Differences of opinion have become more common among groups of NGOs. The organisations tend to consult independently, developing policy interventions reflecting their particular perspectives. Regional caucuses have become a permanent feature. The two-day special NGO meetings prior to each COP, intended to prepare for the conference's substance, also provide an introduction for new participants. The tradition of regular reporting on a region's progress in relation to NGO and civil society activities, perceived by many observers as an essential accountability function, is being increasingly abandoned. A number of participants and other UNCCD partners have expressed a desire to refocus the NGO delegates' full attention to the COP and UNCCD processes.

To advance the consultation on RIOD's future, Solidarité Canada Sahel (SCS), in its role as RIOD Facilitation Unit, commissioned a performance evaluation of the network facilitation function to be discussed at COP-6. The evaluation strongly recommended the clarification and renewal of mandates for all focal points. It contended that the RIOD structure needed to be adapted to the increasingly complex responsibilities of NGOs in the context of implementing the UNCCD (Knabe 2003). Following these recommendations, network members present at COP-6 decided to embark on a bottom-up renewal process to return focal points to their national bases.

Solidarité Canada Sahel was asked to continue as RIOD Facilitation Unit until COP-7 in October 2005 to support the renewal process and co-ordinate activities. The subsequent mandate review, which was undertaken by SCS in consultation with RIOD members, confirmed the Facilitation Unit as the primary communicator within the network and with the wider UNCCD community (Knabe 2004). The continuation of the mandate depended on ongoing funding for the network facilitation provided by external sources. The search for institutional funds, however, was not successful, and SCS had to terminate its mandate in the summer of 2004. The impact of suspending the global facilitation function on NGO participation cannot yet be fully gauged. National and regional focal points are continuing their activities of co-ordination. But, overall, a weakening of the international NGO community in terms of networking and preparation for the next COP can be expected.

Participating in National Action Programmes

The UNCCD requires that NAPs be built on multi-layered partnerships. Co-operation among national authorities, civil society representatives — including NGOs and grassroots organisations and research institutions — and representatives from the international community including donors is fundamental for the successful implementation of the programmes.

A full analysis of civil society involvement in these processes would go beyond the scope of this chapter. Nonetheless, several independent studies on the first round of NAPs seem to conclude that, in a general sense, the design phase has been more inclusive of NGOs and other civil society actors than the implementation phase was (see, in particular, UNSO 1999).

UNSO's preliminary overview of NAP processes conducted in 1998 found that there was 'broad interest and commitment to the CCD [Convention To Combat Desertification] implementation process, and most affected countries [had] initiated measures and activities to combat desertification at national level' (UNSO 1999, 4). NGOs involvement, the study noted, had been instrumental in some countries in ensuring their government's commitment to launching the NAP process. While NGOs and other stakeholders had been actively involved in formulating NAPs, their influence was unequal and a clearer delineation of roles would have facilitated interaction.

In their preliminary analysis of NAP implementation by the time of COP-2 in 1998, NGOs concurred with the UNSO assessments. They found that the level of local community participation in introducing NAPs in their respective countries had varied enormously. The better civil society was organised, the more successful was the mobilisation of the population for the struggle against desertification. In many affected countries, NGOs contributed to the process by establishing national focal points and co-ordination committees, reaching beyond their own groups, in order to ensure wider and more effective partnerships with government institutions. NGOs concluded that despite difficulties, in part due to competition among players, the innovative character and inclusiveness of some NAP processes needed to be preserved and implanted into the anti-desertification strategies.[22]

Mali presents an interesting example of the complexities involved in the NAP processes. The government started its NAP early, conscious of its particular importance for the country's population (Republic of Mali 1998). Linked to the National Environment Action Plan, the NAP consultation process established a co-ordination committee to which NGOs and professionals from socioeconomic organisations and development partners were invited. Special strategy sessions were organised with the NGOs to ensure the widest possible involvement. With more than 800 active NGOs, regional and local consultations involved some 3000 people.[23] Mali's report to COP-3 stressed that in response to the diverse and independent programmes being implemented in a decentralised fashion, a national monitoring committee that included NGOs was needed (Republic of Mali 1998). Furthermore, the report underlined the vital importance of NGOs in programme delivery and fund development.

Anne Mossige, Yonis Berkole, and Sina Maiga (2001) reached a more cautious conclusion in their evaluation of civil society involvement in the NAP processes in Mali and Ethiopia. They suggested that success in raising awareness and involving local populations had been limited, and that authorities and NGOs need to make a major effort to moving toward full co-operation. One of the constraints identified in their evaluation was the lack of financial resources and organisational expertise delaying the NAP processes. These findings suggested that local communities required better knowledge of the NAP and technical assistance in designing, implementing, monitoring, and evaluating projects, as well as financial support. The evaluation found that NGOs and community-based organisations often had relevant experience in assisting local communities and that their efforts could complement government programmes, particularly in remote rural areas.

The lack of substantive civil society involvement in the NAPs was reiterated at COP-4 in December 2000. The opening NGO statement expressed concern that civil society participation was not taken sufficiently into account in the design and implementation of NAPs. Delegates were reminded that parties had accepted the obligation of fully integrating civil society including local communities in the UNCCD processes. It concluded that 'we all are aware that there is no alternative to the participation of civil society for the successful implementation of the Convention to Combat Desertification. We therefore urge the parties to allow for the establishment of indicators for assessing the participation of civil society in the design and implementation of the convention, as well as for the mainstreaming of gender' (Gliese 2000).

With the COP's establishment of the Committee for the Review of the Implementation of the Convention (CRIC), the review of country progress reports submitted by parties moved to a new level. Country reports need, *inter alia*, to 'reflect the views of a wider set of actors and stakeholders, in particular those of relevant ministries, public agencies, non-governmental organizations (NGOs), community-based organizations (CBOs), academic institutions, the private sector, regional and local authorities and the media' (UNCCD 2003b, para. 25). This opens up a new opportunity for NGOs to see their views reflected in NAP processes.

Strengthening Capacities

It has often been said in UNCCD forums that NGOs and community-based organisations, like other national stakeholders, require support and advice in training and capacity building. The early assessments of NAPs highlighted the need for specific technical and organisational expertise to complement the existing programme knowledge of communities and their supporting organisations. To address capacity-building needs, in 2001 the Global Mechanism (GM) established the Community Exchange and Training Programme (CETP), which encourages the establishment of systematic dialogue and exchanging experiences between natural resource users as a basis for building long-term partnerships (GM 2005; see also Chapter 11). By 2003, the programme had developed 19 community-based projects in 17 countries.

Many civil society groups concentrate their activities at the local and national levels. Local NGOs and community-based organisations, in particular, may not necessarily be well informed about the broader role that they could play in the UNCCD. Conversely, NGOs participating in the UNCCD carry the explicit responsibility for maintaining communication among the local, national, regional, and international levels. In some countries, national co-ordination bodies have assisted in developing these important links, helping to strengthen the capacity of local communities to address desertification and related challenges.

Strengthening North-South Partnerships

NGO networking represents an important component of civil society engagement in the UNCCD. It applies at both the local and the national levels as much as at the level of North-South partnerships and of the international level. The absence of Northern and major international NGOs from this convention has already been pointed out. Its consequences are especially felt at the local programming level. Northern NGOs, often unaware of the UNCCD and its importance for their Southern partners, can undermine the implementation of NAPs by conducting desertification activities outside the framework of the convention. Insufficient awareness of the UNCCD among Northern NGOs also leads to the loss of funding opportunities for Southern organisations.

In 1998, the German Working Group on Desertification solicited views and experiences from 650 organisations from Southern, Eastern, and Northern countries, with interest in the UNCCD or in international programme partnerships (Arbeitsgruppe Desertifikation 1998). Most of the 140 respondents (77 percent) were based in the South. In the context of implementing the convention, 85 percent of participants stressed the importance of North-South NGO partnerships. The enthusiasm for such partnerships was more evident among Southern NGOs (more than 90 percent) than Northern NGOs (60 percent). Southern NGOs placed greater emphasis on operational co-operation, such as training, capacity building, and funding, while Northern NGOs stressed the importance of advocacy and outreach. The authors state that, overall, the response level from Northern NGOs was disappointing. Non-response was sometimes explained by the 'lack of relevance' of the UNCCD to current activities. This lack of response reinforces the perception of Northern NGOs disinterest in the UNCCD.

Despite growing activity by networks such as the European Networking Initiative Desertification (eniD), which was established in 2001 and supported by national networks in participating countries, the number of Northern NGOs that participate in UNCCD processes and are accredited has not substantially increased. Moreover, NGOs have been not effective in attracting major international NGOs to the UNCCD. Northern developmental NGOs tend to operate primarily through bilateral partnerships rather than through international forums.

Divisions often exist between environmental and development NGOs, and between large, financially secure NGOs and small groups operating with limited staff and modest resources. While funding constraints may prevent some NGOs from participating, large NGOs tend to focus on what they perceive as policy interests of their constituents. As Helmut Breitmeier and Volker Rittberger (1998) write, 'environmental NGOs in industrialized countries attribute to climate change whereas other issues such as soil conservation or desertification tend to be neglected'.

Conclusion

The more than ten-year partnership between civil society and the other stakeholders in the UNCCD has evolved in a multitude of ways. It is worth recalling the three phases of NGO involvement in the convention. In 1993, at the beginning of INCD negotiations, the principle of full involvement of NGOs as civil society's representatives was established. NGOs pressed for a bottom-up approach and the full incorporation of needs and aspirations of local communities, community-based organisations, and NGOs into the text of the convention. Speaking with a single voice on these key concerns contributed to positive outcomes, and the influence of NGOs on the proceedings was recognised by all stakeholders.

During the second phase of the convention process, NGO contributions in the COPs shifted the focus to relevant UNCCD provisions, in particular to the operation of the CST and NAPs. NGOs successfully insisted that their voice be heard clearly at the table. The recognition of traditional knowledge and the inclusion of civil society participation into benchmarks and indicators were both major negotiation points. With perseverance and constructive contributions, NGOs reached their objectives with regard to the CST. The acceptance of NGO expertise, reflected in the revised criteria for the roster of experts, has enabled local know-how and NGO competence to enrich the committee's work.

In the third and current phase, NGOs' responsibilities have broadened and become more complex. While their contribution to promoting the UNCCD has been an aspect of their engagement from the outset, the establishment of NAPs has brought additional tasks. Supplying expertise and experience has also meant seeking new partnerships at the national and local levels. NGOs must be fully engaged in the evaluation of NAPs and must contribute to the critical assessment of strengths and weaknesses of implementation.

The growing recognition of their role as partners in the COPs and other consultations has brought new responsibilities for NGOs. In order to continue and strengthen their contribution, NGOs must develop their capacities. Funding of their participation in the UNCCD remains an ongoing concern in that regard, at the international level and at other levels. Adequate resources are required to strengthen NGO capacities to fulfil the important functions that flow from the partnership with governments and other stakeholders.

In relation to the NAPs, NGOs must expand their reach into affected communities and increase communication with local people and community-based organisations to hear their demands and support their capacity building. National RIOD focal points may need to attract additional members as well as adequate financial resources in order to operate as communication hubs to link local populations with the wider international network and feed information back to the communities.

At present, the UNCCD structures and implementation bodies do not have a mechanism to financially or technically support NGO co-ordination efforts, which are essential between sessions. Moreover, weaknesses in national and regional co-ordination structures also have become apparent in reduced NGO participation, co-ordination, and quality of contributions at recent COPs, in contrast to their initial goal of equal participation. Although NGOs have made important gains in enhancing their formal status within the UNCCD framework, financial and structural weaknesses prevent them from fully playing this role.

Notes

1 Many scholars have explored history and trends in defining civil society: see Clark (2003).
2 'Civil society in general and NGOs in particular are very much key cooperating partners, which have shown commitment to the CCD [Convention to Combat Desertification]

throughout the negotiation process. This exemplary cooperation between the UNCCD and NGOs contributes significantly to the successful implementation of the Convention' (UNCCD 2005a).
3. The UNCCD glossary defines 'NGO' only as 'non-profit', covering research institutions, environmental groups, and business associations (UNCCD 2005b).
4. For a detailed narrative account of the INCD process and its salient issues, see Chapter 5.
5. There were six references in the Regional Implementation Annex for Africa, two in each of the other regional annexes (Asia, the Northern Mediterranean, and Central and Eastern Europe) except for the Regional Implementation Annex for Latin America and the Caribbean, which includes only one reference to 'affected communities and society in general' (UNCCD 1994).
6. The last session of the INCD prior to the first Conference of the Parties (COP) was the resumed tenth session in Geneva, 18–22 August 1997.
7. The full statement is published in *Eco* no. 9 (17 January 1997), the bulletin produced daily during the INCD sessions by RIOD.
8. For a detailed comparative analysis of the MEAs, see Oberthür (2002).
9. Based on discussions by the author with participants and on personal research.
10. RIOD was established at an international NGO meeting on desertification in Ouagadougou, Burkina Faso, that was attended by approximately 50 NGOs from around the world. The United Nations Environment Programme (UNEP), UNSO, and the governments of France, Switzerland, and the Netherlands supported the meeting.
11. Three types of activities were identified in order to meet RIOD's objectives: the core local activities, lobby activities at the international level, and joint network activities.
12. The NGO position that was finally agreed upon sidestepped the main question and focussed instead on the need for predictability, transparency, and co-ordination of funding mechanisms at all levels.
13. *Eco* no. 9 (17 January 1997) provides further details on the round tables.
14. A recent study by Friederike Knabe and Lene Poulsen (2004) offers a thorough analysis and review of advances and constraints experienced in 'mainstreaming gender' into the UNCCD processes.
15. *Eco* special edition (6 October 1997) contains a report of the first COP.
16. *Eco* no. 5 (2001), personal collection.
17. Calculations based on the number of participants and the NGO accreditation lists.
18. RIOD Global Meeting, Dakar, 23–28 August 1999.
19. Brown, M. (1997). 'Strengthening NGO Networking for Sustainable Dryland Development'. Internal RIOD document.
20. Some 100 NGOs and community-based organisations from all regions attended the Second Global RIOD Meeting in Dakar. They discussed the challenges facing the network in light of the increasing demands placed on them by the UNCCD implementation processes.
21. This function was fulfilled by ENDA-TM until 2000, when Solidarité Canada Sahel (SCS) was elected at COP-4 to take on the responsibility for the Facilitation Unit. SCS and the unit were supported by the Canadian International Development Agency (CIDA) between 2000 and 2004.
22. *Eco* COP-2 #9 10 December 1998 <www.enda.sn/cdp2/ECO9.htm> (December 2005).
23. *Eco* COP-2 #9 10 December 1998 <www.enda.sn/cdp2/ECO9.htm> (December 2005).

References

Arbeitsgruppe Desertifikation (1998). 'The Combat against Desertification in Global Partnership: A Challenge for Civil Society'. Offprint.

Betsill, Michele M. and Elisabeth Corell (2001a). 'NGO Influence in International Environmental Negotiations: A Framework for Analysis'. *Global Environmental Politics* vol. 1, no. 4, pp. 65–85.

Betsill, Michele M. and Elisabeth Corell (2001b). 'A Comparative Look at NGO Influence in International Environmental Negotiations: Desertification and Climate Change'. *Global Environmental Politics* vol. 1, no. 4, pp. 86–107.

Breitmeier, Helmut and Volker Rittberger (1998). 'Environmental NGOs in an Emerging Global Society'. Tübinger Arbeitspapiere zur Internationalen Politik und Friedensforschung No. 32. <www.uni-tuebingen.de/uni/spi/taps/tap32.htm> (December 2005).

Canadian International Development Agency (1998). 'Update from Dakar: Second Session of the Conference of the Parties (COP2) to the United Nations Convention to Combat Desertification'. <www.acdi-cida.gc.ca/cida_ind.nsf/0/a7ba07ff629071c885256761006030e3?OpenDocument> (December 2005).

Clark, John D. (2003). *Worlds Apart: Civil Society and the Battle for Ethical Globalization*. Kumarian Press, Bloomfield CT.

Corell, Elisabeth (1999). 'The Negotiable Desert: Expert Knowledge in the Negotiations of the Convention to Combat Desertification'. Ph.D. thesis. Linköping University, Linköping, Sweden.

Environment and Development Action in the Third World (1999). 'Proposal for the Organization of a RIOD General Meeting'. ENDA TM. <www.enda.sn/energie/desertif/gm.htm> (December 2005).

Environmental Liaison Centre International (1994a). 'Réseau International d'ONG sur la Désertification — RIOD: General Framework for Operation'. NGO Planning Meeting. Nairobi.

Environmental Liaison Centre International (1994b). 'The NGO Action Plan on Desertification: Facilitating the Voice of the Grassroots'. Réseau international des ONG pour la desertification.

Gadgil, Madhav (undated). 'Desertification'. <ces.iisc.ernet.in/hpg/cesmg/2-6.html> (December 2005).

Gliese, Jürgen (2000). 'COP 4 NGO Opening Statement'. 11 December 2000. <www.forumue.de> (December 2005).

Global Mechanism (2005). 'SADC RIOD Community Exchange and Training Programme'. <www.gm-unccd.org/FIELD/Multi/GM/FR_SADC.htm> (December 2005).

Hill, Tony (2004). 'Three Generations of UN-Civil Society Relations: A Quick Sketch'. UN Non-Governmental Liaison Service. <www.globalpolicy.org/ngos/ngo-un/gen/2004/0404generation.htm> (December 2005).

Holmén, Hans and Magnus Jirström (1997). 'Strengthening NGO Networking for Sustainable Development in Drylands'. Report to the United Nations Development Programme and the United Nations Office for Project Services.

International Institute for Sustainable Development (1993). 'Elaboration of an International Convention to Combat Desertification'. *Earth Negotiations Bulletin* vol. 4, no. 8. <www.iisd.ca/vol04/0408001e.html> (December 2005).

International Institute for Sustainable Development (1997). 'Summary of the Resumed Tenth Session of the INC for the Convention to Combat Desertification: 18–22 August 1997'. *Earth Negotiations Bulletin* vol. 4, no. 106. <www.iisd.ca/vol04/enb04106e.html> (December 2005).

International Institute for Sustainable Development (1998). 'CCD COP-2 Highlights'. *Earth Negotiations Bulletin* vol. 4, no. 121. <www.iisd.ca/vol04/enb04121e.html> (December 2005).

Keane, John (2003). *Global Civil Society?* Cambridge University Press, Cambridge.

Knabe, Friederike (2003). 'Evaluation of the RIOD Facilitation Unit during Solidarité Canada Sahel's Mandate December 2000–August 2003'. Solidarité Canada Sahel.

Knabe, Friederike (2004). 'Final Report on the FU-RIOD/SCS Transition Mandate'. Solidarité Canada Sahel.

Knabe, Friederike and Lene Poulsen (2004). 'Promoting Equality between Men and Women through Multilateral Environmental Agreements: How the UN Convention to Combat Desertification and Drought Has Promoted the Role of Women in Decision-Making'. Paper prepared for IUCN, 18 May. <www.iucn.org/themes/cem/documents/drylands/unccdgender.pdf> (December 2005).

Ledgar, Richard (1997). 'NGO Report on the Convention to Combat Desertification for Rio + 5'. Prepared on behalf of the RIOD Steering Committee, Réseau International d'ONG sur la Désertification, January. <www.ecouncil.ac.cr/rio/focus/report/english/riod.htm> (December 2005).

Mossige, Anne, Yonis Berkele, and Sina Maiga (2001). 'Participation of Civil Society in the National Action Programs of the United Nation's Convention to Combat Desertification: Synthesis of an Assessment in Ethiopia and Mali'. DCG Report 16A. Drylands Coordination Group. <www.drylands-group.org/Articles/305.html> (December 2005).

Oberthür, Sebastian, Matthias Buck, Sebastian Müller, et al. (2002). 'Participation of Non-Governmental Organisations in International Environmental Governance: Legal Basis and Practice Experience'. Ecologic. <www.ecologic.de/download/projekte/18501899/1890/-report_ngos_en.pdf> (December 2005).

Paoletto, Glen and Heike Schroeder (1997). 'Enhancing Participation of NGOs in the UNFCCC Process'. Global Environment Information Centre. <www.geic.or.jp/ngo-bonn.html> (December 2005).

Republic of Mali (1998). 'Résumé du PNAE/PAN-CID'. Ministère de l'environnement, Secretariat permanent du PNAE-CID. Bamako. <www.gm-unccd.org/FIELD/Countries/Mali/nap.htm> (December 2005).

Scherl, Lea M. (1996). 'Relationships and Partnerships among Governments, NGOs, CBOs, and Indigenous Groups in the Context of the Convention to Combat Desertification and Drought: An Analysis of Progress'. Environment Liaison Centre International.

Sokona, Youba and Emmanuel S. Seck (2000). 'Report on the Consideration of Local Knowledge by the Action Programmes, Networks, and Mechanisms Set Up by the CCD Secretariat to Promote Programmes Combatting Desertification on the Regional and National Scales'. ENDA TM. <www.enda.sn/energie/desertif/local-knowl.htm> (December 2005).

United Nations Convention to Combat Desertification (1993). 'Elaboration of an International Convention to Combat Desertification in Countries Experiencing Serious Drought and/or Desertification, Particularly in Africa'. A/AC.241/12, 23 August. <www.unccd.int/cop/officialdocs/incd/pdf/24112eng.pdf> (December 2005).

United Nations Convention to Combat Desertification (1994). 'Text of the United Nations Convention to Combat Desertification'. <www.unccd.int/convention/text/convention.php> (December 2005).
United Nations Convention to Combat Desertification (1997). 'Report of the Conference of the Parties on Its First Session, Held in Rome from 29 September to 10 October 1997: Part Two — Actions Taken by the Conference of the Parties at Its First Session'. ICCD/COP(1)/11, 29 December. <www.unccd.int/cop/officialdocs/cop1/pdf/11add1eng.pdf> (2005 December).
United Nations Convention to Combat Desertification (2003a). 'Report of the Conference of the Parties on Its Sixth Session, Held in Havana from 25 August to 5 September 2003: Part Two — Action Taken by the Conference of the Parties at Its Sixth Session'. ICCD/COP(6)/11/Add.1, 7 November. <www.unccd.int/cop/officialdocs/cop6/pdf/11add1eng.pdf> (December 2005).
United Nations Convention to Combat Desertification (2003b). 'National Reporting Process of Affected Country Parties: Explanatory Note and Help Guide'. ICCD/CRIC(3)/INF.3, 19 November. <www.unccd.int/cop/officialdocs/cric3/pdf/inf3eng.pdf> (December 2005).
United Nations Convention to Combat Desertification (2005a). 'The Bottom-Up Approach'. <www.unccd.int/ngo/menu.php> (December 2005).
United Nations Convention to Combat Desertification (2005b). 'Glossary'. <www.unccd.int//knowledge/glossary-eng.html> (December 2005).
United Nations Development Programme's Office to Combat Desertification and Drought (UNSO) (1999). 'A Preliminary Overview of National Action Programme Processes of the United Nations Convention to Combat Desertification and Drought'. Report prepared as a contribution to the deliberations of COP-2. <www.undp.org/seed/unso/lessons/trad/documemts/nap-eng.pdf> (December 2005).
United Nations General Assembly (1992). 'Establishment of an Intergovernmental Negotiating Committee for the Elaboration of an International Convention to Combat Desertification in Those Countries Experiencing Serious Drought and/or Desertification, particularly in Africa'. A/RES/47/188, 22 December. <www.un.org/documents/ga/res/47/a47r188.htm> (December 2005).
United Nations General Assembly (1996). 'Organization of Scientific and Technological Cooperation'. Intergovernmental Negotiating Committee for the Elaboration of an International Convention to Combat Desertification in Those Countries Experiencing Serious Drought and/or Desertification, Particularly in Africa, Ninth Session, 3–13 September, A/AC.241/57. <www.unccd.int/cop/officialdocs/incd/pdf/24157eng.pdf> (December 2005).

Chapter 8

Promoting Good Governance through the Implementation of the UNCCD

Lene Poulsen and Masse Lo

Agenda 21, which was created at the United Nations Conference on Environment and Development (UNCED) in Rio de Janeiro in 1992, marked a global commitment to apply the basic elements of what is now known as 'good governance', including accountability, participation, predictability, and transparency in order to ensure the best use of resources in a sustainable manner.[1] These principles were further developed in the Millennium Development Goals (MDGs) and during the World Summit on Sustainable Development (WSSD), which called for good governance with the participation of civil society, the private sector, and governments in partnerships to fight poverty and promote sustainable development.[2] At the national level, the WSSD Plan of Implementation states that good governance should be based on democratic institutions responsive to the needs of the people, rule of law, anticorruption measures, gender equality, an enabling environment for investment, sound environmental, social, and economic policies (United Nations 2002b).

Based on Agenda 21's recommendations, the UN General Assembly (UNGA) established the Intergovernmental Negotiating Committee (INCD) to draft the UN Convention to Combat Desertification and Drought (UNCCD) in 1992.[3] The INCD process was in itself a critical development in the approach to negotiating international agreements and the sessions included not only the parties to the convention (that is, the governments), but also accredited nongovernmental organisations (NGOs) as well as intergovernmental and international institutions, each playing important and often complementary roles in the negotiating process. Although NGOs were not official members of the INCD and therefore did not have a vote, they did have a voice in the discussions and were often consulted both as a group and individually (see Chapter 7). Indeed, several official delegations included representatives from NGOs and the private sector. The UNCCD bears the marks of these diverse influences, and it is probably the global environmental treaty that gives the greatest importance to partnership building among government and nongovernmental institutions at all levels. Furthermore, the UNCCD specifically obliges the parties to integrate all key stakeholder groups in its implementation.[4]

Good governance principles within the UNCCD are established in various provisions for natural resource management and people-centred development. One critical commitment to good governance is strengthening the role of women in decision making and implementation; another is the continuous response to drought and desertification challenges through iterative planning processes based on institutionalised partnerships, with proper representation of all stakeholders and facilitated by effective communication structures. As such, the UNCCD uses good governance as a means to combat desertification and as an end to ensure sustainable dryland development. The understanding of this complex is extremely important for any appreciation of the UNCCD's effectiveness.

The following chapter briefly introduce the concept of good governance, describes how the UNCCD provides for good governance, and shows examples of responses and challenges to implement the good governance principles of the UNCCD.

The Meaning of Good Governance

Although the concept of governance is used widely within the development community, there is still no single definition of the term. Often, governance is referred to as a buzzword for different kinds of development processes. When the concept is qualified by adding 'good', there seems to be even more ambiguity about the exact meaning apart from alluding to something positive. Without judging about the appropriateness of different definitions of good governance for different contexts, this chapter applies a working definition for the analysis here based on the suggestion from the Global Drylands Partnership that was presented at the Interactive Dialogue Session at the fifth session of the UNCCD Conference of the Parties (COP[5]):

> Good Governance is a social contract between the government and the people in a nation based on a mutual understanding of the rights and responsibilities of all partners.

While the primary focus of this chapter is the national implementation of the UNCCD, it should be noted that good governance at the national level is defined *inter alia* by agreements made by governments at regional and global levels.

In order to adhere to the social contract, a certain number of basic elements need to be in place, particularly:

- *Participation*: civil society involvement and structures to ensure appropriate representation such as decentralised organisations, partnership-building mechanisms, and consensus orientation in decision making;
- *Equality and equity*[6]: equal opportunities for all stakeholder groups to participate in the design and implementation of development programmes;
- *Openness and transparency:* interactive platforms and multi-directional communication structures for information, feedback, easily identifiable structures, and debate;
- *Responsiveness*: the ability of the institutional set-up to respond to the needs of the local populations;
- *Effectiveness and efficiency*: processes and institutions that make the best use of human, institutional, technical, and financial resources;

- *Predictability*: strategic vision and the leadership to implement it, and a certain degree of institutionalisation of the organisational set-up, including procedures;
- *Coherence*: policy alignment, for instance with instruments such as the Beijing Platform for Action (see UN 1995);
- *Legitimacy and accountability*: the trust of stakeholders and their consent to the manner in which power is exercised and resources managed.

As shown by many multilateral and bilateral development agencies, as well as declarations at international conferences such as the WSSD and the International Conference on Financing for Development, there is a broad understanding and recognition within the development community that these eight elements are central and key to good governance (see, for example, UN 2002a). It is furthermore recognised that the elements are mutually supportive and reinforcing in practice and the categorisation can therefore be seen as abstract and conceptual in nature, which might cause some difficulties when applied to an analytical framework. However, a general assessment of the way these eight basic elements are addressed and implemented will define the level and quality of the adherence to good governance.

Provisions for Good Governance in the UNCCD

Although good governance as a concept was referred to explicitly during the INCD process, the final text only refers to good governance implicitly. According to Ravi Sharma (1993), good governance issues were integrated into the UNCCD partly to please donors contrary to the position of some international NGOs that found that the UNCCD should rather focus on global political issues. Moreover, many African NGOs found that if donors had already decided not to invest new development funds in the implementation of the convention and as such to promote global environmental governance, they should not dictate special clauses on good governance within countries affected by drought and desertification. Still, good governance elements were not only promoted by donors during the negotiation but also by affected developing countries.[7] Madagascar, Cape Verde, and Burkina Faso, for instance, highlighted the central role played by women in natural resource management and hence the need for specific attention to their role in the UNCCD and an overall gender sensitive approach (Poulsen 2003). Furthermore, NGOs made a joint statement to the INCD that emphasised the centrality of community-led activities in combating desertification and lobbied for emphasis on implementation of the convention at the local level.

The very objective of the UNCCD outlines that activities to combat desertification and drought will be based on the principles of Agenda 21, which in turn emphasise partnership building, accountability, legal regulatory frameworks, gender equality, co-ordination and co-operation, and transparency and openness through effective information structures (UNCCD 1994).

At the centre of the implementation of the UNCCD are the action programmes, primarily at the national level (NAPs) and supported by sub-regional and regional action programmes (SRAPs and RAPs).[8] The action programmes are therefore the reference points for how well the convention's objective and provisions are being met and how well good governance principles are being put into practice. In fact, articles 9 to 15 of the UNCCD reflect a series of good governance principles, stating that the action programmes should be:

- made public;
- built on existing efforts;
- harmonised, co-ordinated, and integrated with other major and relevant frameworks;
- participatory with effective participation of both men and women;
- all inclusive with multi-stakeholder participation and partnership building as major elements;
- clear on the specific role of each stakeholder;
- flexible with iterative planning, allowing for necessary modifications according to a changing environment;
- regularly monitored and evaluated; and
- predictable and based on long-term planning with a certain security regarding the resources necessary for implementation.

Moreover, the action programmes should include legal frameworks and actions for institutional strengthening and capacity development, including capacities for assessment and systematic observations.

An analysis of the UNCCD text shows that all major good governance elements are taken into consideration in the various provisions of the convention as can be seen in the following.

Participation

The UNCCD calls for active participation of a number of different stakeholder groups, particularly governments at all levels, donors, NGOs, populations in affected areas including landholders and users, girls and women and youth, as well as multilateral and intergovernmental organisations, scientific institutions, and the private sector.[9] In order to ensure the full and effective involvement of stakeholders at all levels, governments — as the overall responsible partners of the implementation of the UNCCD — are requested to create an enabling environment that will include awareness raising, open and transparent planning, participatory implementation and monitoring with representatives from all stakeholder groups, and promotion of partnerships among stakeholders. In fact, the UNCCD includes more than 50 references to local populations, communities, and NGOs in conjunction with participation and participatory activities.

Equality and Equity

The UNCCD has often been highlighted for the committed effort to strengthen women's role at all levels of its implementation — an element that is lacking in most other multilateral environmental agreements (MEAs) such as the Convention on Biological Diversity (CBD) and the UN Framework Convention on Climate Change (UNFCCC) (see, for example, Poulsen 2003). The UNCCD's preamble recognises that women play an important role in dryland development but are often absent in planning and monitoring activities. Special activities are therefore required to promote equality among men and women, and the UNCCD calls for special attention to women when designing and implementing awareness raising campaigns and participatory structures (UNCCD 1994, art. 5[d]). Moreover, the NAPs shall 'provide for effective participation at the local, national and regional levels of non-governmental organizations and local populations, both women and men, particularly resource users, including farmers and pastoralists and their representative organizations, in policy planning, decision-making, and implementation and review of national action programmes' (art. 10.2[f]). Furthermore, the convention provides for special support measures such as capacity building to enhance women's active participation (art. 19.3).

Openness and Transparency

Information collection and exchange are called for as a crucial means to achieve most of the intentions stated in the UNCCD, such as harmonisation with other planning instruments, avoidance of duplication, effective partnerships at all levels building on trust and co-operation, informed decision making, use of appropriate technologies, the inclusion of traditional knowledge, public awareness raising, effective resource mobilisation, and monitoring and evaluation. The parties shall 'exchange and make fully, openly and promptly available information from all publicly available sources relevant to combating desertification and mitigating the effects of drought' using existing information systems when possible (UNCCD 1994, art. 6[f], 18.1[a]). Moreover, the parties shall 'organize awareness raising campaigns for the general public' on the importance of the objectives of the UNCCD (art. 19.3). The UNCCD also calls for measures to inform about potential funding mechanisms for dryland development (art. 21).

Responsiveness

In general, the participatory and decentralised structures promoted by the UNCCD, combined with iterative and flexible planning processes, aim at increasing responsiveness to the needs of the dryland populations.[10] Furthermore, the UNCCD promotes increased knowledge and understanding of barriers and threats for dryland development; as spelled out in the provisions of articles 17 and 18, for example, which call for research and development and the transfer of know-how, the parties are requested to increase the knowledge of causes leading to desertification and vulnerability to drought, taking into account the needs and capacities of different stakeholder groups. In addition, affected countries shall allocate adequate resources to dryland development and establish national desertification funds (NDFs) to ensure rapid and efficient responses to local needs (UNCCD 1994, art. 5[a], 21.1[d]). The UNCCD also promotes a number of specific measures to increase responsiveness such as the establishment of early warning systems.

Effectiveness and Efficiency

The parties shall 'better focus financial, human, organizational, and technical resources where they are needed' (UNCCD 1994, art. 3[b]). This shall be done *inter alia* through co-ordination and integration with other planning instruments (such as art. 9) and budgetary and administrative procedures to increase efficiency (art. 13.1[d]). Furthermore, the parties shall establish capacity building programmes to allow all stakeholders to be active partners in the planning and implementation of dryland development activities(art. 19). In order to optimise resources, the parties are requested to make full use of NGOs and other relevant organisations and to ensure effective co-operation, which could involve promotion of legal frameworks (art. 19.1[j]).

Predictability

The action programmes as described in articles 10 and 11 are intended to be institutionalised as active and iterative planning and implementation processes which by themselves should ensure predictability.[11] Furthermore, according to the general principles of the convention, the parties shall develop a coherent long-term strategy for its implementation (UNCCD 1994, art. 4). /But the UNCCD also recognises that financing is crucial for ensuring its implementation (art. 20.1). Its financial provisions therefore become critical for the predictability of its implementation. As such, action programmes shall be supported by 'financial cooperation to provide predictability' and allow for long term planning (art. 13.1). In addition, developed countries shall provide financial resources, including new and additional funding and funding from the private sector and NGOs (art. 6). Moreover, the COP is requested to ensure regular information on available funding for affected countries (art. 21.1[c]; see also Chapter 9).

Coherence

The parties shall ensure that dryland development concerns such as drought vulnerability and land degradation be mainstreamed into poverty eradication efforts and sustainable development policies (UNCCD 1994, art. 4, 9). Furthermore, synergy with other relevant planning instruments, particularly the CBD and the UNFCCC, shall be promoted to maximise the use of human, institutional, and financial resources (for example, art. 8).

Legitimacy and Accountability

The COP shall monitor the implementation of the UNCCD, including review the appropriateness of the institutional arrangements. Furthermore, according to article 22 the COP and its subsidiary bodies shall make recommendations to enhance the implementation of the UNCCD at all levels, for example in response to national reports that are now presented to the Committee for the Review of the Implementation of the Convention (CRIC), and action programmes shall be reviewed regularly as part of the iterative and participatory planning process promoted by the convention (UNCCD 1994, art. 10.3, 11; see also UNCCD 2001).[12] Again, the relatively strong emphasis given by the UNCCD to establishment of effective information systems as described above should also lead to increased accountability regarding the implementation of the UNCCD. Finally, the parties shall ensure the existence of legal frameworks that will enable the implementation of the UNCCD and its stated objectives.

Regional Annexes

In addition, the regional annexes — by themselves an expression of the importance of subsidiarity and potential mechanisms for promoting good governance — stress issues such as participation, information, coherence, and harmonisation. This is further explained in Ethiopia's NAP from 1998, which states that the call for public participation in articles 8 to 10 of the Regional Implementation Annex for Africa implies the need to develop decentralised and democratic governance structures (Federal Democratic Republic of Ethiopia 1999).

Information

As mentioned above, good governance principles such as transparency, participation, equality, capacity building, and accountability all depend on effective and multi-directional information structures. Although the UNCCD includes a number of provisions for smooth communication and information systems such as awareness-raising campaigns made available in local languages and prompt availability of all publicly accessible sources of information relevant for dryland development, the measures are mostly one-way communication and little is provided to ensure institutional set-ups that will allow all partners to be in constant communication (UNCCD 1994, art. 19.3[d], 16[f]).

Different Scales

As should be expected from a global legal framework, the review of good governance principles contained in the UNCCD shows that the text offers a general description of what should be done and not an exact manual on the 'how to'. Furthermore, the UNCCD led to common definitions for many terms that hitherto had numerous interpretations such as the very term of 'desertification' (see Chapter 2), but concepts related to governance such as participation, representation, and equality are not defined in the UNCCD. What is meant, for instance, when the convention states that 'all stakeholders' shall be represented? Does the concept of 'local population' have the same connotation in China, with more than 1.6 billion people, as in Nauru, with a population of 13 000 people?[13] What is meant by 'participation'?[14] What does it mean when some of the regional annexes use the term 'consultation', which, for many sociologists, implies that stakeholders can express their opinion, which might or might not be taken into account in the final decision making — or is 'consultation' just used as a synonym for 'participation'?

Overall Framework for Good Governance

The convention should thus be seen as an agenda for action programmes to be developed at local, national, sub-regional, and regional levels with proper development of methodologies and operational guidelines that take into account local conditions. In support of this principle, a number of UN agencies and NGOs took the initiative to develop a series of operational guidelines before the UNCCD entered into force in December 1996. These guidelines put particular emphasis on the institutional aspects and good governance principles of the UNCCD. The UNCCD secretariat, for instance, prepared documentation on awareness-raising campaigns, while the United Nations Development Programme (UNDP), in collaboration with NGOs, developed specific guidelines for partnership building, participatory approaches, gender mainstreaming, NGO networking, and NDFs.[15]

Still, the text of the UNCCD is more prescriptive on a number of good governance issues than other global environmental governance instruments are. For example, it calls for educational and adult literacy programmes and opportunities for all, particularly for girls and women, on the identification, conservation and sustainable use and management of the natural resources of affected areas (UNCCD 1994, art. 19.3[e]). In comparison, the CBD does not include any special provisions for promoting gender-sensitive approaches, although its preamble recognises that women should participate fully at all levels of policy making and implementation (Poulsen 2003).

The convention contains several rather specific suggestions for more technical solutions to mitigate the impact of drought and combat land degradation, such as enhance water availability through cloud seeding (UNCCD 1994, art. 17.1[g]). This might reflect the political sensitivity to good governance principles as described at the beginning of this chapter. In fact, while good governance is increasingly integrated in international environmental governance documents, it remains a controversial subject. Part of the controversy arises from diverging assessments of the appropriateness of importing international good governance principles and practices into local development.

Another controversy arises from the concurring objectives of promoting good governance at the international and national levels. For instance, while both the 2000 Millennium Declaration and the 2002 WSSD Plan of Implementation confirm the importance of good governance at national and international levels, the discussions during the WSSD in Johannesburg on the possible inclusions of good governance principles in its chapter entitled 'Institutional Framework for Sustainable Development' were particularly contentious (United Nations 2002b; UNGA 2000). The European Union strongly promoted binding guidelines for good governance principles at national level, a position opposed by the G77, which insisted on international governance measures such as free and equitable trade rather than national governance structures.

Good Governance in the Implementation of the UNCCD

Donors' Response to Good Governance in Support of the UNCCD

By ratifying the UNCCD, donor countries have committed to promote good governance principles through partnerships with developing countries affected by drought and desertification (UNCCD 1994, art. 6, 20). The following assessment is based mainly on donors' reports on the implementation of their obligations (art. 26).[16]

Most donors highlight aspects of coherence such as policy alignment and harmonisation as a basis for their support to the UNCCD. As such, the majority of donors to the UNCCD have adopted the MDGs as the overall framework for official development assistance (ODA). The United Kingdom's Department for International Development (DFID) states for instance that the support to the UNCCD is provided through MDG-1, which aims to halve extreme poverty by 2015, and through MDG-7, which aims at environmental sustainability (UK 2004). Similarly 'new' donor countries such as Poland identify the MDGs as the overall framework for ODA, including support to the UNCCD (Republic of Poland 2004). Likewise, donors also give particular attention to mainstreaming of the UNCCD within overall development efforts, particularly poverty eradication (see also chapters 9 and 10). Sweden (2004) states that the goal for Swedish support to the UNCCD is the integration of dryland issues into bilateral and regional development co-operation programmes.

Many donor reports recognise the special nature of the UNCCD with its strong emphasis on sustainable livelihoods and poverty eradication and therefore also the need for special attention to institutional arrangements that will promote issues such as equality and participation. For instance, according to Sweden (2004), the challenges for achieving the objectives of the convention do not lie in the technical solutions alone but also include the efficiency of institutions and the capacity for stakeholder participation.

In line with the increasing reference to good governance in the development discourse, the concept is also increasingly highlighted in donor reports to the COP/CRIC. New Zealand (2002) describes the country's guiding principles for development assistance as including equity and participation through good governance. However, the exact role of good governance as a means for dryland development or as an end is not clear. The United States (2002) confirms that the UNCCD's development orientation corresponds with the mandate of the U.S. Agency for International Development (USAID) not only to help protect the world's environment but also to spur economic growth and good governance. USAID supports the participation of civil society, the private sector, and other major groups as 'engines of growth and as essential for good governance' (8). However, Canada (2005) highlights in its International Policy Statement that considering the complexity of desertification special emphasis is given to formation of enabling environments, including good governance.

In its 2003 report Sweden (2004, 4) emphasises good governance as a means to sustainable development, stating that 'decentralisation combined with empowerment of the local population is conducive to sustainable use of natural resources, if supported by proper regulatory frameworks, sound economic policies and good governance'. This statement also shows one of the ambiguities in the understanding of the concept of good governance and its components, which in theory include decentralisation and the participation of local stakeholders. In fact, many donor countries make similar development policy statements. For instance, in 2002 Finland (2002) states that the country's ODA is based on equality, good governance, democracy, and the respect for human rights.

The ambiguity and confusion about the understanding of the exact meaning of the good governance concept, as illustrated by these examples, has further implications for the partnerships between donors and affected developing countries inasmuch as many donors insist that their ODA is based on the partner countries' commitment to good governance (see, for example, the Netherlands 2004; Finland 2004; Kingdom of Belgium 2002). But as stated by David Brown, Kathrin Schreckenberg, Gill Shepherd, and Adrian Wells (2003, 2), 'good governance is a powerful concept but it could easily become no more than trendy jargon if steps are not taken to tie down its meaning'.

Still, it is important to note that the review of donors' dryland development policies and concrete support to the implementation of the UNCCD shows a commitment to promote good governance. Indeed, some donor reports actually recognise certain limitations in their dryland development support that have an impact on the implementation of good governance principles. For instance, while recognising that good governance is required to counteract the malfunctioning of the land use systems, Sweden (2002) also acknowledges some limitations in its support to the UNCCD, including the lack of a coherent long-term strategy for dryland development and the limited attention to promoting synergies between the different conventions.

The approval in 2002 of land degradation as an additional focal area of the Global Environmental Facility (GEF) and the subsequent allocation of US$500 million for projects to promote sustainable land management in developing countries between 2003 and 2006, constitute the largest financial commitment to UNCCD-related activities (see Chapter 9). It is therefore also important that the framework that establishes the criteria for GEF funding for sustainable land management calls for coherence through mainstreaming of land degradation into development frameworks, partnership building, and fund allocation to promote effective stakeholder participation in project design, implementation, monitoring, and evaluation, thereby increasing the legitimacy and accountability of the activities (GEF 2003). Furthermore, the GEF will fund information management systems to support decision making at national and local levels on integrated land-use planning and management.

Affected Developing Countries' Integration of Good Governance in the Implementation of the UNCCD

The following review of the application of good governance principles in the implementation of the UNCCD in countries affected by drought and desertification is mainly based on the NAPs prepared by affected countries as well as on national reports prepared by the parties for the COP and the CRIC. The national reports follow a set of guidelines that require parties to review, among other things, the participatory approaches in the structures and mechanisms established for implementing the convention with a special focus on national co-ordination bodies or steering committees, including the means to identify stakeholders.[17] In terms of equality, the guidelines suggest information on the number of men and women participating in different institutions as well as specific activities to promote the role of women.[18] Furthermore, the guidelines request information on communication structures, efficiency, legal measures, and the predictability in form of long-term strategies and funding prospects.

As of January 2005, more than 70 affected developing countries have developed long-term strategies for dryland development through the preparation of NAPs. Initially, awareness-raising campaigns were organised in most affected countries, establishing a good basis for public information and participation in the development of NAPs.

Most countries have established national steering committees, typically with the participation of various line ministries and representatives from NGOs and other civil society organisations with the potential for promoting a series of the good governance principles, including openness and transparency, responsiveness, predictability, and accountability. Several countries, such as Swaziland (2000), include representatives from the international community in their steering committees, which in principle should allow for better co-ordination and coherence as well as planning based on realistic funding expectations — and thereby increased predictability. Furthermore, some countries such as Eritrea (2002) have established a system of regional and local coordinators for the NAP process, thereby allowing for improved responsiveness.

As part of the initial NAP preparation activities, most countries organise multi-stakeholder workshops to identify major dryland problems and possible solutions. The identification of participants for these workshops has often included detailed background analysis and public information campaigns. Argentina's NAP, for instance, was developed through discussions with major stakeholder groups in 1995 identified by the provisional steering committee, and included NGOs, producer associations, women's groups, and representatives from local governments (Republic of Argentina 1997). The discussions with the stakeholder groups focussed on drought and desertification challenges, networking, education, capacity building and awareness raising, institutional and legal aspects, and potential funding mechanisms, and formed the basis for a draft NAP that was later presented to the stakeholders for further consultation.

Still, a number of national reports and action programmes recognise the limitations to the participatory approaches to implementing the UNCCD. This is summed up well in the SRAP West Asia, which states that the limited participation of civil society is one of the major challenges in the region aggravated by 'the weakness — or in some cases absence — of national legislation, (especially laws on transparency and accountability of the governance system), [that] has impeded effective participation and hence, compliance' (UNCCD 2000, 10).

In terms of equality among men and women, many countries plan concrete activities to promote women's participation. Meanwhile, reports and NAPs do not provide any specific information as to how countries will promote equality between men and women in dryland development (Poulsen 2003). In terms of concrete implementation, many countries report specifically on the involvement of women in the NAP process (for example, Mongolia and Barbados) and some have taken special measures to counteract cultural challenges for women and men to participate in the same meetings. Yemen, for instance, has established separate male and female commissions for its NAP. Some countries, including Zambia, Tunisia, and Kenya, have established special budget lines in their NAPs for activities to strengthen the role of women and gender mainstreaming. Generally speaking, these budget allocations represent less than 3 percent of the overall NAP budget.

Many countries have paid special attention to coherence and effectiveness in the design of their NAP processes, particularly alignment of the NAP with national programmes to eradicate poverty, which is highlighted in the majority of country reports submitted between 2002 and 2004. Uganda, for instance, has launched activities to mainstream its NAP with the national development framework, the Poverty Eradication Action Plan, which focusses on accountability, transparency, participation, and decentralised administration (Republic of Uganda 2004). Burkina Faso (2004) realised in 2002 that its NAP process had been carried out in isolation and it became clear to its national steering committee that the approach was not sustainable because funding could not be mobilised. Consequently, a study on synergies was undertaken and formed the basis for a mainstreaming strategy of the NAP into the country's poverty reduction strategy.

Similarly, many countries have harmonised their NAP processes with National Environmental Action Plans (NEAPs) to allow for better use of limited resources and thereby increase efficiency and effectiveness. In Bangladesh, for instance, the NEAP preparation is based on an extensive eight-month consultative process at community level, which fed into the NAP process (People's Republic of Bangladesh 2002). However, normally the NEAP does not aim at continuous and iterative planning processes as the NAPs do, and the lack of sufficient additional organisational measures limit the effectiveness of NEAPs in terms of promoting the UNCCD's principles of institutionalised good governance structures.

Algeria states that regardless of the decentralisation of sustainable development activities, actions at local levels are still inadequate due to limited capacity and lack of financial resources (Republic of Algeria 2004). Local workshops, a newsletter prepared by NGOs, and the UNCCD focal point for all stakeholders have been discontinued.[19] Similarly, both Burkina Faso and Mali report that despite decentralised rural development policies, community-based organisations still lack the capacity to participate fully in the NAP process partly due to the slow transfer of resources and power to the local level (Burkina Faso 2004; Republic of Mali 2004).

Egypt states that although technical solutions to drought and desertification are well developed, it encounters capacity problems in responding to the requests of the UNCCD regarding participatory approaches (Arab Republic of Egypt 2004). Consequently, its NAP is still under development. Similarly, newly independent states such as Georgia (2002) states that the anti-desertification efforts during the Soviet era were based on technical solutions and that the restructured administrations need capacity building for local governance.

A large part of the information and communication in the NAP processes is based on single events and often one-way communication such as awareness-raising campaigns, consultations, and websites reaching a limited numbers of stakeholders. Several of the 2004 CRIC reports show that communication is a limiting factor for a participatory and representative NAP process. Namibia (2004), for instance, states that communication between the civil society representatives on its steering committee and the rest of the communities is insufficient; Senegal states that the legal framework for its NAP process is still unknown to most stakeholders and concludes that the framework for civil society harmonisation and co-ordination does not function (Republic of Senegal 2004). Other communication problems are linked to dependence on ordinary mail and the limited opportunities for organising representative meetings due to transportation difficulties. Even within the steering committees, many countries encounter severe communication limitations such as reported by Tajikistan in 2002 (Republic of Tajikistan 2002). The Solomon Islands (2002) have established a special government information service that in principle would be responsible for promoting effective information and communication on UNCCD-related issues. However, lack of resources has rendered the service ineffective, leaving the information management for dryland development deficient.

The Aarhus Convention on access to information, public participation in decision making, and access to justice in environmental matters, which came into effect in 2001, emphasises the need for citizen participation in environmental issues (UN Economic Commission for Europe 1998). Hence, environmental information held by the public authorities should be made accessible to all citizens, including information related to drought and desertification. According to Belarus's national report, a national co-ordination body with participation of NGOs and the environmental ministry was established in 2001 to implement the Aarhus Convention; this body is expected to play a facilitating role for the implementation of the UNCCD (Republic of Belarus

2002). On the other hand, Georgia (2002) reports that the structures established for implementing the Aarhus Convention do not involve local organisations due to lack of resources. In general, although the provisions of the Aarhus Convention in principle should further instruments such as the UNCCD, the actual implementation is still limited and no communication strategies have been established on drought and desertification.

Monitoring and evaluation systems have been established in many countries but some national reports highlight the deficiencies of the present systems and consequently the limited accountability and transparency (see, for example, Republic of Djibouti 2004; Burkina Faso 2004; Republic of Algeria 2004). Likewise, activities for the development of indicators have been launched in several countries, but socioeconomic and process indicators are still rather weak as reported by, for instance, Egypt and Burkina Faso, who suggest that issues such as transparency and capacity for participation are still not covered satisfactorily by the indicators (Arab Republic of Egypt 2004; Burkina Faso 2004).

A number of the problems encountered in applying the principles of good governance to the implementation of the UNCCD are linked to the challenges for designing an effective organisational set-up to accommodate the cross-sectoral characteristic of the convention (see, for example, Republic of Algeria 2004). Likewise, Jordan reports that the overlap of UNCCD-related responsibilities among various government bodies has created inertia and delay in the NAP process (Hashemite Kingdom of Jordan 2002). Part of the problem is the location of the UNCCD within the ministries of environment or agriculture as well as insufficient legal measures to provide the focal point and steering committee with sufficient authority to undertake effective interdisciplinary activities. In Burkino Faso, during a recent administrative reorganisation it was suggested to move the national steering committee to the office of the prime minister. However, the move was halted and the steering committee is now within the National Council for the Environment and Sustainable Development (Burkina Faso 2004).

Other countries such as Cape Verde report about severe limitations to the efficiency of the steering committees due to lack of resources and capacity in integrated natural resource management (Republic of Cape Verde 2004). Similarly, Namibia (2004) reports the discontinuation of its NAP steering committee as a result of the conclusion of funding by the Deutsche Gesellschaft für Technische Zusammenarbeit (GTZ, or German Technical Development Co-operation) after almost ten years of close collaboration. At the time of the Namibian reporting in 2004, the future of the NAP and related activities remained unclear. Likewise, Mali's steering committee did not meet at all in 2002 and 2003 due to a decline in external funding (Republic of Mali 2004).

In some countries, the national legal frameworks actually promote good governance according to the national report. Ethiopia's Transitional Charter and Constitution, for instance, includes provisions for decentralisation and local participation that create the necessary structure for good governance (Federal Democratic Republic of Ethiopia 2004). On the other hand, countries such as Uganda describe that the different laws affecting decentralisation are incompatible (Republic of Uganda 2004).

Although the UNCCD does not specifically address countries with complex emergencies, sustainable natural resource management based on good governance principles can play a special role in post-conflict situations.[20] In fact, countries such as Liberia (2002), Comoros, (2004), and Haiti (2002) highlight in their national reports the role of good governance programmes in the reconstruction phases, and Burundi (2004) reports that, in spite of some NAP preparatory activities organised by an NGO, the complex emergency in that country implies that there is still no NAP process (Republic of Liberia 2002; Comores 2004; Republic of Haïti 2004; Burundi 2004).

In general, affected developing countries are committed to applying the principles of good governance in implementing the UNCCD, and a number of good practices in the preparatory phases of the NAP can be observed. However, they also recognise the severe limitations that exist in putting good governance into practice due to budgetary constraints, dependence on outside funding, inappropriate organisational arrangements, and the lack of capacity at various levels for particularly participatory approaches.

Nonetheless, it is important to note that the NAP process has shown strength in a number of contexts as an innovative and participatory mechanism. As such, the participatory process developed through the implementation of the UNCCD has been used for other planning instruments, including a revision of the poverty reduction strategy in Burkina Faso and Cape Verde's preparation of the National Environmental Action Plan 2004–14 (Burkina Faso 2004; Republic of Cape Verde 2004).

Conclusion

Several conclusions can be drawn from this review of the principles of good governance promoted by the UNCCD, the actual implementation of those principles by countries affected by drought and desertification, and the specific support promoted by donors.

First, the institutional framework suggested in the UNCCD has the potential to promote good governance through the implementation of NAP processes. Most donors have declared their support to good governance and affected countries have established structures to promote participation and equality, particularly in the design of the NAP processes. Furthermore, many affected countries have realised that effective dryland development requires aligning the NAPs with major planning instruments such as poverty reduction plans. However, even the best planned NAP processes can be effective only if designed according to a sustainable funding structure that can ensure the continuity of participatory approaches, information and communication systems, and so on, and the lack of sustainable funding mechanisms has shown to be a major obstacle to implementing good governance principles.

Second, the review process established by the COP, which is based on regular national reports, is an important instrument for strengthening the openness, transparency, accountability, and efficiency of the UNCCD, particularly the NAP processes. As the review process follows the guidelines developed by the CRIC, the explicit integration of the principles of good governance is seen as essential. From the perspective of good governance, it is therefore very positive that the guidelines focus on issues such as participatory approaches, monitoring and evaluation, communication structures, co-ordination and harmonisation, legal measures, and funding prospects.

Third, one can choose to take an optimistic point of view and recognise that many principles of good governance are reflected in the implementation of the UNCCD. As such, it is positive, for instance, that more than 2000 stakeholders could participate in preparing the NAP document in Argentina, 925 in Uganda, and 1700 in Iran. However, compared to the total dryland population, these numbers are insignificant, particularly if no efficient communication structure is in place to ensure the proper and equitable representation of all stakeholders.

Finally, it is important not to lose sight of the innovative aspect of the convention, which calls for flexibility and the need to develop and try new methods and approaches. Furthermore, this implies the need for effective knowledge management with the sharing of lessons learned among the various parties to the UNCCD and, in fact, the acceptance that not all new methods will be effective or replicable. This is particularly true for the implementation of the principles of good governance.

Notes

1 Agenda 21 was the major outcome of the UNCED, and sets out how to ensure economic, social, and environmental sustainability in a number of different fields. See UNCED (1992).
2 The MDGs set clear targets on, *inter alia*, poverty, hunger, disease, illiteracy, environmental degradation, and discrimination against women by 2015 and promote partnership building between developed and developing countries (United Nations General Assembly [UNGA] 2001; see also UN Millennium Project 2005).
3 UNGA created the INCD in response to Agenda 21's call for the preparation of a UN convention to combat desertification and drought. INCD was organised in January 1993 and held five substantial sessions before the adoption of the final text of the UNCCD on 17 June 1994. During the interim period between the adoption of the convention and its entry into force in December 1996 (with the ratification of 50 countries), the INCD met six times.
4 The implementation of the UNCCD includes concrete actions in the field as well as the creation of an enabling environment through policies, mechanisms, structures, and institutions to promote concrete dryland development activities.
5 The Global Drylands Partnership is a joint effort to address dryland issues, and includes the Canadian International Development Agency (CIDA), UN agencies, foundations, and NGOs.
6 Equality and equity are often defined as synonymous, but for the purposes of this discussion they refer to equal opportunities for various stakeholder groups to participate. This does not necessarily translate into equal participation with, for instance, equal numbers of women and men participating in meetings. While equality and equity in principle refer

to equal opportunities for a number of groups — such as young and old, rich and poor, rural and urban — the term here is mainly used to refer to equal opportunities for men and women.

7   Affected developing countries refer to developing countries as defined by the UN and that are affected by drought or desertification, or both.

8   For a detailed analysis of NAPs and their implementation in West Africa, see Chapter 10 by Richard Pearce.

9   The use of the term 'landholders' is unexpected considering the overall spirit of the UNCCD, and seems to refer to land users in general; it might be the result of a oversight when the text was being finished in the wee hours of 18 June after two weeks of intense final negotiations (the text was officially finished on 17 June 1994). The reference to 'women and girls' as a stakeholder group can be misleading. Apart from an overall division of all people into two stakeholder groups, namely men and women, women are not a stakeholder group as such — there are women farmers, women ministers, and so on — and the lumping of women and girls together can also cause some problems as the two groups generally have very different needs and priorities. While there is a reference to youth as a stakeholder group, their special needs or interests in relation to dryland development are not spelled out in the UNCCD.

10   Subsidiarity, which is the principle that activities should be implemented at the lowest level possible, is closely linked to decentralisation, improved possibilities for openness, and transparency and participation, as well as responsiveness. As such, many authors include subsidiarity as a basic principle for good governance (see, for instance, UN Human Settlements Programme 2002). Subsidiarity seems to be an unwritten but overall accepted principle in the implementation of the UNCCD in the sense that the convention promotes the principle of doing things at the lowest possible level.

11   'Institutionalise' is used here as in sociology to describe well-established and structured patterns of behaviour that are approved and accepted by all stakeholders because they believe that they are important and as such will sustain them. In another words, the action programmes will never become effective if seen as an obligation imposed from outside and from which the actual players do not see any benefits. Concepts such as ownership and sustainability are closely linked to the concept of institutionalisation.

12   The CRIC was established by COP-5 in 2001 as a subsidiary body to review reports on implementing the UNCCD prepared by countries and other partners such as UN agencies.

13   It should be noted that Nauru has not developed a NAP *per se*, but the National Land Use Committee with government representatives and local leaders is responsible for the country's implementation of the UNCCD.

14   Here it should be remembered that 'assistance' does not necessarily mean 'active participation'.

15   This initiative was taken up by the UNDP's Office to Combat Desertification and Drought (UNSO), which has now been transformed into the Dryland Development Centre.

16   In accordance with article 26 of the UNCCD, each party is required to report regularly on actions taken to implement the UNCCD. The first comprehensive reports were submitted to COP-3 in 1999 for review at the following COP in 2000). To ensure that proper attention was given to the assessment process, COP-3 also established an Ad Hoc Working Group (ADWG), followed by the establishment of the CRIC. Reports have been prepared in 1999 (Africa and donors), 2000 (Asia, Latin America and the Caribbean, Northern Mediterranean, Central America, and donors), 2002 (all parties), and 2004 (Africa and donors).

17 The guidelines are available from the UNCCD website at <www.unccd.int/cop/cric3/helpguides.php> (December 2005). National co-ordination bodies and steering committees are partnerships established to implement the UNCCD, and normally advise the national body responsible for implementing the UNCCD, which is typically the party's ministry of the environment.
18 This information does not necessarily show whether men and women have equal opportunities to participate.
19 There were 48 workshops organised locally between 2001 and 2003. The newsletter was issued regularly from 1996 to 2001.
20 Complex emergencies are defined as humanitarian crises that require international responses that go beyond the mandate or capacity of any single agency (Inter-Agency Standing Committee 1994).

References

Arab Republic of Egypt (2004). 'National Report for Combating Desertification'. Report prepared for the Third Session of the Committee for the Review of the Implementation, 2–11 May. Desert Research Center, Cairo. <www.unccd.int/cop/reports/africa/national/2004/egypt-eng.pdf> (December 2005).

Brown, David, Kathrin Schreckenberg, Gill Shepherd, et al. (2003). 'Good Governance: What Can We Learn from the Forest Sector?' Forest Policy and Environment Programme, Overseas Development Institute. <www.odifpeg.org.uk/activities/environmental_governance/IN1/good%20gov_group_paper.pdf> (December 2005).

Burkina Faso (2004). 'Troisième rapport national du Burkina Faso sur la mise en oeuvre de la Convention des Nations Unies sur la lutte contre la désertification'. Report prepared for the Third Session of the Committee for the Review of the Implementation, 2–11 May. Ouagadougou. <www.unccd.int/cop/reports/africa/national/2004/burkina_faso-fre.pdf> (December 2005).

Burundi (2004). 'Troisième rapport national sur la mise en oeuvre de la Convention des Nations Unies sur la lutte contre la désertification'. Report prepared for the Third Session of the Committee for the Review of the Implementation, 2–11 May. Ministere de l'Aménagement du territoire, de l'Environnement et du Tourisme, Bujumbura. <www.unccd.int/cop/reports/africa/national/2004/burundi-fre.pdf> (December 2005).

Canada (2005). 'Canada's International Policy Statement: A Role of Pride and influence in the World — Development'. April. <www.acdi-cida.gc.ca/IPS> (December 2005).

Comores (2004). 'Rapport national sur les mesures prises en vue de lutter contre la désertification dans le cadre de la Convention des Nations Unies sur la lutte contre la désertification'. Report prepared for the Third Session of the Committee for the Review of the Implementation, 2–11 May. <www.unccd.int/cop/reports/africa/national/2004/comoros-fre.pdf> (December 2005).

Eritrea (2002). 'The National Action Programme for Eritrea to Combat Desertficiation and Mitigate the Effects of Drought (NAP)'. Ministry of Agriculture, Asmara. <www.unccd.int/actionprogrammes/africa/national/2002/eritrea-eng.pdf> (December 2005).

Federal Democratic Republic of Ethiopia (1999). 'National Report on the Implementation of the United Nations Convention to Combat Desertification'. Environmental Protection Authority, Addis Ababa. <www.unccd.int/cop/reports/africa/national/1999/ethiopia-eng.pdf> (December 2005).

Federal Democratic Republic of Ethiopia (2004). 'The 3rd National Report on the Implementation of the UNCCD/NAP in Ethiopia'. Report prepared for the Third Session of the Committee for the Review of the Implementation, 2–11 May. Environmental Protection Authority, Addis Ababa. <www.unccd.int/cop/reports/africa/national/2004/ethiopia-eng.pdf> (December 2005).

Finland (2002). 'Report by Finland'. Report prepared for the First Session of the Committee for the Review of the Implementation, 11–22 November. Helsinki. <www.unccd.int/cop/reports/developed/2002/finland-eng.pdf> (December 2005).

Finland (2004). 'National Report by Finland for the Committee for the Review of Implementation of the United Nations Convention to Combat Desertification'. Report prepared for the Third Session of the Committee for the Review of the Implementation, 2–11 May. Ministry for Foreign Affairs, Helsinki. <www.unccd.int/cop/reports/africa/developed/2004/finland-eng.pdf> (December 2005).

Georgia (2002). 'Second National Report of Georgia on the Implementation of the United Nations Convention to Combat Desertification'. Report prepared for the First Session of the Committee for the Review of the Implementation, 11–22 November. Ministry of Environment and Natural Resources Protection, Tblisi. <www.unccd.int/cop/reports/centraleu/national/2002/georgia-eng.pdf> (December 2005).

Global Environmental Facility (2003). 'Operational Program on Sustainable Land Management'. Operational Program No. 15. <www.gefweb.org/Operational_Policies/Operational_Programs/OP_15_English.pdf> (December 2005).

Hashemite Kingdom of Jordan (2002). 'The Hashemite Kingdom of Jordan National Report on the UNCCD Impelementation'. Report prepared for the First Session of the Committee for the Review of the Implementation, 11–22 November. General Corporation for the Environment Protection, Amman. <www.unccd.int/cop/reports/asia/national/2002/jordan-eng.pdf> (December 2005).

Inter-Agency Standing Committee (1994). 'Working Paper on the Definition of Complex Emergency'. Geneva.

Kingdom of Belgium (2002). '2002 Report to the United Nations Convention to Combat Desertification (UNCCD)'. Report prepared for the First Session of the Committee for the Review of the Implementation, 11–22 November. Ministry of Foreign Affairs, Brussels. <www.unccd.int/cop/reports/developed/2002/belgium-eng.pdf> (December 2005).

Namibia (2004). 'Namibia's Third National Report on the Implementation of the United Nations Convention to Combat Desertification'. Report prepared for the Third Session of the Committee for the Review of the Implementation, 2–11 May. Desert Research Foundation of Namibia, Windhoek. <www.unccd.int/cop/reports/africa/national/2004/namibia-eng.pdf> (December 2005).

Netherlands (2004). 'Netherlands Communication to the CRIC on the Convention to Combat Desertification (CCD): Policy and Programmes in Support of the African Region'. Report prepared for the Third Session of the Committee for the Review of the Implementation, 2–11 May. Ministry for Foreign Affairs, The Hague. <www.unccd.int/cop/reports/africa/developed/2004/netherlands-eng.pdf> (December 2005).

New Zealand (2002). 'Report prepared for the First Session of the Committee for the Review of the Implementation, 11–22 November'. Report prepared for the First Session of the Committee for the Review of the Implementation, 11–22 November. Wellington. <www.unccd.int/cop/reports/developed/2002/new_zealand-eng.pdf> (December 2005).

People's Republic of Bangladesh (2002). 'Second National Report on Implementation of United Nations Convention to Combat Desertification: Final Draft'. Report prepared for the First Session of the Committee for the Review of the Implementation, 11–22 November. Ministry of Environment and Forests, Dhaka. <www.unccd.int/cop/reports/asia/national/2002/bangladesh-eng.pdf> (December 2005).

Poulsen, Lene (2003). 'Promoting the Role of Women in the Implementation of the UNCCD'. Introductory Presentation to the Global Biodiversity Forum on the Ecosystem Approach to Dryland Management: Integrating Biodiversity Conservation and Livelihood Security, Sixth Meeting of the Conference of the Parties to the UN Convention to Combat Desertification and Mitigate the Impacts of Drought, 30–31 August. Havana. <www.gbf.ch/Session_Administration/upload/paper%20Lene%20Poulsen.doc> (December 2005).

Republic of Algeria (2004). 'Rapport national de l'Algériel sur la mise en oeuvre de la Convention des Nations Unies de Lutte Contre la Désertification'. Report prepared for the Third Session of the Committee for the Review of the Implementation, 2–11 May. Ministère de l'Agriculture et du Développement rural, Alger. <www.unccd.int/cop/reports/africa/national/2004/algeria-fre.pdf> (December 2005).

Republic of Argentina (1997). 'Programa de Acción Nacional de Lucha contra la Desertificatión: Documento de Base'. Ministerio de Desarrolo Social, Buenos Aires. <www.unccd.int/actionprogrammes/lac/national/1997/argentina-spa.pdf> (December 2005).

Republic of Belarus (2002). 'National Report of the Republic of Belarus on Implementation of the UN Convention to Combat Desertficiation'. Report prepared for the First Session of the Committee for the Review of the Implementation, 11–22 November. Ministry of Natural Resources and Environmental Protection, Minsk. <www.unccd.int/cop/reports/centraleu/national/2002/belarus-eng.pdf> (December 2005).

Republic of Cape Verde (2004). 'Mis en oeuvre de la Convention des Nations Unies de Lutte contre la Désertification'. Report prepared for the Third Session of the Committee for the Review of the Implementation, 2–11 May. Ministère de l'Environnement, de l'Agriculture et des Pêches, Praia. <www.unccd.int/cop/reports/africa/national/2004/cape_verde-fre.pdf> (December 2005).

Republic of Djibouti (2004). 'Troisième rapport national sur la mise en oeuvre de la Convention des Nations Unies sur la lutte contre la désertification'. Report prepared for the Third Session of the Committee for the Review of the Implementation, 2–11 May. Ministere de l'Agriculture de l'élevage et de la mer chargé des ressources hydrauliques de Djibouti. <www.unccd.int/cop/reports/africa/national/2004/djibouti-fre.pdf> (December 2005).

Republic of Haïti (2004). 'Rapport national de la République d'Haïti sur la mise en oeuvre de la Convention sur la Lutte contre la Désertification'. Report prepared for the Third Session of the Committee for the Review of the Implementation, 2–11 May. Ministère de l'Environnement, Port-au-Prince. <www.unccd.int/cop/reports/lac/national/2002/haiti-fre.pdf> (December 2005).

Republic of Liberia (2002). 'First National Report of the Implementation of the United Nations Convention to Combat Desertification'. Report prepared for the First Session of the Committee for the Review of the Implementation, 11–22 November. Liberian National Coordinating Committee to Combat Desertification, Monrovia. <www.unccd.int/cop/reports/africa/national/2002/liberia-eng.pdf> (December 2005).

Republic of Mali (2004). 'Rapport national du Mali'. Report prepared for the Third Session of the Committee for the Review of the Implementation, 2–11 May. Ministère de l'Environnement et de l'Assainissement, Bamako. <www.unccd.int/cop/reports/africa/national/2004/mali-fre.pdf> (December 2005).

Republic of Poland (2004). 'The National Report of the Republic of Poland on the Implementation of the United Nations Convention to Combat Desertification'. Report prepared for the Third Session of the Committee for the Review of the Implementation, 2–11 May. Warszawa. <www.unccd.int/cop/reports/africa/developed/2004/poland-eng.pdf> (December 2005).

Republic of Senegal (2004). 'Troisième rapport national sur la mise en oeuvre de la Convention des Nations Unies sur la Lutte contre la Désertification'. Report prepared for the Third Session of the Committee for the Review of the Implementation, 2–11 May. Ministère de l'Environnement et de la Protection de la nature, Dakkar. <www.unccd.int/cop/reports/africa/national/2004/senegal-fre.pdf> (December 2005).

Republic of Tajikistan (2002). 'National Report of the Republic of Tajikistan to Combat Desertification'. Report prepared for the First Session of the Committee for the Review of the Implementation, 11–22 November. State Committee of the Republic of Tajikistan for Land Management, Dushanbe. <www.unccd.int/cop/reports/asia/national/2002/tajikistan-eng.pdf> (December 2005).

Republic of Uganda (2004). 'Third National Report to the Conference of the Parties on the Implementation of the United Nations Convention to Combat Desertification in Uganda'. Report prepared for the Third Session of the Committee for the Review of the Implementation, 2–11 May. Ministry of Agriculture, Animal Industry, and Fisheries, Entebbe. <www.unccd.int/cop/reports/africa/national/2004/uganda-eng.pdf> (December 2005).

Sharma, Ravi (1993). 'Desertification: NGOs Reject Cautious Approach'. *Eco* vol. 2005, no. 85, pp. 10 (27 August). <www.climatenetwork.org/eco/INCs/Eco10_0893.html> (December 2005).

Solomon Islands (2002). 'National Report to the United Nations Convention to Combat Desertification'. Report prepared for the First Session of the Committee for the Review of the Implementation, 11–22 November. <www.unccd.int/cop/reports/asia/national/2002/solomon_islands-eng.pdf> (December 2005).

Swaziland (2000). 'National Action Programme'. <www.unccd.int/actionprogrammes/africa/national/2000/swaziland-eng.pdf> (December 2005).

Sweden (2002). 'Implementation of the UN Convention to Combat Desertification: Report by Sweden'. Report prepared for the First Session of the Committee for the Review of the Implementation, 11–22 November. Swedish International Development Agency, Stockholm. <www.unccd.int/cop/reports/developed/2002/sweden-eng.pdf> (December 2005).

Sweden (2004). 'Implementation of the UN Convention to Combat Desertification'. Report prepared for the Third Session of the Committee for the Review of the Implementation, 2–11 May. Swedish International Development Agency, Stockholm. <www.unccd.int/cop/reports/africa/developed/2004/sweden-eng.pdf> (December 2005).

United Kingdom (2004). 'Report by the United Kingdom of Great Britain and Northern Ireland on Measures Taken to Support the Implementation of the United Nations Convention to Combat Desertification with a Particular Focus on Affected Developing Country Parties in Africa'. Report prepared for the Third Session of the Committee for the Review of the Implementation, 2–11 May. Department for International Development, London. <www.unccd.int/cop/reports/developed/2004/united_kingdom_of_great_britain_and_northern_ireland-eng.pdf> (December 2005).

United Nations (1995). 'Report of the Fourth World Conference on Women'. A/CONF.177/20, 17 October. Beijing. <www.un.org/documents/ga/confl77/aconfl77-20en.htm> (December 2005).

United Nations (2002a). 'Report of the International Conference on Financing for Development'. A/CONF.198/11, 18–22 March. Monterrey. <www.un.org/esa/ffd/aconf198-11.pdf> (December 2005).

United Nations (2002b). 'Plan of Implementation of the World Summit on Sustainable Development'. Johannesburg. <www.un.org/esa/sustdev/documents/WSSD_POI_PD/English/WSSD_PlanImpl.pdf> (December 2005).

United Nations Conference on Environment and Development (1992). 'Agenda 21'. 14 June. Rio. <www.un.org/esa/sustdev/documents/agenda21/english/agenda21toc.htm> (December 2005).

United Nations Convention to Combat Desertification (1994). 'Text of the United Nations Convention to Combat Desertification'. <www.unccd.int/convention/text/convention.php> (December 2005).

United Nations Convention to Combat Desertification (2000). 'Sub-regional Action Programme (SRAP) to Combat Desertification and Drought in West Asia'. Meeting on the Endorsement of the Sub-regional Action Program to Combat Desertification in West Asia, 16 February. Dubai. <www.unccd.int//actionprogrammes/asia/subregional/2000/westernasia-eng.pdf> (December 2005).

United Nations Convention to Combat Desertification (2001). 'Implementation of the Convention: Conference of the Parties Fifth Session, Geneva, 1–12 October'. ICCD/COP(5)/3/Add.1, 29 August. <www.unccd.int/cop/officialdocs/cop5/pdf/3eng.pdf> (December 2005).

United Nations Economic Commission for Europe (1998). 'Convention on Access to information, Public Participation in Decision-Making and Access to Justice in Environmental Matters'. Aahrus Convention, 25 June. Aahrus. <www.unece.org/env/pp/documents/cep43e.pdf> (December 2005).

United Nations General Assembly (2000). 'Resolution Adopted by the General Assembly: 55/2. United Nations Millennium Declaration'. A/RES/55/2, 8 September. <www.un.org/millennium/declaration/ares552e.htm> (December 2005).

United Nations General Assembly (2001). 'Road Map towards the Implementation of the United Nations Millennium Declaration: Report of the Secretary General'. A/56/150, 6 September. <www.un.org/documents/ga/docs/56/a56326.pdf> (December 2005).

United Nations Human Settlements Programme (2002). 'International Legal Instruments Addressing Good Governance'. Nairobi. <www.unhabitat.org/campaigns/governance/documents/Intl%20legal%20instruments%20addressing%20good%20gov.pdf> (December 2005).

United Nations Millennium Project (2005). 'About the Goals'. <www.unmillenniumproject.org/goals> (December 2005).

United States (2002). 'United States Activities in Support of the UN Convention to Combat Desertification'. Report prepared for the First Session of the Committee for the Review of the Implementation, 11–22 November. Washington DC. <www.unccd.int/cop/reports/africa/developed/2002/united_states_of_america-eng.pdf> (December 2005).

Chapter 9

The Global Mechanism and UNCCD Financing: Constraints and Opportunities

François Falloux, Susan Tressler, and Karel Mayrand

The United Nations Convention to Combat Desertification (UNCCD) negotiations concluded without a strong commitment by developed country parties to contribute new and additional resources to its implementation and without a consensus on the nature of the financial mechanisms that would support its implementation. This outcome differed considerably from that of the fellow conventions, the Convention on Biological Diversity (CBD) and the UN Framework Convention on Climate Change (UNFCCC), which both benefited from well-defined issues for which the global political and scientific communities had created the urgency and direction needed to instigate an influx of financing. The UNCCD's financial provisions, including the Global Mechanism (GM), were instead designed to mobilise, channel, and co-ordinate financial flows to fight poverty and land degradation in drylands.

In 1998, when the GM was established, both it and the parties to the UNCCD faced the challenge of financing a convention with a broad mandate and stagnating funding for official development assistance (ODA). At the time this chapter was written, traditional sources of funding and institutional roles in securing them had still not met expectations. Financing the implementation of the UNCCD has therefore proven to be a challenge. The reasons are complex, and lie among a number of constraints inherent to the UNCCD context. Drawing from evaluations of the GM presented to the second session of the Committee for the Review of the Implementation of the Convention (CRIC), the sixth Conference of the Parties (COP) to the UNCCD, and to the World Bank's Development Grant Facility (DGF), this chapter examines these constraints and suggests solutions to address them and maximise potential for financing UNCCD implementation.

UNCCD Financial Provisions

Articles 20–21 of the UNCCD provide for the substantive and institutional issues related to mobilising resources to implement the convention (UNCCD 1994). Article 20.1 states that 'given the central importance of financing to the achievement of the objective of the Convention, the Parties, taking into account their capabilities, shall make every effort to ensure that adequate financial resources are available for programmes to combat desertification and mitigate the effects of drought'. Moreover, under article 20.2, the parties commit to 'facilitate through international cooperation the transfer of technology, knowledge and know-how' and 'explore, in cooperation with affected developing country Parties, innovative methods and incentives for mobilizing and channelling resources, including those of foundations, non-governmental organizations and other private sector entities, particularly debt swaps and other innovative means which increase financing by reducing the external debt burden of affected developing country Parties, particularly those in Africa'.

Article 21 contains provisions for the financial mechanisms of the UNCCD. These focus on the mobilisation and improved co-ordination of existing sources with a view to maximise their impact on UNCCD implementation. Article 21(4) establishes the GM, which aims to mobilise and channel existing financial resources: 'In order to increase the effectiveness and efficiency of existing financial mechanisms, a Global Mechanism to promote actions leading to the mobilization and channelling of substantial financial resources, including for the transfer of technology, on a grant basis, and/or on concessional or other terms, to affected developing country Parties, is hereby established.'

Pursuant to article 21.5, the functions of the GM are to '[identify and draw] up an inventory of relevant bilateral and multilateral cooperation programmes that are available to implement the Convention; [provide] advice, on request, to Parties on innovative methods of financing and sources of financial assistance and on improving the coordination of cooperation activities at the national level; [provide] interested Parties and relevant intergovernmental and non-governmental organizations with information on available sources of funds and on funding patterns in order to facilitate coordination among them'. Following decisions by the COP at its first session, the GM was formally established late in 1998, more than two years after the UNCCD came into force. It was decided that the GM would be hosted by the International Fund for Agricultural Development (IFAD) (UNCCD 1997).

The GM is not an autonomous institution, but is rather a brokering process established within the legal framework of the convention. It is accountable to the COP and its decisions. It works closely with its three founding members (its hosting institution — IFAD — and the United Nations Development Programme [UNDP] and the World Bank) and the other members of its Facilitation Committee.[1] The GM's governance system therefore consists of three interlinked elements: the COP, to which the GM is accountable; IFAD, whose president has management responsibility for the GM between COP sessions; and the Facilitation Committee, which is assigned an advisory and support role.

Financing Development in Rural Drylands: Sources and Trends

Development financing was allocated in the rural drylands well before the UNCCD came in force. Beginning in the late 1960s, a major portion of this multilateral and bilateral financing went to humanitarian assistance and food aid. A significant share also went to agriculture and land management. The UNCCD's financial provisions were conceived to facilitate the mobilisation of resources to combat desertification and concurrently to identify and channel existing resources through a coherent and integrated framework. This section looks at trends in resources mobilisation and co-ordination in the context of the UNCCD.

Trends in Public Financing and Development Assistance Support

ODA, including allocations to desertification-affected countries, stagnated in real terms between 1992 and 2003. ODA flows reached US$69 billion in 2003, or 0.25 percent of the gross national income (GNI) of the members of the Development Assistance Committee (DAC) of the Organisation for Economic Co-operation and Development (OECD) (Development Co-operation Directorate 2005).[2] Although at the 2002 Monterrey Conference on Financing for Development, there were international calls to increase this financing to 0.30 percent of DAC members' GNI, it has not yet materialised. The overall share of ODA for desertification-affected countries has not increased, nor have the affected countries realigned their borrowing priorities to reflect UNCCD objectives. Although the UNCCD was designed as an instrument to channel existing funds and mobilise new ones for fighting desertification, it seems to have had little impact on ODA flows so far.

Moreover, as the international community was turning its focus to poverty alleviation, resources available for agriculture and rural development declined by nearly 50 percent (from US$4.9 billion in 1988 to US$2.5 billion in 1999) (IFAD 2002, 5). The decrease in financing to rural development includes those contributions from the major lending institutions with the exception of IFAD, which focusses exclusively on agriculture and rural development. At the same time, domestic resources allocated to agriculture and rural development fell in several affected developing countries. According to the International Monetary Fund (IMF), between 1990 and 1998 the share of agriculture in government expenditures fell from 6.2 percent to 3.9 percent in sub-Saharan Africa, from 8.4 percent to 5.4 percent in South Asia, from 3.2 percent to 1.9 percent in Latin America, and from 4.1 percent to 1.1 percent in the Near East and North Africa (IMF Government Financial Statistics, 2000, cited in IFAD 2002, 5).

Since its establishment in 1991, the Global Environment Facility (GEF) has been providing incremental grant financing to assist developing countries and countries with economies in transition in addressing environmental issues of global importance. The GEF has proved to be a useful granting scheme for financing multilateral environmental agreements (MEAs), but total investment remains relatively small (about 2 percent of total ODA). In addition, the leveraging of GEF resources has been less than expected (GEF 2002). Under article 20.2(b) of the UNCCD, developed countries are committed to 'promote the mobilization of adequate, timely and predictable financial resources, including new and additional funding from the Global Environment Facility of the agreed incremental costs of those activities concerning desertification that relate to its four focal areas, in conformity with the relevant provisions of the Instrument establishing the Global Environment Facility'.

Until October 2002, support from the GEF for land degradation issues was only available under the four focal areas recognised at that time: biological diversity, climate change, international waters, and ozone layer depletion. In October 2002, the GEF's general assembly designated land degradation as a new focal area, with the primary focus on desertification and deforestation. The new focal area enabled the GEF to enhance its support to the implementation of the UNCCD. It was further decided in 2003 that the GEF would become a financial mechanism of the UNCCD. In 2003, the GEF adopted the Operational Program on Sustainable Land Management (OP-15) to operationalise financing of the land degradation focal area. The objective of the programme is 'to mitigate the causes and negative impacts of land degradation on the structure and functional integrity of ecosystems through sustainable land management practices as a contribution to improving people's livelihoods and economic well-being' (GEF 2003, para. 31).

The opening of a GEF window for desertification and the decision to make it a financial mechanism for the UNCCD have provided a new opportunity for financing. However, the operational linkages between the OP-15 delivery mechanism and the UNCCD process and objectives must still be formalised in order to ensure the coherence of resource allocation with UNCCD objectives. In addition, although the allocation for OP-15 of US$250 million over a three- to five-year period under the GEF was perceived as considerable progress, this amount was soon exhausted, and many observers suggest that this sum should be doubled in the fourth GEF replenishment.

Trends in Private Financial Flows

In 1990, private investment capital flows to developing countries were at about the same level as total ODA (about US$50 billion). They are now more than 2.5 times greater and growing. Indeed, flows of foreign direct investment (FDI) to developing countries reached US$172 billion in 2003. However, they were highly concentrated in only a few countries, with China, Hong Kong, Singapore, Mexico, and Brazil accounting for 56 percent of total FDI to developing countries. In 2003, Africa received 8 percent of FDI, or US$15 billion, mostly concentrated in the natural resource sector (UN Conference on Trade and Development [UNCTAD] 2004). FDI flows are close to nonexistent in several regions, especially in sub-Saharan Africa. Recent research reveals the existence of a poverty trap in remote rural areas, including rural drylands where poverty and environmental degradation combines with the lack of infrastructure and human capital to create conditions that are not favourable to private investment (Sachs et al. 2004).

The near-total absence of private financial flows in rural drylands should not discourage attempts to mobilise some of these resources for improving soil and land management, especially in Africa. Private financial flows could be mobilised through the emerging carbon market, in the context of the UNFCCC. There has been more than US$0.5 billion in transactions to date. The financing figure could reach US$10 billion annually by 2008, the start of the first commitment period of the Kyoto Protocol. Out of this total, 20 percent to 30 percent could be transacted under the UNFCCC's Clean Development Mechanism (CDM), which allows developed countries to obtain carbon credits for investment made in developing countries. However, reaching the full potential of the CDM in drylands would require the inclusion of agriculture in eligible activities under the CDM. The elaboration of the post-Kyoto regime could provide an opportunity to reform the CDM's eligibility criteria to that effect.

The Global Mechanism and the UNCCD Financing Record

Ten years after the UNCCD was signed, the expectation that funding would be channelled to rural drylands within the convention framework has not been met. Multilateral and bilateral donors still do not have a common interpretation of what constitutes direct support to the UNCCD process and what supports the objectives of the convention indirectly. Consequently, there is insufficient information to assess the level of funding directed at combating land degradation and poverty in the rural drylands.

However, funding for UNCCD implementation raised by the GM was documented in the evaluation process undertaken for the second CRIC session and COP-6, and for the DGF, on which the following analysis is based.[3] The evaluations showed a predominant focus on public financing for process-related activities rather than on investments in the implementation of National Action Programmes (NAPs). In addition, the evaluation found that the financing process had not yet adequately addressed the need to diversify funding, in particular bringing public and private sources together. Between 1998 and 2003, the GM pursued its mission by focussing on three areas: securing direct funding, catalysing funds through small investments, and providing information to support national-level financing efforts.

Direct Funding

The 2003 Independent Evaluation of the Global Mechanism categorised two types of resources secured by the GM: process funding and investment funding. Process funding addresses short-term actions within planning processes. It is untargeted when the funded actions lead only to further planning activities. Process funding is appropriate and necessary when it is targeted to secure investment funding. Investment funding leads directly to on-the-ground actions that address land degradation, where the ultimate beneficiaries (namely, the rural people) are directly involved.

From 1998 through 2002, GM revenues totalled US$16.6 million. Sources included bilateral and multilateral agencies, regional development banks, and the COP's core allocation. Of this total, 68 percent (US$11.3 million) was granted to client countries both directly and indirectly (through technical assistance), while 32 percent covered GM administrative costs. GM fundraising focussed primarily on ODA; however, the list of contributing donors is short.[4] This suggests that a significant number of OECD countries have not been sufficiently engaged.

About US$2.1 million (19 percent) of the US$11.3 million was allocated to targeted process or pilot investment financing. This proportion was not sufficient to stimulate more substantial investment financing to implement the convention. Although some allocation of funding to support planning processes was warranted to jump-start national processes in the early stages of the UNCCD, the evaluation supported the view that the focus should now be almost fully on investment in implementation.

Of the US$11.3 million granted, 49 percent (US$5.5 million) was allocated in small grants to affected countries, 4 percent to strategic initiatives (for example, carbon sequestration and methodologies for desertification monitoring) and the remaining 47 percent for technical advice (such as consultancies to support NAP elaboration). The GM has launched a small grants programme with the assistance of its Technical Advisory Group (TAG).[5] Through this programme, US$5.6 million has been allocated to affected countries, with Africa receiving 48 percent of TAG grants, Latin America and the Caribbean receiving 23 percent, Asia receiving 18 percent, and 11 percent allocated on a global basis. Moreover, nearly US$1.5 million was raised to support civil society participation in implementing the convention.[6]

Catalysing Funds

Catalytic activities aim to trigger investment financing through the strategic placement of small grants and expert advice to support, for example, capacity building, the mainstreaming of UNCCD action programmes into national development planning, and partnership building to finance the implementation of NAPs. The concept is illustrated by the examples that follow.

In Tunisia, US$80 000 was invested to assist in bridging key ministries, raising awareness about the country's obligations under the UNCCD and building understanding about the connection between desertification, sustainable development, and poverty alleviation. NAP priorities are now integrated into the country's socioeconomic plans, and Tunisia has committed US$18 million to address these issues in priority projects — a 23 percent increase in domestic resources dedicated to land degradation issues until now. The increase will be pursued during the five-year socioeconomic development plan as decided by the government. Bilateral agencies from Japan, France, and Germany have expressed an interest in matching this governmental effort through increased participation in project development.

In China, the GM helped to streamline two parallel processes: NAP development at the national level and a concurrent but separate initiative involving the Chinese government in the western region of the country, the GEF, and the Asian Development Bank (ADB). A US$8000 grant was allocated to study root causes of desertification in China (funded by the DGF) and a US$50 000 grant (from IFAD's Community

Exchange and Training Programme) was allocated to test models for action and recruit additional donors. The German and Italian governments, the European Union, and the Food and Agriculture Organization (FAO) expressed interest in participating; the GM and Chinese representatives agreed to develop a private sector component.

The GM has also provided funds to initiatives to develop projects for GEF funding. In addition, efforts to stimulate innovative approaches to financing include initial assessment of potential of carbon markets to support technical aspects of implementing the convention. More general opportunities with the private sector and debt swaps have also been pursued. These include some steps toward forging relationships with the private sector and brokering debt swaps. The GM, the FAO, IFAD, the World Food Programme (WFP), and the Italian government are involved in debt swaps with Ecuador and Peru, where small grants were awarded to develop NAP-based projects and efforts have been made to channel debt-swap proceeds to areas of high priority for UNCCD.

Providing Information

Several funders and UNCCD stakeholders consider that the lack of knowledge and reliable information on desertification constitutes a hurdle in the process of mobilising and co-ordinating resources for UNCCD implementation. In order to resolve this knowledge gap, the GM has developed a comprehensive source of baseline information on desertification-related funding sources, trends in financing, projects, background information on the convention and its operations, and a variety of related technical and policy reports. This information is made available to UNCCD parties and others through the Financial Information Engine on Land Degradation (FIELD), an on-line and CD-ROM resource.[7]

FIELD benefits from collaboration with other institutions involved in gathering and disseminating information on similar or complementary subjects. Collaboration with the OECD's DAC began when the resource was first conceived, and enabled the GM to build on existing DAC standards and definitions. More recently, the GM has worked with the DAC to assess the quality of official data on UNCCD-related aid through the OECD/DAC Working Party on Statistics. This should help align the large discrepancies between what OECD countries report to the DAC compared to what they report to the UNCCD secretariat.[8]

An Assessment of the UNCCD Effectiveness in the Context of Financing

The process intended to channel existing development financing toward implementing the UNCCD has not yet succeeded, nor has the UNCCD had an impact on the declining trend in ODA flows toward rural drylands. Donors remain reluctant to finance NAPs and any financing that does exist is heavily oriented toward process-related activities. Two factors analysed below explain this situation.

The UNCCD: A Convention for Environment or Development?

The UNCCD is perceived as an environment convention, even though its primary focus is to fight land degradation through sustainable rural development while reducing poverty. In the aftermath of the 1992 UN Conference on Environment and Development (UNCED) at Rio, it was initially thought that the environment label would provide higher visibility and better support to the UNCCD. Unfortunately, this has not materialised. As a consequence, national UNCCD focal points are generally situated within ministries of environment without adequate links to ministries of finance, agriculture, and other departments positioned to influence policies and budgets pertinent to UNCCD implementation (such as rural development or agriculture). Other factors that contribute to the weakness of UNCCD focal points are insufficient human and financial resources, high staff turnover, lack of outreach capacity, and general inactivity.[9] As a result, desertification is rarely cited among the priorities put forward by developing countries in their discussions with donor agencies.

However, in developed countries focal points are located within development agencies. Although developed countries have committed to support the UNCCD, they have so far failed to take the lead in helping their developing country counterparts to raise the profile of desertification into development planning. Nationally and internationally, the result is that the money and actions are focussed elsewhere, with inadequate commitment to the development objectives of the UNCCD. Recognition of the convention's focus on development is growing (see, for example, the Plan of Implementation of the World Summit on Sustainable Development UN 2002a); however, subsequent action has not materialised, and the UNCCD is still perceived as an environmental convention even though it deals with poverty in rural drylands (see Chapter 8).

The Financing Structure: Can It Attract Adequate Financing?

After an examination of funding patterns and sources, and a discussion of the need for more and different sources of funding, the question becomes whether or not the structure of the UNCCD is positioned to attract adequate financing. The 2003 Independent Evaluation of the Global Mechanism found that the roles of the entities within the GM structure were not working to their potential (UNCCD 2003a; 2003b). The report found that IFAD, despite incorporating the GM in its administrative framework in terms of personnel and financial accounting, has not been sufficiently engaged in the management of the GM.

The GM's facilitation committee provides a structure that supports co-operation and the exploitation of comparative advantages, and its members represent a diversified financial and technical potential for implementation of the UNCCD. But the evaluation suggested that these relationships had not yet moved beyond *ad hoc* co-operation at the regional and country levels. At the policy level, the lending practices of neither IFAD nor the World Bank had shifted as a result of UNCCD objectives. One result of these findings was an initiative to mainstream support of the GM process, and ultimately UNCCD implementation, more effectively among the founding member institutions and members of the facilitation committee.

The evaluation's review of COP directives showed that, although the COP has clearly interpreted the GM's overall mandate in the text of the convention, a number of its subsequent decisions have resulted in confusion over the role that the GM should play in carrying it out. The COP is best positioned to provide broad strategic direction. By focussing at a level of micro-management, it may have diluted the focus of the GM by diminishing flexibility and the ability to explore new directions and opportunities. By concentrating on the demand-driven aspects of the GM's mission, the COP set a context within which it is difficult to prioritise activities and focus on mobilisation of substantial resources. Not surprisingly, the 2003 evaluation process revealed considerable criticism about the GM's lack of prioritisation, and that it spreads itself too thin. Decision 9 from COP-3 suggests that the COP was concerned as well, as it calls for the GM to set clear priorities.[10]

Furthermore, NAPs, which were conceived to serve as UNCCD financing instruments, were expected to serve dual roles, acting as a framework for participation and a tool for resource mobilisation. In some countries, NAPs have served well in the first role. However, experience to date shows that they do not serve well as financing instruments. Among the members of the facilitation committee and donors interviewed for the 2003 evaluation, there was broad consensus that NAPs are inconsistent in quality, lack scientific foundation, and are without the baselines and indicators needed for monitoring. Frequently NAPs are presented simply as 'shopping lists' in which donors have little interest (see Chapter 10). In addition, NAPs drown in a plethora of plans and strategy papers required by other conventions (namely, the UNFCCC and CBD), poverty reduction strategy plans, and a variety of development processes and policies.

The 2003 evaluation reported that while NAP processes serve a strategic role in national-level planning, NAPs are often inadequate instruments for attracting investment financing. Existing NAPs should be used to identify key priorities for project development and funding or, if feasible, transformed into crisp, strategic documents. These views are not shared by all affected developing countries. There is a basic misunderstanding between donors and these countries about the role of NAPs, which seems to emanate from a differing view of the pace and process needed for both effective national policy development and sound financing operations.

Within this overall context, the GM focussed in its first few years on the demand side, in essence to help prepare affected countries for funding opportunities. This approach led to the GM's extensive involvement in NAP elaboration and mainstreaming, a decision that has prompted controversy about GM operations. Focus on the demand side has been to the detriment of building support on the supply side, among the bilateral and multilateral agencies.

In its report to COP-5, the GM states its case for taking this direction, arguing that integration of NAPs into broader development policies and frameworks is essential for successful resource mobilisation.[11] The argument is not without merit.

The GM has therefore focussed on process, in particular on enhancing quality of NAPs, the Sub-regional Action Programmes (SRAPs), and the Regional Actions Programmes (RAPs), and on mainstreaming the objectives of the UNCCD into the larger development policy frameworks that will make them more relevant politically and financially.

While the GM's small granting activities respond to the COP mandate and provide important support to the desertification-affected parties, substantial investment funding for land degradation issues has not yet been realised. This circumstance is due, in part, to the GM's efforts to address issues that hamper funding potential (such as mainstreaming UNCCD objectives into policy frameworks), even though they fall outside of its core mandate. Although these activities are important for financing success, other GM partner institutions are better equipped to carry them out. For example, the World Bank and, to a lesser extent, the UNDP are lead agencies to assist countries in formulating their poverty alleviation and development strategies and policies. However, this assistance is not systematically co-ordinated with the GM, the result being that the UNCCD objectives are not always effectively integrated into these strategies and policies. The potential for a strong partnership in this regard has not been realised.

Harnessing Synergies with the Climate Change Convention

Limited resources suggest not only the need for an entrepreneurial approach, but also the need to converge efforts amongst complementary environmental conventions. The UNFCCC in particular shares objectives with those of the UNCCD, and provides a political context within which emergence of carbon markets has flourished. Moreover, the entry into force of the Kyoto Protocol in February 2005 has accelerated the growth of carbon markets.

Links between land degradation and carbon sequestration lead to synergies between the UNCCD and UNFCCC, as land degradation can be translated into loss of soil fertility. This corresponds to decreasing soil organic matter that in turn corresponds to soil carbon de-stocking. Reversing the downward spiral of land degradation is essentially re-stocking organic matter — that is, carbon sequestration. Likewise, the decreasing dry forest and range cover correspond to another form of carbon de-stocking, which can be resolved by agro-forestry and improved range management. This means, again, carbon sequestration. Limiting deforestation is also linked to improving the sustainability of fuelwood production and increasing biomass energy efficiency, which results in an increasing carbon content and decreasing emissions of greenhouse gases (GHGs). The potential for carbon sequestration and GHG emission reduction may be lower per hectare in desertification-affected countries than in the humid tropics, but the areas may be much larger.

Carbon markets offer a new type of investment financing that corresponds to the sale of an environmental service involving the reduction of GHG emissions and carbon sequestration. While carbon-financing opportunities are growing, it will be a challenge for desertification-affected countries to access those funds. For one thing, as mentioned earlier, the exclusion of agriculture from CDM-eligible activities constitutes a major impediment to the channelling of carbon-related investment in rural drylands.

In addition, the lack of accurate data and scientific knowledge on carbon sequestration in the drylands, as well as scarce expertise and information channels to market drylands environmental services, are impediments to the development of such schemes in regions affected by desertification. Capacity building will be needed in this new field, including for developing pilot projects, attracting public and private carbon purchasers, negotiating complex contracts, and monitoring actual GHG emission reduction and carbon sequestration. In particular, African countries must be prepared first to ratify the Kyoto Protocol and then to vigorously pursue those opportunities presented by the Protocol and carbon markets. IFAD has agreed in principle to participate in the World Bank Bio Carbon Fund. The GM and IFAD could work together in this domain of opportunity to facilitate transactions in the rural drylands by serving as brokers.

Adaptation measures to climate change are another synergistic area between the UNCCD and the UNFCCC. Such measures aim to reduce vulnerability particularly in the desertification-affected countries. It is expected that they will be financed out of 2 percent of the value of carbon transacted. This could amount to about US$200 million annually.[12] In brief, the UNFCCC could dramatically contribute to implementing the UNCCD by promoting investment projects that would combine activities such as soil improvement and forest activities (carbon sequestration), more efficient biomass energy (GHG emission reduction) with adaptation measures to climate change. National-level programmes and policies would do well to explore potential for targeting implementation of both conventions and potential collaboration that might benefit both.

Overcoming Constraints and Seizing New Opportunities for UNCCD Financing

The context within which the GM works therefore presents a mix of opportunities and constraints (see Table 9-1).

Additional constraints have emanated from the initial conception of the UNCCD itself:

- The UNCCD's initial position as an environment convention has overshadowed its key link to agriculture and rural development;
- NAPs, expected to be key for financing the convention, have not been adequate for resource mobilisation, even though some have played a strategic role in framing national actions to fight land degradation;
- Parties that are OECD members, expected to increase their support to the convention, have not fully met their financing obligations (Article 20), nor have practices of major lending institutions been influenced toward convention objectives; and

- Responsibility for implementing the convention, expected to involve key country decision makers, has usually fallen outside national financing and development policy streams.

On the positive side, although funding has not yet been adequate, the creation of a GEF focal area on desertification and its 2003 designation as UNCCD financial mechanism constitute important contributions to UNCCD financing and implementation.

Enhancing the success of UNCCD financing will require several adjustments to the current financing environment. These include refocussing the convention on poverty and development, in the first instance by promoting its links to sustainable development and moving it from the periphery of development policy to the centre stage. In

Table 9-1   Opportunities and Constraints of the Global Mechanism

Opportunities	Constraints
Increasing private investment capital flows to developing countries; nearly six times ODA	Private investment mainly flows to larger, medium-income developing countries, much less to Africa where it focusses on oil- and mineral-producing countries
Strong synergy between UNFCCC and UNCCD with potential to stimulate investment through emerging carbon markets	The CDM's current eligibility criteria exclude agriculture, thereby diminishing synergies with UNCCD
Declarations from the WSSD, the Millennium Summit, and the Millennium Task Force call for an increase in ODA financing and priority attention to rural development and UNCCD	Agriculture-related ODA is decreasing and a new mobilisation by donor agencies toward UNCCD implementation is not yet evident
The GM is aligned with powerful and well-positioned agencies, primarily IFAD and the facilitation committee	Partner institutions are not sufficiently aligning practices and policies in support of UNCCD
The opening of a new GEF focal area on land degradation has led to the mobilisation of new funds for the UNCCD	Funds available under OP-15 remain low compared to other lending windows; operational coherence with UNCCD objectives has not been fully achieved

Notes:
CDM = Clean Development Mechanism
GM = Global Mechanism
IFAD = International Fund for Agricultural Development
ODA = official development assistance
UNCCD = United Nations Convention to Combat Desertification
UNFCCC = United Nations Framework Convention on Climate Change
WSSD = World Summit on Sustainable Development

line with the declaration issued at the 2002 World Summit on Sustainable Development (WSSD), the UNCCD should focus more on promoting sustainable agriculture and rural development by engaging a wide range of decision makers in efforts to fight desertification (UN 2002b). Attaining such an environment could lead to, for example, joint ventures and networks among the bilateral and multilateral financial agencies to promote integration of UNCCD issues into development assistance and loan programmes.

At the national level, UNCCD focal points should involve key decision makers in the process of UNCCD implementation. NAPs, which have proven inadequate for financing, should be recognised more as strategic documents highlighting priority actions to pave the way to the development and implementation of investment projects. Consequently, more attention should be paid to supporting the formulation of these projects and financing their implementation. Furthermore, efforts to direct domestic resources toward land degradation and rural poverty should be increased.

Within the GM structure, a system is needed between the GM, the World Bank, and the UNDP to promote mainstreaming UNCCD objectives systematically and co-operatively into development and policy frameworks. IFAD, the World Bank, and other financial agencies should more effectively incorporate UNCCD objectives into their lending programmes. Finally, all members of the GM's facilitation committee should become more fully engaged in information gathering and dissemination activities, particularly IFAD, the World Bank, and the UNDP.

Notes

1 At its first session, the COP endorsed collaborative institutional arrangements with IFAD, the UNDP, and the World Bank, establishing the GM Facilitation Committee (UNCCD 1997, decision 25). In addition, the Facilitation Committee includes the UNCCD secretariat, the secretariat of the Global Environmental Facility (GEF), the Food and Agriculture Organization (FAO), the United Nations Environment Programme (UNEP), the African Development Bank (AfDB), the Asian Development Bank (ADB), and the Inter-American Development Bank (IADB). At the suggestion of the World Bank, the Consultative Group on International Agricultural Research (CGIAR) has been closely associated with the GM's efforts, as are nongovernmental organisations (NGOs).
2 The DAC regroups the OECD's 23 major bilateral donors. Its members are Australia, Austria, Belgium, Canada, Denmark, Finland, France, Germany, Greece, Ireland, Italy, Japan, Luxembourg, the Netherlands, New Zealand, Norway, Portugal, Spain, Sweden, Switzerland, the United Kingdom, the United States, and the Commission of the European Communities.
3 This section is based on the independent evaluation of the GM conducted for sixth COP and the second session of the CRIC, in which the authors of this chapter took part (see UNCCD 2003a; 2003b).
4 Donors to the GM are the Arab Fund for Social and Economic Development (AFSED), the Canadian International Development Agency (CIDA), the International Development Research Centre (IDRC), Denmark, Finland, IFAD, Italy, the Islamic Development Bank (IsDB), Germany, the Netherlands, Norway, the Organization of Petroleum Exporting

Countries (OPEC), Portugal, Sweden, Switzerland, the U.S. Congressional Hunger Center, and the World Bank.
5   Some donors have chosen to establish granting funds within the GM. The TAG, which comprises representation from the GM, IFAD, the FAO, and the UNCCD secretariat, then allocates small grants to the countries (to governments and civil society).
6   As mandated by the second COP (UNCCD 1999a, decision 18). The NGO funds are managed through three granting facilities, one managed through the TAG, one managed by the GM, and the third managed by IFAD.
7   The third version of FIELD, published early in 2003, contains 5600 cross-referenced records, including 1300 documents, reports, and publications related to the implementation of the UNCCD, as well as thousands of links to external websites and sources of information. The information covers six continental regions, 185 countries, 28 bilateral donors, 69 multilateral donors, 98 public and private foundations, 145 NGOs, 90 private sector companies, and 52 research and academic institutions and other donors. See the GM's website at <www.gm-unccd.org/English/Field> (December 2005).
8   According to a DAC study on bilateral aid reporting between 1998 and 2000, only in 3 percent of the cases were bilateral aid activities reported consistently to both the OECD/DAC and the UNCCD secretariat (OECD 2002).
9   See chapter 10 by Richard Pearce, who points out that the national focal points have never been really been functional in many countries of West Africa.
10 Indeed, the report from COP-3 stressed that 'the Global Mechanism shall clearly identify priority activities and submit them to its fourth session for consideration and adoption by the Conference of the Parties in order to ensure effectiveness and impact of its activities as well as to avoid overlaps with the activities of existing institutions and organizations, and therefore achieve concrete results, in the shortest possible time, in mobilizing and channelling financial resources to affected developing country Parties, particularly those in Africa, for the implementation of the Convention' (UNCCD 1999b, 35).
11 Experience has shown that the prospects for effective resource mobilisation are contingent upon, *inter alia,* the ability of the concerned governments to mainstream the NAPs and Sub-regional Action Programmes (SRAPs) into national and sector planning frameworks, and to create synergies between conventions (see UNCCD 2001).
12 This amount refers to the US$10 billion estimate of the total carbon market by 2008, the starting year of the implementation period for the Kyoto Protocol.

References

Development Co-operation Directorate (2005). 'The DAC Journal: Development Co-operation Report 2004'. Organisation for Economic Co-operation and Development. Paris.
Global Environmental Facility (2002). 'Focusing on the Global Environment: The First Decade of the GEF'. Second Overall Performance Study (OPS2). <www.gefweb.org/1Full_Report-FINAL-2-26-02.pdf> (December 2005).
Global Environmental Facility (2003). 'Operational Program on Sustainable Land Management'. Operational Program No. 15. <www.gefweb.org/Operational_Policies/Operational_Programs/OP_15_English.pdf> (December 2005).
International Fund for Agricultural Development (2002). 'Theme Paper for the Twenty-Fifth Session of the Governing Council: Financing Development — The Rural Dimension'.

GC 25/L.11, 19–20 February. Rome. <www.ifad.org/events/gc/25/e/GC-25-L-11.pdf> (December 2005).
Organisation for Economic Co-operation and Development (2002). *Creditor Reporting System on Aid Activities: Aid Targeting the Objectives of the Rio Conventions 1998/2000*. Organisation for Economic Co-operation and Development, Paris.
Sachs, Jeffrey, John W. McArthur, Guido Schmidt-Traub, et al. (2004). 'Ending Africa's Poverty Trap'. Brookings Papers on Economic Activity 1. <www.earthinstitute.columbia.edu/about/director/documents/bpea0104.pdf> (December 2005).
United Nations (2002a). 'Plan of Implementation of the World Summit on Sustainable Development'. Johannesburg. <www.un.org/esa/sustdev/documents/WSSD_POI_PD/English/WSSD_PlanImpl.pdf> (December 2005).
United Nations (2002b). 'Johannesburg Declaration on Sustainable Development: From Our Origins to the Future'. World Summit on Sustainable Development, 4 September. <www.un.org/esa/sustdev/documents/WSSD_POI_PD/English/POI_PD.htm> (December 2005).
United Nations Conference on Trade and Development (2004). 'World Investment Report 2004: The Shift Towards Services'. Geneva.
United Nations Convention to Combat Desertification (1994). 'Text of the United Nations Convention to Combat Desertification'. <www.unccd.int/convention/text/convention.php> (December 2005).
United Nations Convention to Combat Desertification (1997). 'Report of the Conference of the Parties on Its First Session, Held in Rome from 29 September to 10 October 1997: Part Two — Actions Taken by the Conference of the Parties at Its First Session'. ICCD/COP(1)/11, 29 December. <www.unccd.int/cop/officialdocs/cop1/pdf/11add1eng.pdf> (2005 December).
United Nations Convention to Combat Desertification (1999a). 'Report of the Conference of the Parties on Its Second Session, Held in Dakar from 30 November to 11 December 1998: Part Two — Action Taken by the Conference of the Parties at Its Second Session'. ICCD/COP(2)/14/Add.1, 5 February. <www.unccd.int/cop/officialdocs/cop2/pdf/14add1eng.pdf> (December 2005).
United Nations Convention to Combat Desertification (1999b). 'Report of the Conference of the Parties on Its Third Session, Held in Recife from 15 to 26 November 1999'. ICCD/COP(3)/20/Add.1, 31 December 1999. <www.unccd.int/cop/officialdocs/cop3/pdf/20add1eng.pdf> (December 2005).
United Nations Convention to Combat Desertification (2001). 'Global Mechanism: Review, Pursuant to Article 21, Paragraph 5(d), of the Convention of the Report on the Activities of the Global Mechanism, and the Provision of Guidance to It'. ICCD/COP(5)/4, 11 September. <www.unccd.int/cop/officialdocs/cop5/pdf/4eng.pdf> (December 2005).
United Nations Convention to Combat Desertification (2003a). 'Global Mechanism: Review, Pursuant to Article 21, Paragraph 7, of the Convention of the Policies, Operational Modaliites and Activities of the Global Mechanism, and the Provision of Guidance to It'. ICCD/CRIC(2)/5, 20 June. <www.unccd.int/cop/officialdocs/cric2/pdf/5eng.pdf> (December 2005).
United Nations Convention to Combat Desertification (2003b). 'Independent Evaluation of the Global Mechanism'. Final Report to the World Bank Development Grant Facility, ICCD/COP(6)/MISC.1, 24 August. <www.unccd.int/cop/officialdocs/cop6/pdf/misc1eng.pdf> (December 2005).

Chapter 10

Decentralisation and Sustainable Resources Management in West Africa: A Line of Action for Revisiting National Action Programmes

Richard Pearce

The United Nations Convention to Combat Desertification (UNCCD) places desertification control in a holistic approach of environmental and development issues.[1] For this reason, the 2002 World Summit on Sustainable Development (WSSD) adopted the convention as a relevant tool for curbing poverty and promoting sustainable development in countries affected by desertification and drought (UN 2002, para. 7, 41).

The countries affected by drought that signed the UNCCD are committed to developing and implementing National Action Programmes (NAPs). According to the UNCCD (2005),

> these programmes must adopt a democratic, bottom-up approach. They should emphasize popular participation and the creation of an 'enabling environment' designed to enable local people to reverse land degradation through self-help. Of course, governments remain responsible for creating this enabling environment. They must make politically sensitive changes, such as decentralizing authority, improving land tenure systems, and empowering women, farmers and pastoralists. ... these action programmes are to be fully integrated into other national policies for sustainable development. They should be flexible and adopted to changed circumstances.

Furthermore, they 'will improve coordination and channel development assistance to where it can be most effective [and] will also produce partnership agreements that spell out the respective contributions of affected states, donor states and international organizations' (UNCCD 2005). Thus, NAPs are both participatory tools for strategic planning and tools for co-ordinating and mobilising national and international resources to combat desertification.

In accordance with the commitment made to the convention and with the financial support from foreign partners, West African countries have begun developing their NAPs. National co-ordinating bodies (NCBs) responsible for steering the process were established, stakeholders information and sensitisation campaigns were carried out, wide consultations were undertaken, national forums were organised to ensure participatory validation and adoption of these programmes, and round tables were organised to introduce NAPs to technical and financial partners.[2] In most countries in the sub-region, NAPs have been developed and adopted, whereas the operational stage of the process was launched several years ago (for example, in 1998 in Mali and Senegal and in 2000 in Burkina Faso and Niger).

Unfortunately, many observers acknowledge that after more than five years of NAPs in some countries, the implementation of the UNCCD in West Africa is at a standstill (Toulmin 2001). Very few partnership agreements have been signed. The working relations established between NCBs and development partners in view of developing NAPs as well as the participatory processes initiated with various representatives of civil society — nongovernmental organisations (NGOs), elected community representatives, associations, and so on — have tailed off. Although the UNCCD is perceived as an African countries' convention, the general impression is increasingly that it is a debate on desertification control that no longer interests the countries in West Africa.

This chapter is based on the author's experience as an observer of the UNCCD's implementation over the last six years. It attempts in the first instance to explain why, at present, development actors (technical services, civil society, development partners) are unwilling to invest in NAPs in West Africa. Then it suggests main lines of reflection for revising the NAPs – a process that is provided for by the UNCCD although no such review has been undertaken in any country in the sub-region.

Why Do Development Actors Not Invest in National Action Programmes or Desertification Control?

No critical analysis of UNCCD implementation in West Africa can be undertaken without referring to the work done in the framework of the UNCCD's Committee for the Review of the Implementation of the Convention and of the Ad Hoc Working Group (AHWG).[3] These reviews reveal a number of causes that block UNCCD implementation. The three causes most frequently mentioned relate to:

- Funding problems: there have been very few partnership agreements between donor countries and affected states for the implementation of the NAPs (see Chapter 9);
- Lack of political will: affected states barely invest in the implementation of their NAPs and rarely mention them as priority programmes during negotiations of co-operation agreements; and
- Poor coherence and integration of NAPs with other development strategies and policies of governments of affected countries: most often, NAPs are only a mere programme among others in the development strategy of affected countries.

These AHWG and CRIC review processes also lead to some recommendations:

- Funding must be increased (which affected countries would like to hear);
- The UNCCD should become a priority both for governments of affected countries and for development partners (which the UNCCD secretariat would like to hear); and
- NAPs should be mainstreamed in other development frameworks, strategies, and programmes of affected countries (which developed countries and donor agencies would like to hear).

But the situation in the field is different. Although these recommendations have been endorsed by the Conference of the Parties (COP), they could not trigger the relaunching of UNCCD implementation process, at least not in West Africa.

Relaunching the implementation process requires an analysis of NAPs in order to define whether they are currently in line with some of the key principles of the convention. Based on this analysis, it is possible to formulate recommendations to improve NAPs through possible review processes. The key principles on which such analysis should be grounded are the following: NAPs should define long-term strategies and specify the roles of state, local communities, and farmers (UNCCD 1994, art. 10); NAPs should be integrated into national policies (art. 10); and NAPs should strengthen the participation of populations and local communities in developing and implementing programmes to control desertification (art. 3).

The first part of this chapter seeks to understand why some of these principles have been implemented incorrectly in the field, and to propose concrete alternatives to facilitate their real implementation. This analysis will outline some of the current weaknesses in the NAPs.

A Field of Activity Yet to Be Specified

> National action programmes shall specify the respective roles of government, local communities and land users and ... incorporate long-term strategies to combat desertification (UNCCD 1994, art. 10.2).

One thing is crucial in explaining why UNCCD implementation is at deadlock in West Africa: in general, the stakeholders involved in desertification control (such as technical services, development partners, and representatives of the civil society) do not share a common vision of what can or should be done in the specific framework of desertification control.[4]

This situation can be explained by the fact that the UNCCD undoubtedly represents an international consensus on the desertification phenomenon (causes and consequences) and on the principles of an efficient control — which have something in common with the principles of sustainable development — but it lacks genuine operational guidelines to guide actors in their actions to combat desertification.[5] The priority areas of intervention set by the UNCCD for NAPs are too wide. Some of these areas are the following:

- Promoting new livelihoods and improving the national economic environment to reinforce poverty elimination and food security programmes;
- Sustainable management of natural resources;
- Development of ecologically sustainable farming practices;
- Development and efficient use of various sources of energy;
- Reforming institutional and legal frameworks;
- Strengthening the means for systematic evaluation and monitoring, especially by creating hydrology and meteorology services; and
- Capacity building, education, and raising public awareness (UNCCD 1994, art. 10.4).

Furthermore, opinions diverge on concrete actions to be conducted on the field to enhance desertification control.[6] A succinct analysis of NAPs developed by affected West African countries and of the reactions they arouse, particularly among development partners, shows that the contents of the various NAPs in the sub-region differ depending on how the NCBs interpret the UNCCD. However, NAPs fall into wo main categories: 'shopping list' NAPs and 'strategic' NAPs.

The first category of NAPs consists of shopping lists of projects that plan for implementing integrated projects that take into account the social, economic, and environmental dimensions of development.[7] These projects generally give greater importance to building the self-management capacity of local populations, developing socioeconomic opportunities, and establishing environment conservation activities (which are often limited to technical operations of reforestation, land reclamation, water flow control, and so on) in limited geographical zones.

One recurrent criticism of such NAPs is their lack of long-term vision for the 'enabling environment' that they are expected to create in order to improve the exploitation of natural resource at the national level. Moreover, development partners think they are already providing large support to this type of projects under various names and umbrellas (for example, forest management, farmland management, local development projects under the umbrella of the ministries responsible for the economy and finance, or agriculture, animal breeding, forestry and water resources). Hence, the problem that frequently appears when signing partnership agreements to implement NAPs is that development partners want to determine the additional value that NAPs bring to what is being already done. The development partners reproach their counterparts for not taking into account the experiences they are supporting and for not providing for the required mechanisms to enhance the value of these experiences and to institutionalise them. The NAPs are considered additional programmes with questionable capacity to effect the changes recommended by the UNCCD.

The second category includes a few 'strategic' NAPs. These try to define long-term objectives likely to create the enabling environment recommended by the

UNCCD. The problem with these strategic NAPs is their failure to comply with all the UNCCD principles. As a consequence, they look similar to strategies for sustainable rural development that require the intervention of several actors, and especially of several ministries.[8] Therefore, these NAPs often do not translate into proposals with concrete, structured actions that can convince their partners. (This point is discussed further on.)

As stated above, the UNCCD rightly underscores the link between development and the depletion of natural resources. The UNCCD further recommends comprehensive approaches to integrated development that require actions in various domains. Unfortunately, activities for controlling desertification are not neatly delimited, and in the field both shopping-list NAPs and strategic NAPs still raise many questions. What concrete actions should be taken in the field? What are the roles and responsibilities of the various stakeholders in the specific framework of controlling desertification? What is the added value of NAPs?

If NAPs in West Africa are unable to play a strategic role in promoting sustainable development, this is mainly because they lack either relevant objectives relating to the enabling environment they are expected to create or clear action plans to attain these objectives. This situation does not allow governments and their partners to come together for common projects, and therefore does not facilitate the partnership agreements as recommended by the convention.

In 2003, the UNCCD's Global Mechanism (GM) sponsored an interesting study in Burkina Faso to capitalise on efforts conducted to control desertification (Conseil National pour l'Environnement et le Développement Durable [CONEDD] 2003a). This study met a key requirement of development partners who believed that the experiences they supported were inadequately taken into account in NAPs. The study also shed more light on the field actions that were best in line with the UNCCD principles. However, it did not help define the role that NAPs could play in improving UNCCD implementation in the country. Indeed, this study revealed that more than one hundred projects classified under a variety of denominations and conducted by various national structures complied with UNCCD principles. The study provides development partners with arguments to assert that they are already strongly backing up desertification control in Burkina Faso, but it does not clarify how NAPs could improve the impact of these interventions.

Integrated National Action Programmes and Desertification Control: Controversial Solutions

> National Action Programmes shall ... incorporate long-term strategies to combat desertification and ... emphasize implementation and be integrated with national policies for sustainable development (UNCCD 1994, art. 10.2).

One of the key observations made by the CRIC and AHWG review processes concerned the fact that NAPs are neither coherent enough nor integrated sufficiently into other government development strategies and policies in affected countries.

To solve this and to facilitate the integration of NAPs into a wider strategic framework that relates to sustainable development, the fifth COP recommended the establishment of an appropriate consultation mechanism in each country.

Indeed, in most West African countries, NCBs have never been truly functional and have rarely played the role of consultation mechanism aiming at favouring the integration of environment issues into sector-wide policies, plans, and programmes.[9] An argument mostly mentioned to explain why strategic NAPs were unable to propose relevant operational action plans with clear roles and responsibilities for the various stakeholders, in particular the various technical ministries, to combat desertification efficiently, is the light political weight of NCBs, which are regarded as belonging to the ministries responsible for the environment. Therefore, NAPs have no influence on the activities of other technical ministries. Most countries held several discussions on the status of the NCB and its institutional umbrella in order to ensure that it runs efficiently. But these discussions and initiatives did not yield the expected results and, unfortunately, there is almost no doubt about the NCBs' inability to become functional. The two main reasons are described below.

The first reason is that, in these countries, several strategies, plans, and programmes have been developed, not only in the area of environment (for instance national action plans for biodiversity, climate change, or forest programmes), but also in the domain of development (poverty reduction strategy papers [PRSPs], for example, or programmes for supporting decentralisation, sustainable development in rural sectors, education, or health). All these programmes, strategies, and action plans recommend crosscutting and integrated approaches, and most have interdepartmental orientation and steering and monitoring committees similar to NCBs.

The second and complementary reason is that affected governments, development partners, civil society organisations (such as local communities, professional organisations, associations, and even NGOs pay less and less political attention to desertification control. This can be explained, on the one hand, by the poor impact of NAPs, especially regarding funding mobilisation, and, on the other, by the arrival of new priority strategic frameworks on the political scene. This is particularly the case for PRSPs, which have become the reference strategic frameworks for all development actors in West Africa.

Thus an affected country's NCB is unable to impose any kind of control on desertification as a priority intervention in its various development programmes or policies. This can lead to two types of reactions.

First, to comply with the recommendation that requires the integration of the existing various programmes and plans into an overall development policy, affected countries tend to favour the approach consisting of integrating their NAPs, as they stand, together with their action programmes for biodiversity and climate change into a single broad NAP for the environment that will then be integrated into the priority development framework called the PRSP, alongside other plans on education, health, or infrastructure.[10] This produces a set of nested Russian dolls, with the various programmes fitting into one another, often ranging from the weakest to the strongest. In this agglomerate, the NAP and desertification control are the most often mentioned, but little attention is paid anymore to what it is supposed to bring to the country's sustainable development policy.

Since the NAP and desertification control are regarded as a sub-component of the priority development programme of these countries, the other typical reaction has been increasing the amount of time devoted by those in charge of implementing the UNCCD to developing integrated field projects to be submitted to the Global Environment Facility (GEF) since it released the Operational Program on Sustainable Land Management (OP-15) in 2003 (GEF 2003; see also Chapter 9). In this context, existing resources are not channelled to revisiting NAPs.

It is regrettable that in the process of revisiting the NAPs there is no emphasis on defining the genuine specific interventions to combat desertification that are likely to bring about gains to other development programmes in the country.

Community Participation: A Challenge Yet to Be Taken Up

> Parties should ensure that decisions on the design and implementation of programmes to combat desertification and/or mitigate the effects of drought are taken with the participation of populations and local communities and that an enabling environment is created at higher levels to facilitate action at national and local levels (UNCCD 1994, art. 3[a]).

> African country Parties undertake to ... sustain and strengthen reforms currently in progress toward greater decentralization and resource tenure as well as reinforce participation of local populations and communities (annex 1, art. 4.2[b]).

One of the main differences between the UNCCD and previous action plans or declarations on drought or desertification is that it places community participation at its very core and sets it among the quality criteria for desertification control activities (United Nations General Assembly [UNGA] 1977; Organization for African Unity [OAU] 1985). Community participation is also a central concern of most conventions and agreements relating to environment and sustainable development (see, for example, the CBD, Agenda 21, the Millennium Development Goals [MDGs], and the WSSD).

However, in the framework of the UNCCD, the meaning of the term 'community participation' has changed through time. Initially, in view of various experiences in combating desertification in the preceding two decades, community participation was understood as involvement in the identification and implementation of desertification control actions.[11]

In the field, following the signature of the UNCCD and in the framework of the development of their NAPs, all affected countries in West Africa engaged in community information, public awareness campaigns, and consultations. These processes were clearly flagged by affected countries as responses to the UNCCD requirements for 'participation'.[12] But the effects of these consultation processes are questionable, as most of them appear to lead to shopping-list NAPs that often limit popular participation to their physical involvement in the realisation of activities that were already identified and planned (see Chapter 8).

Currently, in most West African countries, civil society participation in implementing the UNCCD is limited to the mere presence of some NGOs at workshops in order to validate reflections of the NCB. The consultation process has been maintained to seek adherence of local community representatives to predefined objectives or initiatives. However, the conditions for a genuine investment of local communities in actions to control desertification have not yet been created. In fact, civil society interest in this issue has faded.

For some time now, it is more and more understood that local participation should be characterised by people's own investment in desertification control actions. The leaflet available on the UNCCD website says that NAPs 'must adopt a democratic, bottom-up approach' and 'should emphasize popular participation and the creation of an "enabling environment" designed to enable local people to reverse land degradation through self-help' (UNCCCD 2005). This implies that as a national programme, the NAP should identify and implement the means required to promote changes in practice and behaviour in those who use natural resources so as to ensure sustainable management of these resources. This distinction between the two meanings given to community participation is crucial and has been the source of much confusion.

Recommendation to Overcome Current Weaknesses in NAP

Ideally, depending on the socioeconomic and political contexts of a country, each NAP should try and define more explicitly the exact responsibilities of the various actors in desertification control, especially at local level, because actions to combat desertification must be undertaken in the field. Who can or must do what to combat desertification? This question includes defining the roles of the public and private sectors, communes and local communities, NGOs, decentralised technical services, and also researchers and development partners. But, in order to achieve this, it is necessary to define specifically what a NAP can accomplish.

To promote the integration of NAPs and desertification control into development policies and strategies in affected countries, efforts should be put into defining the scope of their activity. To this effect, the NAP and desertification control are not the only sustainable development plan, but rather only one programme among others from the ministries responsible for the environment.[13] Therefore, it is important to specify the gains that this programme, can bring to other programmes within the environmental ministries (for example, programmes for forests or biodiversity) and to the various development plans and strategies considered priorities by the governments of these countries (such as decentralisation policy, rural sector development, and PRSPs).

In order to define the main areas that NAPs should attack — which is an absolute prerequisite in order to clarify the responsibilities of the various actors — it is imperative to consider what is being done in these countries regarding sustainable development. Indeed, in the affected countries in West Africa, irrespective of NAPs, several major political reforms are underway (in particular those relating to decentralisation) and contribute toward UNCCD implementation. Most development partners provide institutional and field-level support to these policies through various projects (for instance, village land-management projects, local development projects, and community forest development projects).

In this changing political context, it is necessary to seek the added value of NAPs. Efforts should aim at delimiting the field of activities of NAPs against other plans, strategies, and national programmes. The UNCCD recommends integrating — or mainstreaming — NAPs into national policies to contribute to the delimitation of the field of action for controlling desertification (UNCCD 1994, art. 10). The next section analyses how this UNCCD principle is implemented on the field.

In the framework of decentralisation, West African affected countries have undertaken several legal and institutional reforms (for example, on forestry codes, pasture land charters, water codes, and land reforms) in order to hand the responsibility of developing and managing natural resources over to local communities. These reforms also aim at revisiting the distribution of benefits derived from the exploitation of natural resources between the state and local communities. Through these reforms, populations are equipped with the legal, technical, and financial means to involve themselves further in desertification control. Moreover, several field initiatives supported by many development partners contribute to strengthening the participation of local populations in managing natural resources with interesting tools, methods, and results.

All these reforms and initiatives contribute to the implementation of the UNCCD. Unfortunately, in the sub-region, the contribution of NAPs to these processes is limited or even non-existent. NAPs should help local populations take charge of desertification control activities. In the context of decentralisation, a certain number of activities could be proposed for implementation by the environmental ministries to promote local populations' involvement and responsibility in issues relating to natural resource development and management. The second part of this chapter elaborates further on this line of propositions.

Possible Lines of Reflection for Revisiting National Action Programmes

This section builds on the main issues identified and recommendations made above in order to offer operational proposals for revisiting NAPs in West African countries. First, some basic principles will be reviewed before proposing a specific objective for NAPs in West Africa.

The Three Priority Lines of Intervention in Desertification Control in West Africa

The main stake in controlling desertification is to assist the societies that live in arid, semi-arid, and sub-humid dry zones so that, with current contemporary data on their environment (covering climate, population, economy, policies, and environment), they can change their practices and develop new ways to enhance the value of natural resources that further provide long-term security.

According to Antoine Cornet (2002), chair of the Comité scientifique français de la désertification, three priority areas of intervention for desertification control can be distinguished:

- Intervention Area 1: The development of techniques enabling improved use of resources, increased productivity, or regeneration. The objective is to develop, with local people, improved and adapted practices in agriculture, livestock rearing, and the use of biomass and soils. These research and extension-based actions depend institutionally on different technical departments (such as agriculture, livestock, and environment).
- Intervention Area 2: The development of sustainable resource management systems. This intervention aims at developing effective systems to control access to natural resources and their use. With the failure of state-directed resource management systems and the advent of decentralisation in West African countries, the policies, strategies, and action plans (for forest, fauna, water, and so on) are advocating the development and creation of such management systems, the establishment at various levels (from local to sub-regional) of a process for consultation and negotiation among the various natural resource users. At the institutional level, this support for the development of sustainable natural resource management systems is almost exclusively the responsibility of the ministries in charge of environmental issues.
- Intervention Area 3: The establishment of institutional mechanisms and policies conducive to economic development and natural resource conservation. The objective is to establish legislations and regulations, put in place economic and fiscal incentives, develop infrastructures, and strengthen human resources. At the institutional level, the creation of such mechanisms is the responsibility of various technical ministries (responsible, for example, for economy and finance, territorial administration, infrastructures, agriculture, livestock, or the environment).

The proposed line of action for NAPs described below is only partial, as it only focusses on the second priority area — the implementation of which depends institutionally more directly on the ministries responsible for the environment.

In reference to the facts and the recommendations noted above, in order to play a strategic role in the promotion of sustainable development in affected countries in West Africa, NAPs should:

- facilitate the commitment of the local people to actions to combat desertification;
- seek the added value that they may bring, as programmes of the environmental ministries, to the other development programmes, plans, and strategies of the governments; and
- identify specific operational objectives that are shared by all stakeholders (including technical services, civil society, and development partners).

Therefore, there is a need first to try to delineate the potential scope of NAPs by identifying specific operational objectives for the ministries in charge of the environment and desertification control. To provide a concrete illustration, the following section focusses on what could be one line of action taken on by West African NAPs.

Strengthening the Role of the Ministry of Environment: A Possible Line of Action for Revisiting National Action Programmes

As described above, NAPs should facilitate the empowerment of local populations in their efforts to conserve and manage of natural resources. In West Africa, many projects supported by funding partners are backing the empowerment process of local populations in taking charge of development actions. These projects are commonly known as 'local development' or 'decentralisation support', depending on the level of commitment of the country to the decentralisation process; they aim mainly at building the capacity of populations for self-management, socioeconomic development, and environmental conservation.[14]

Such projects contribute to desertification control as underlined in a 2003 study in Burkina Faso, which concludes that 'the procedures of the local development type of projects are interventions that best agree with the principles of the UNCCD' (CONEDD 2003b). Yet, it is also noted that such projects, as currently implemented in most countries in the sub-region, tend to overlook the issue of sustainable natural resource management to the benefit of economic and social development issues.

One may thus wonder about the possibilities of improving the inclusion of environment and promotion of sustainable natural resource management in the efforts to involve and empower the local communities that were adopted in these interventions.

While many ministries are involved in implementing local development projects as well as in defining modalities for setting up and supporting decentralisation processes (for example, the ministries of territorial administration, finance, and agriculture), the role of the ministry responsible for the environment could thus be to bring in proposals for solutions (tools and methods) to enable environmental conservation in the current move toward decentralised development.

In order to facilitate the empowerment of local populations in natural resource preservation and sustainable management actions, NAPs should support the decentralisation process and notably endeavour to develop the local management of natural resources. As discussed above, desertification control involves not only the development of technical solutions, such as improving soil productivity (see intervention area 1) and institutional and policy measures (for instance, developing economic or legislative incentives — see intervention area 3), but also the implementation of a sustainable system to manage natural resources so as to maintain the rate of use of these resources below their regeneration rate (see intervention area 2).

Such management systems apply mainly to the so-called shared natural resources (pastoral, forest, fish, and fauna resources), which are characterised by the fact that they are non-produced and tapped by different users. For several years, and notably since the failure of state-management systems of natural resources, it has been agreed that such systems cannot be effective unless they are legitimate in the eyes of the users of the target resources, which implies that the latter are closely involved in developing the rules set by these management systems to govern the use of those resources.

The current decentralisation processes generally provide for the transfer of responsibility for natural resource development and management to the local communities. The development of natural resource management systems at the local level, where stakeholders may contribute to the negotiation of modalities for access to and use of target resources, is therefore a major challenge to preserving natural resources and hence to controlling desertification.

The ministries responsible for environmental issues should therefore promote the development of such management systems at the local level for the control of shared natural resources use. NAPs should support this approach by enhancing the value of existing experiences and build the capacity of local actors (such as technical departments and local populations) so that they develop such systems by themselves. The regional conference on land tenure and decentralisation in the Sahel, organised in June 1994 in Praia in Cape Verde under the aegis of the Permanent Inter-State Committee on Drought Control (CILSS) and the Club du Sahel (1994), recommended that Sahelian states develop participatory and decentralised management of natural resources. In particular, it made two suggestions: they should focus their efforts on preparing legislation that defines, at the national level, the general orientation toward sustainable natural resource management while leaving implementation measures to be defined at the local level; and they should focus on building capacity in local players in the realm of sustainable natural resource management.

In response to this and to the principles of most conventions and agreements relating to the environment and sustainable development (*inter alia* Agenda 21, the CBD, and the UNCCD), West African governments undertook many policy and legislative reforms aimed at empowering local communities in natural resource management (with regard to, for example, land tenure, decentralisation, rural development policy, forest codes, pastoral charters, water codes, and plans of action to combat desertification). Furthermore, experiences in developing and implementing local management systems of shared natural resources are contributing interesting results, methods, and tools.

The different institutional and legal reforms are not often very successful, and the respective actors that play a role in the local management of natural resources (such as technical government departments, associations, and other civil society representatives) lack information about current policy developments, their new responsibilities in the framework of these reforms, and practical modalities for their implementation.

Therefore, a specific objective for the NAPs in West Africa should be that the ministries in charge of the environment strive to enhance the value of experiences in developing and establishing local systems for managing natural resources and in building the capacity of local actors (particularly technical departments and local populations) so that they can develop such systems by themselves.

Conclusion

Apart from the legislative reforms on forests, water, hunting, and so on, yet to be further strengthened, the ministries in charge of the environment and desertification — in the framework of NAPs and also as part of their contribution to the process of decentralisation — should contribute to building the capacity of development actors (such as government workers, local populations, and their representatives) so as to prepare them for their role in the local management of natural resources.

This proposal for West African NAPs is fully in line with the COP, which promotes the new measures to be taken within the framework of the implementation of the convention. In fact, the COP-6 urges all affected developing countries, acting with the assistance of the developed parties and institutions concerned, to promote measures aimed at building capacity and participatory processes in the area of natural resource management (UNCCD 2003a).

The stakes of these NAPs, which would facilitate the promotion of natural resources through capacity building, are numerous and in line with sustainable development stakes of these countries: the long-term impact of natural resource preservation; the medium-term impacts of decreased conflicts arising from natural resource use, reduced vulnerability of populations, and the development of grassroots democracy; and of course the short-term impacts of the implementation of government policies on decentralisation, sustainable rural development, poverty reduction, and desertification control.

This proposal is, however, partial because it covers only the first of the three priority intervention areas identified. Obviously, as part of the full review process of NAPs, a similar exercise should be undertaken for the two remaining intervention areas, namely enabling the development of more sustainable resource management techniques and establishing institutional and policy mechanisms that would promote natural resource preservation. For these two areas, there is also the need to ensure that the operational objectives included in NAPs are adequately specific and limited in relation to what was planned in other existing development programmes and plans, and that they contribute to facilitating the commitment of local communities.

Notes

1 See, for example, the speech by Kofi Annan on World Day 2003 (United Nations 2003). Indeed, the lack of capital and economic opportunities are key factors in the 'mining' of resources and environment, whereas the reduction of land productivity can generate malnutrition, famine, conflicts, and migration.

2 In most West African countries, the NCBs — known by various names depending on the country — have been entrusted with the role of developing not only the NAP but also all the other environment programmes. More generally, the NCB is entrusted with the task of ensuring that environmental issues are taken into account in other sector policies. The NCB is a political, interministerial organ that endorses environmental policies and monitors their implementation through development policies. It also has a technical body often known as 'permanent secretariat', which is responsible for preparing plans, strategies, and programmes and submitting them to political decision makers. The NCB and its permanent secretariat are often considered a 'department' in the ministry responsible for environment in these countries.

3 In March 2001, the AHWG met to review the NAPs and the first national progress reports on the implementation of the convention in parties. A number of conclusions were derived from this analysis and were validated by Decision 3 taken at the fifth Conference of the Parties (COP) (UNCCD 2001, 10). Later on, the UNCCD established the CRIC, which is responsible for regular monitoring of convention progress in parties. At its first meeting in Rome in 2002 and its second meeting in Havana in 2003, CRIC came up with the same conclusions as the AHWG (UNCCD 2003b; 2003c).

4 This was noted [in three countries (Burkina Faso, Mali, and Mauritania)] during the consultation process initiated in 2002 by the Union Mondiale pour la Nature (known in English as the World Conservation Union, or IUCN) to share problems encountered in implementing the UNCCD with the various actors involved in desertification control (Union Mondiale pour la Nature and Bureau Régional de l'Afrique de l'Ouest 2004)

5 As Antoine Cornet (2002) discusses, the UNCCD states a number of guidelines including particularly the following ones: desertification control requires a long-term commitment of parties through international co-operation and partnership agreements; affected populations should be at the centre of the desertification control process and therefore their participation should be reinforced; governments of affected countries have a new role to play mainly in initiating co-ordination and establishing the legal and regulatory frameworks conducive to desertification control and sustainable management of their natural resources; science and technology are essential tools that should be mobilised, focussing on technology exchange; the value of traditional knowledge should be enhanced; and there need to be integrated approaches to natural resource conservation and socioeconomic development.

6 These diverging views are fuelled by the differing definitions of desertification and land degradation used by development agencies. As noted in Chapter 2, the lack of consensus on the definition of desertification adds to the confusion about the scope and focus of the UNCCD.

7 These NAPs seem to derive their legitimacy on the one hand from Article 4 of the UNCCD on 'general obligations', which states that 'in pursuing the objective of this Convention, the Parties shall ... adopt an integrated approach addressing the physical, biological and socio-economic aspects of the processes of desertification and drought'. On the other hand, this legitimacy derives from a partial reading of Article 10 on the content of NAPs:

'The purpose of national action programmes is to identify the factors contributing to desertification and ... emphasize implementation and be integrated with national policies for sustainable development.'
8   For example, the strategy proposed by Burkina Faso's NAP is quite similar to its policy paper on decentralised rural development.
9   See Burkina Faso's Decree No. 98-337/PRES/PM/MEE of 30 July 1998 on the organisation, responsibilities, and functioning of the Conseil National pour la Gestion de l'Environnement (CONAGESE), which is the country's NCB.
10 The UNCCD is not the only convention or institution that recommends the integration of development plans, programmes, and policies into a more coherent and larger framework. The WSSD declaration and the CBD, as well as the World Bank and various development partners, also favour this approach.
11 Examples are the National Action Plans to combat desertification that were adopted in the early 1980s following the 1977 United Nations Conference on Desertification (UNCOD) in Nairobi.
12 See the first reports to the COP from affected West African countries on activities undertaken to implement the convention, which are available on the UNCCD's website at <www.unccd.int/cop/reports/africa/africa.php> (December 2005). These reports were reviewed by the UNCCD group of independent experts (at the 2001 meeting of the AHWG in Bonn).
13 Recommendations such as 'desertification should be made a political priority' have no operational value. If something coherent is to be suggested to relaunch UNCCD implementation in West Africa, it must take into account the fact that desertification control in these countries comes under the responsibility of ministries in charge of environment.
14 In Mali, for example, the harmonisation of methods and tools for local development projects enabled the government to put in place a national mechanism to support decentralisation, which was largely supported by its development partners.

References

Conseil National pour l'Environnement et le Développement Durable (2003a). 'Revue des projets et programmes en cours d'exécution en vue de l'évaluation des ressources mobilisées pouvant entrer dans le cadre de la mise en oeuvre du Programme d'Action National de lutte contre la désertification au Burkina Faso'. Ouagadougou.

Conseil National pour l'Environnement et le Développement Durable (2003b). 'Funding Orientations in Desertification Control'. Ouagadougou.

Cornet, Antoine (2002). 'La désertification à la croisée du développement et de l'environnement: un problème qui nous concerne'. In R. Barbault, A. Cornet, J. Jouzel et al., eds., *Johannesburg: Sommet Mondial du Développement Durable 2002 — Quels enjoux? Quelle contribution des scientifiques?*, pp. 93–129. Ministère des Affaires étrangères, Paris. <www.adpf.asso.fr/adpf-publi/folio/johannesburg/pdf/joburg-desert-4.pdf> (December 2005).

Global Environment Facility (2003). 'Operational Program on Sustainable Land Management'. Operational Program No. 15. <www.gefweb.org/Operational_Policies/Operational_Programs/OP_15_English.pdf> (December 2005).

Organization of African Unity (1985). 'Declarations and Resolutions as Adopted by the Twenty-First Ordinary Session of the Assembly of Heads of State and Government'. 18–20 July.

Addis Ababa. <www.africa-union.org/Official_documents/Heads%20of%20State%20Summits/hog/uHoGAssembly1985.pdf> (December 2005).

Permanent Inter-State Committee on Drought Control in the Sahel and Club de Sahel (1994). 'Report on the CILSS/Club de Sahel Regional Conference on Land Tenure and Decentralisation'. 20–25 June. Praia, Cape Verde.

Toulmin, Camilla (2001). 'Lessons from the Theatre: Should This Be the Final Curtain Call for the Convention to Combat Desertification'. World Summit on Sustainable Development Briefing Papers. International Institute for Environment and Development, London. <www.iied.org/pubs/pdf/full/11017IIED.pdf> (December 2005).

Union Mondiale pour la Nature and Bureau Régional de l'Afrique de l'Ouest (2004). 'Rapport d'activités du projet: «Appui à la mise en oeuvre de la CCD en Afrique de l'Ouest»'. January. <www.iucn.org/brao/articles/0402/appuiCCD.pdf> (December 2005).

United Nations (2002). 'Report of the World Summit on Sustainable Development'. A/CONF.199/20 and A/CONF.199/20/Corr.1, 26 August–4 September. Johannesburg. <www.unctad.org/en/docs/aconf199d20&c1_en.pdf> (December 2005).

United Nations (2003). '"Desertification Is Both a Cause and a Consequence of Poverty", Secretary General Says in Message for International Day'. Press Release SG/SM/8750 OBV/355, 12 June. <www.un.org/News/Press/docs/2003/sgsm8750.doc.htm> (December 2005).

United Nations Convention to Combat Desertification (1994). 'Text of the United Nations Convention to Combat Desertification'. <www.unccd.int/convention/text/convention.php> (December 2005).

United Nations Convention to Combat Desertification (2001). 'Report of the Conference of the Parties on Its Fifth Session, Held in Geneva from 1 to 12 October 2001: Part Two — Action Taken by the Conference of the Parties on Its Fifth Session'. ICCD/COP(5)/11/Add.1, 7 November. <www.gm-unccd.org/English/DOCS/COP5Dec.pdf> (December 2005).

United Nations Convention to Combat Desertification (2003a). 'Report of the Conference of the Parties on Its Sixth Session, Held in Havana from 25 August to 5 September 2003: Part Two — Action Taken by the Conference of the Parties at Its Sixth Session'. ICCD/COP(6)/11/Add.1, 7 November. <www.unccd.int/cop/officialdocs/cop6/pdf/11add1eng.pdf> (December 2005).

United Nations Convention to Combat Desertification (2003b). 'Report of the Committee on Its First Session, Held in Rome from 11 to 22 November 2002'. ICCD/CRIC(1)/10, 17 January. <www.unccd.int/cop/officialdocs/cric1/pdf/10eng.pdf> (December 2005).

United Nations Convention to Combat Desertification (2003c). 'Review of the Implementation of the Convention and of Its Institutional Arrangements, Pursuant to Article 22, Paragraphs 2(a) and (b), and Article 26 of the Convention: Review of the Report on Enhanced Implementation of the Obligations of the Convention'. ICCD/CRIC(2)/3, 2 July. <www.unccd.int/cop/officialdocs/cric2/pdf/3eng.pdf> (December 2005).

United Nations Convention to Combat Desertification (2005). 'Explanatory Leaflet'. <www.unccd.int/convention/text/leaflet.php> (December 2005).

United Nations General Assembly (1977). 'Plan of Action to Combat Desertification'. Resolution 32/172, 19 December.

Chapter 11

Knowledge and the UNCCD: The Community Exchange and Training Programme

Noel Oettlé

The single greatest challenge facing humanity is the management of natural resources in a manner that will prove to be sustainable in a rapidly changing global environment.

In 1992 the seminal United Nations Conference on Environment and Development (UNCED) in Rio de Janeiro provided the first comprehensive overview of the problem. Chapter 12, 'Managing Fragile Ecosystems: Combating Desertification and Drought', recognised the need 'to support local communities in their own efforts in combating desertification, and to draw on the knowledge and experience of the populations concerned, ensuring the full participation of women and indigenous populations' (United Nations General Assembly [UNGA] 1992, art. 12.56[d]).

The thinking embodied in this statement was subsequently elaborated in the United Nations Convention to Combat Desertification (UNCCD), an instrument designed to achieve sustainable development by implementing strategies that focus on 'sustainable management of land and water resources' (UNCCD 1994, art. 2). With its emphasis on information provision and participation, the UNCCD recognises the need for knowledge-based approaches that will 'integrate, enhance and validate traditional and local knowledge, know-how and practices' (art. 17[c]).

Knowledge alone is insufficient to bring about the desired result of sustainable resource use; it is also essential to agree on how resources will be managed for the greater common good, and to act. The UNCCD (at the level of a macro-institutional response) is unique in recognising that micro-institutional responses are vital in achieving sustainable management of natural resources. Article 3, on principle, emphasises that 'design and implementation' of measures should be 'taken with the participation of populations and local communities'.

This chapter explores how a knowledge-based approach, the Community Exchange and Training Programme (CETP) of the Global Mechanism (GM), has been elaborated and implemented within the context of the UNCCD, and examines some of the results from the perspective of the field-based practitioner. The final section provides an assessment of the contribution of the CETP to the implementation of the UNCCD, and identifies shortcomings. It examines the way in which knowledge-based approaches achieve sustainable livelihoods at a local level.

Institutions for Sustainable Resource Use

Natural resource management may be seen as involving the interaction of the natural and human domains, and the institutions that structure the transformation when these two domains interact (Maarleveld 2003). As such, the term 'institutions' here appears not in the context of organisations, but rather refers to the 'rules of the game' as regards the management of resources. Institutions can be understood as collective assumptions and convictions about the strategy to be followed in maintaining and improving livelihoods (Maarleveld 2003). This includes policies, laws, regulations, norms, and codes of conduct. In this sense, institutions differ from organisations, which are groupings of people bound together by a common goal. The purpose of institutions in the realm of natural resources is to ensure good governance in the interest of the collective good (see Chapter 8). Strong and effective local institutions are a major prerequisite for sustainable resource management (Narjisse 2000).

In the rapidly evolving global social, economic, and biophysical spheres, institutional weaknesses have been identified as fundamental stumbling blocks to sustainable resource use. The authority and power of most local, traditional institutions have been eroded by diverse forces such as central government intervention, formal education, and subversion by local and other economic elites bent on exploitation of common resources. However, higher order or centralised institutions (national laws and policies, or multilateral environmental agreements [MEAs]) are frequently poorly communicated to resources users, and poorly understood by them. All too often these institutions are either ignored or resisted because enforcement is weak and resource users are not sufficiently conscious (or at least convinced) of the benefit that they might derive from adherence.

Evolving adequate, responsive, flexible, and effective institutions at the local level is the crucial task of all stakeholders in the use and management of natural resources. Knowledge is a key driver of institutional evolution, and patterns of governance increasingly require networks of knowledge, dispute resolution procedures, shared experience, trust building, and agreed predicators for taking action (O'Riordan and Stoll-Kleemann 2002).

Mobilising rural people to contribute to the task of achieving sustainable resource management and livelihoods in the drylands has been a vital role filled by nongovernmental organisations (NGOs). NGOs also have a vital role in supporting the development of local democratic organisations and, through them, sound local institutions for managing resources. This chapter examines how knowledge-based approaches have been explored, developed, systematised, and 'scaled out' in the implementation of the UNCCD at local level through participation by NGO members of the Réseau international des ONG pour la desertification (RIOD, or the International Network on Desertification; see Chapter 7).

The Community Exchange and Training Programme

The CETP was initiated in 1999 following discussions between the UNCCD Global Mechanism and members of RIOD. It was intended to support the 'generation of appropriate knowledge about sustainable natural resource management and production at a local level, and the initiation of actions to capitalise on new opportunities' (GM 1999). The overall goal of the programme was for 'dryland communities [to] improve the management of their natural resources and production systems to better their livelihoods'.

In its design, the drafters of the CETP took account of emerging awareness that mobilisation is most easily achieved when people share common understandings of the challenges faced in managing shared resources and join forces to ensure adequate access to markets and services. The CETP aimed to provide opportunities for the low-cost dissemination of knowledge and mobilisation of rural people, particularly those who were not served by existing development projects. Local ownership of the development process was encouraged by providing only supplemental funds for learning initiatives in which community-based and NGO participants also invested their own resources. The link to the implementation of the UNCCD was enhanced by designing mechanisms of project approval that ensured that national co-ordinating bodies (NCBs) supported and linked with the projects, thereby creating a conscious fit with the UNCCD National Action Programme (NAP) in the specific country.

The knowledge and capacity development aspects of the CETP were expressed in the objectives of the programme, which include the following:

- The natural resource management and production knowledge of communities living in drylands is validated, shared, and further developed.
- The ability of the community members to identify, articulate, and respond to limiting factors and opportunities in natural resources management and production is enhanced.
- The coping strategies and resilience capacities of the communities are enhanced.
- Understanding and local capacities to take action in both the short and long term are promoted.

Piloting the Community Exchange and Training Programme

The outcomes of the transition from concept to action are unpredictable. To minimise risk, the GM invited proposals for viable CETP projects, short-listed these, and sought partnership with other agencies with expertise that could contribute to a successful outcome. A promising proposal from South Africa's Environmental Monitoring Group (EMG) was shared with colleagues from the World Bank's Indigenous Knowledge for Development Program. The project was initiated in mid 2000 as a joint GM/World Bank pilot project.

The proposal was drawn up with members of the Suid Bokkeveld community of the Northern Cape Province. This area is home to a resource-poor, dryland community that depends on sheep production and indigenous rooibos tea cultivation for its livelihood. Community members wanted to establish a range of integrated initiatives to improve their lives and to promote the sustained use of natural and cultural resources.

People expressed the need for greater knowledge and insight into a set of linked problems and issues, ranging from declining productivity to poor market access. The design of the project took into account the new paradigm approaches to new ways of learning about the world (Pretty and Chambers 1994).

A process was planned that involved knowledge exchange, facilitated learning, and planning to be undertaken in partnership with appropriate support agencies (see Figure 11-1). The processes were to be documented to enable learning from them, and so the approach could be replicated. The project involved two journeys, the first focussed on eco-tourism and the second on the production and marketing of indigenous rooibos tea.

The project began with workshops designed to engage a wide range of members of the community and to establish a shared vision for the future. By exchanging their ideas, participants developed a visual depiction of what an improved situation would look like, and then described it in words. In the context of locally available resources, they explored opportunities and identified social, physical, and economic constraints that would need to be addressed. They also set concrete objectives for their development process.

The exchange concept was developed in light of the vision and objectives as well as accessible funding. A concept was developed for sharing what would be learned by the travellers, and criteria were then set for their selection. Through networking with other NGOs and government agencies, a team was formed to facilitate the visits. The team learned the skills necessary to facilitate the visits, and developed a team contract so that it could be self-managing. Community members identified the places to visit from a number of alternatives.

Following a planning and capacity development workshop with the facilitation team, community-based preparatory workshops were held. Delegations for the visits were chosen through a transparent process, and learning objectives for the visits were developed interactively. In a crucial process of 'matchmaking', the selected communities were informed about the delegations and the learning objectives. Plans for the journeys were made on the basis of people's needs and limitations. Practical arrangements were finalised in consideration of the host communities' capacities and suggestions.

There were two exchange visits, focussed on two very different topics. As far as possible the visitors learned experientially and from interaction with their peers. 'Expert' input from outsiders was minimal, although workshops were organised to promote interaction. Facilitation served to maximise participation, to promote learning and reflection, and to allow the travellers to make any necessary adjustments to their programme.

Figure 11-1 Steps Followed in the Implementation of the Pilot Project*

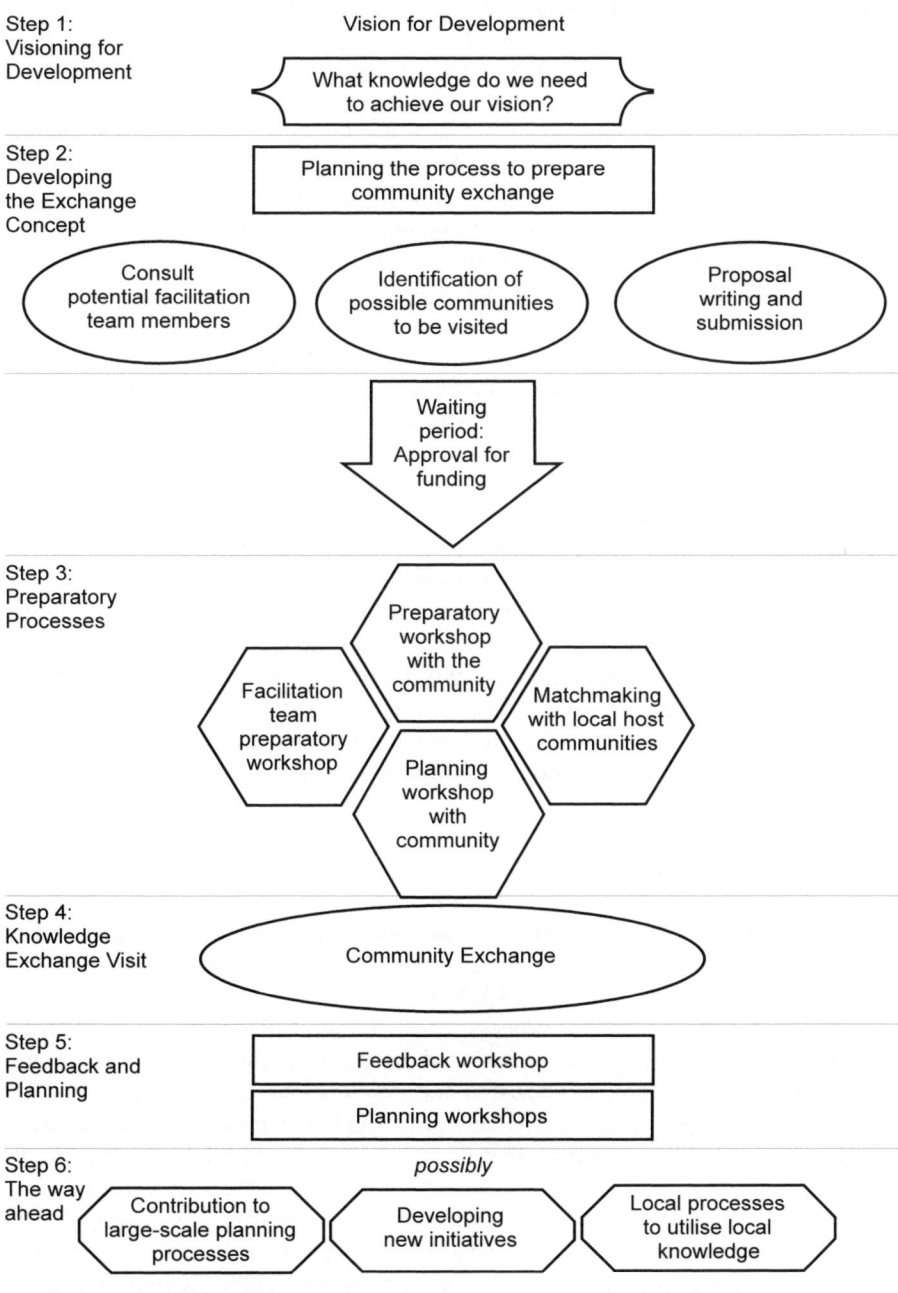

Source: Oettlé and Koelle 2003.

Once the travellers had returned home, two feedback and planning workshops were held for the wider community. Nominated rapporteurs gave feedback on the visits, and participants discussed what they would like to undertake in the light of their experiences. Visual media were used to evoke the actual experiences of the visits. Most importantly, plans were made for new initiatives.

The plans that were made depended on skills and knowledge that were not yet held by members of the community. To address this, follow-up training was provided to develop the skills and competencies needed for the emerging initiatives; it ranged from organic agricultural production techniques to book-keeping and hospitality training. Two enterprises emerged from the exchange visits, and both have since grown and provided enhanced incomes and new economic opportunities. These businesses are self-managed, and, in turn, have sparked the growth of further initiatives. One of them, the Heiveld Co-operative, now exports certified organic products to markets throughout Europe. A local farmer-led initiative that promotes the sustainable use of natural resources was also inspired by the process, and its members collaborate with scientific researchers and NGO partners to enhance their ecological management of their land. Incomes from sustainably produced rooibos tea have more than trebled.

The entire process was recorded on film, and the video title 'Full Circle' was produced as a training tool for community knowledge exchange.

These processes depended on the evolution of local institutions, which have been able to develop and respond to an increasingly complex institutional environment. An example of this is the regime that has been developed by the members of the Heiveld Co-operative in conjunction with scientific researchers for the sustainable harvest of wild stands of rooibos. The wild crop is harvested only every second year, giving it sufficient time to re-grow without losing strength. This, in turn, has contributed to establishing a good market for the sustainably harvested product.

Lessons Learned from the Pilot Project

A number of invaluable lessons were learned from the pilot project, and subsequently influenced the design of a larger, sub-regional programme within the Southern African Development Community (SADC).

The knowledge-exchange process must be embedded in comprehensive community-based development work, and its purpose should arise from other processes in which a collective vision for development is explored. If not, the knowledge exchange is likely to lack any firm foundation and could be experienced simply as an interesting journey that does not provide any clear pivot-point for further action.

The success of the initiative depended greatly on strong ownership by the members of the local community. From its initiation they actively defined the purpose of the exchange visits and established the criteria by which they themselves selected the delegates. Once the delegation had been selected, its members established a code of conduct and agreed on how they would learn together. Local ownership was strengthened because the members of the community who undertook the exchange visit were responsible for reporting back to those who stayed behind.

Skilled external facilitation of the process was important in maximising the learning opportunities by leaving participants unencumbered by logistical and process considerations. Journeys expose people to novel situations and unexpected challenges. Effective facilitation can enable people to focus on the issues that they have collectively identified as being at the core of the enquiry while in an emotionally 'safe' environment, without unnecessary distraction.

Clear learning objectives were established early, and the process was progressively shaped toward achieving them. Communicating the learning objectives to the host communities enabled them to prepare themselves to present appropriate experiences and examples of their successes and frustrations. This resulted in a well-focussed learning experience. However, hosting strangers who are not used to travelling requires more than simply knowing what they want to learn about. It was important to prepare the communities to be visited as well, by briefing them on the background and special needs of the visitors.

Government has a key role in providing ongoing support to local initiatives. In the Suid Bokkeveld experience, involving government departments not only helped to develop the capacity of staff, but also contributed to the sustainability of the initiatives that were subsequently launched. If the knowledge-exchange process is successful, spillover effects will occur that will continue into the distant future and increased ability for self-development will attract other resources to the community. The role of government service providers will evolve in response to these initiatives, and to the enhanced capacities of the community.

Knowledge exchanges should be open-ended learning processes, and should allow for serendipity — positive events that emerge unexpectedly from problematic situations. Sound logistics are essential to ensure the health and comfort of the travellers. However, plans and logistical arrangements should serve the process, and not lead it. Some of the best learning opportunities cannot be anticipated. Things will not go entirely according to plan: one must embrace the unplanned, unexpected learning opportunity that may present itself.

Scaling Out the Community Exchange and Training Programme in the Southern African Development Community

Through the RIOD, other NGOs became aware of the CETP and of the success achieved by the Suid Bokkeveld pilot project. In September 2001, national representatives of the network met in Zambia for a strategic planning workshop, and agreed to launch the CETP as a sub-regional programme within the SADC. A project appraisal committee was established to screen and verify the eligibility of project proposals submitted by various NGOs and community-based organisations, endorse qualified project proposals, provide feedback to project proponents, and seek potential sources for funding.

Four of the UNCCD focal points were elected to serve on the project appraisal committee: the Zambia Alliance of Women (Zambia), the Environmental Monitoring Group (South Africa), the Journalists' Environmental Association (Tanzania), and the Zimbabwe Women's Bureau (Zimbabwe).

In 2002, the committee established procedures and parameters for the programme. These were designed to give more specificity to the broad outlines described in the programme document and to enable wiser, less subjective decisions to be taken as to which projects would be supported. A CETP guide was subsequently published to assist interested organisations to develop and submit appropriate proposals. The guide provided detailed information on the requirements and criteria used to assess each proposal; these were designed to lead to the selection of projects that embodied certain key principles and qualities associated with success (see Table 11-1).

In 2003 five projects were approved and implemented by NGOs in the SADC. The projects involved interaction among rural communities in Zambia, Tanzania, Malawi and South Africa. The implementing organisations were the Zambia Energy and Environment Organisation (ZENGO), the Farmer Support Group, the Zambia Alliance of Women, the Green Living Movement, and the Kinondoni Welfare and Development Association. The projects were supported via the World Bank's Indigenous Knowledge for Development Program, which was able to utilise resources provided by the Swiss Development Corporation Studies Fund.

The focus areas of the exchanges varied greatly, and included wetland management, craft production and marketing, organic and sustainable agriculture, soil fertility management, fuel-efficient cook stoves and business development, and capacity development of farmers.

Facing the Challenges

A knowledge-exchange process involves challenging logistics and complex budgeting. These aspects of the enterprise can all too easily take up most of the

---

Table 11-1 Criteria Applied by SADC/RIOD for the Selection of Knowledge Exchange Projects

- Process orientation (as opposed to technology orientation)
- Strong local ownership of the process on the part of both local communities and implementing agencies, within a broader sustainable development context
- Sufficient formalisation and proven competence on the part of implementers
- The needs of specific target groups should be addressed by the process
- Compatibility between the social and natural ecologies of the groups to be linked
- Relatively local focus (not promoting travel to exotic places)
- Development of local capacities through the knowledge exchange and follow-up processes

available resources, and there is a danger that the learning process may receive little attention. Two key challenges should be recognised and addressed.

First, community exchanges are invariably profoundly stimulating for people from rural communities who rarely have opportunities to travel. There is so much interest and excitement that, in the absence of a well-structured learning process, the visit can lose focus and become just an interesting and exciting journey of general discovery. Whereas participants might enjoy the experience profoundly, it is not likely that the trip will lead to any collective efforts to improve the local situation.

Second, the organisers of an exchange visit bear a huge responsibility for the safety and comfort of the travellers. This aspect is vitally important, but it takes attention away from the equally vital task of designing and facilitating the learning experience.

To address this inherent difficulty and to ensure overall sound design and support of the knowledge exchanges, the EMG undertook to provide 'backstopping' to project implementers. This involved interactive review of proposals and work plans, regular liaison, and a fair amount of hands-on support in planning and implementing the projects, with the aim of developing the capacities of the implementers. EMG advised and guided implementers on methodology, logistics, monitoring and evaluation, follow-up activities, and dissemination of lessons learned and information about the project outcomes. In addition, EMG managed the funds and entered into contractual arrangements with implementers.

The backstopping function played an essential role in ensuring the development of sound, realistic, and well-conceptualised proposals. This included electronic communication about content and concepts, as well as telephone and face-to-face input. Regular interaction in person with the project facilitation team is invaluable, but the great distances involved limited the opportunity to do so. In the case of the Zambia Energy and Environment Organisation and the Farmer Support Group, engagement in the planning phase was useful in developing the capacities of the teams and avoiding pitfalls that were not obvious from the project proposals.

The author of this chapter was involved as a backstopper with the Zambia Energy and Environment Organisation, the Zambia Alliance of Women, and the Green Living Movement. Backstopping provided rich learning opportunities and enabled some mid-stream course corrections, as well as providing the facilitation team members with stimulus and support.

Debriefing sessions were conducted in the case of four of the five projects, and provided a wealth of insight into the process, outcomes, and opportunities. As the knowledge exchanges were designed to stimulate further developmental processes, the follow-up process is critical and is likely to benefit from the reflections that arise in the course of a debriefing.

Views from the Field: Lessons Learned from the Community Exchange and Training Programme in the Southern African Development Community

Knowledge exchanges are most fruitful when 'nested' in ongoing relationships and processes of development. Where previous processes had established clear common principles and direction to development processes, the knowledge-exchange processes added momentum and value. Where little common purpose existed prior to the exercise, it was difficult to maintain momentum after the exchange visit. Simply agreeing on what an improved future situation would look like was a vital point of reference to anchor the entire development process.

Transparent requirements and selection criteria appear to have contributed to sound proposals being submitted. Developing criteria for assessing projects and communicating these in a transparent manner helps project designers to develop more appropriate proposals, which will have a greater chance of successful implementation. The engagement of the SADC RIOD project appraisal committee in the process resulted in a well-balanced set of selection criteria for the project and in transparent and robust decision making.

The entire process of designing and implementing a knowledge exchange is best conceptualised as a learning process. In this context, interaction and discourse with project designers is likely to enhance the quality of proposals and plans. If they have the opportunity to discuss, defend, or adjust the apparent shortcomings of proposals in response to feedback, they will be better able to design effective development processes and anticipate difficulties.

Within the tumult of a journey with its many practical and conceptual challenges, it is important to create the space for serendipitous learning events. The outcomes of knowledge exchanges are not predictable, and many serendipitous learning events are likely to take place in the course of the visits. Tight schedules and close adherence to rigid plans will close off opportunities for these to occur.

It is well known that adults learn most effectively from experience, or learning by doing, as opposed to propositional learning. Participation in simple tasks such as basket making, seed selection, stove manufacture, tree planting, and food preparation not only created bonds of shared experience but also enabled all of people's senses and intelligence to be engaged in the learning process. These experiences were far more memorable than talks and demonstrations by others, and tended to lead to future actions.

Reflection deepens and focusses learning processes. Evaluation played a key role in all of the exchange visits: reflection enabled participants to deepen their learning and share what had excited them, and it also served as a vitally important mechanism for keeping the process focussed on its agreed purpose, adjusting plans if necessary, addressing problematic behaviour, and responding effectively to unexpected opportunities.

Follow-up is essential to realise the potential of knowledge exchanges. Once a knowledge-exchange visit has been completed, implementing agencies should expect to need to respond to a range of new initiatives from people and groups that have been associated with the process. Successful agencies have interfered as little as possible, and created maximum space for people to evolve their ideas and transform these into actions. This being said, appropriate responses included actions such as providing or accessing expert advice (financial, legal, technical), capacity development and training, networking with other support agencies, as well as the facilitation of planning of new initiatives. Outside agencies must provide the necessary stimulus to ongoing processes without becoming the 'owners' of the problems, or the solutions.

Active learning can lead to empowerment, enhanced livelihoods, and improved environmental management. Participants in a number of knowledge exchanges realised that. In the case of the Suid Bokkeveld CETP process, market success and higher prices for agricultural products have been assured by implementing and promoting a conservation-friendly approach, and enhanced production has also been achieved by retaining or reintroducing biodiversity into cultivated lands (Oettlé et al. 2002).

Civil society organisations should strive to engage government agencies in the process. It was very difficult to obtain the endorsement of NCBs for all CETP proposals. In some cases, the NCB was effectively non-functional and did not meet for many months or even years. This created major delays, thus leading to the decision to accept endorsement by the government's national focal point. Nevertheless, seeking endorsement brought the initiatives to the attention of the national government agencies, which do not always have a clear sense of rural realities and processes. As such, opportunity was created for NGOs to contribute to national policy debates.

At the local level, engagement of civil servants was far more effective, reflecting the sound working relationships on the ground. This engagement was worthwhile in all cases, and contributed to building more effective partnerships in facilitation of the development processes with the participating communities, while also ensuring that information about the processes was widely disseminated.

Finally, it was clear that the contributions made (usually in kind) by the implementing NGOs and participating communities contributed to the sustainability of the initiatives. The process of negotiating and agreeing on local contribution made it clear to all that the process would not be a joy ride, based on handouts. Contributing their own resources to knowledge-exchange processes strengthened the ownership of implementers and participants of the process and the outcomes. Despite the challenges involved, most people see making their own contribution as a source of pride, and retain or further develop their sense of independence and self-determination.

Broadening Knowledge, Growing Institutions

The CETP has stimulated project activities in diverse and isolated communities in the SADC, yet has also linked them to the NAPs and the overarching principles of the UNCCD. Even though those principles are part of international law, they too easily remain abstract and do not necessary inform practice. Grassroots CETP projects give form and shape to them. When experiences from CETP projects are shared and disseminated to wider audiences, an excellent opportunity is created to showcase the UNCCD in action.

Commitment of government to CETP projects and their outcomes is notoriously difficult to engage and sustain, but is nevertheless essential if initiatives are to be supported (and not undermined) by government agencies. Engagement with government also enables the lessons learned in the course of the projects to be disseminated widely, and in some cases the CETP projects have been adopted as showcase projects by government.

Because the processes of knowledge exchanges differ so widely from the propositional knowledge-based approaches that are embedded in most agricultural, forestry, or environmental educational and training institutions, the capacity of government officials and even NGO staff is usually not adequate to facilitate a knowledge exchange. Inappropriate leadership can subvert the process, with the result that it will at best be a study tour, in the course of which participants will be told what they should do without any engagement of their own experience and wisdom.

Knowledge-based approaches like the CETP are essentially learner- and process-oriented. They should not be conceived as linear development projects, aimed at a tightly defined set of outcomes. By definition, the participants will open themselves up to new experiences and new ideas from which new initiatives will emerge. The process orientation implies that facilitators should be sufficiently skilled to enable the learning processes to unfold without trying to direct them to a predetermined objective.

Knowledge exchanges are not a panacea for all development problems, and should be used with discretion. They should be regarded as one tool within a larger development toolkit. Nevertheless, they are very valuable interventions at various stages in a development process, including the initiation of new ventures and stimulating people who are already involved in a production or management process. They have also been used successfully by professionals from a particular community of practice whose members lack the stimulus of meeting and exchanging ideas with their peers.

The networks established by knowledge exchanges may prove a long-term resource for both the visitors and the visited. The South African pilot project led to a subsequent process of 'sharing back' with the host community, and resulted in the development of stronger institutional mechanisms for managing wild stands of rooibos. The development of effective local institutions can be immeasurably strengthened when people are able to share their insights and the successes that they have achieved with their peers.

The true impact of a knowledge exchange can only be assessed years later, and the best people to assess the impact are the participants themselves, who will be able to reflect on the ways in which it has influenced them and enabled them to improve their lives and enhance the management of their resources. The development of local organisations is usually a good indicator, but more important will be the effectiveness of local institutions to manage resources in a sustainable manner.

Community-based knowledge exchange processes are an extremely effective way to stimulate and support knowledge-based approaches at local level, and within communities of practice such as farmers, craftspeople, artisans, or honey gatherers. They broaden horizons, stimulate learning, inspire vision, and develop the capacities that people need to address the challenges of sustaining their livelihoods in a fast-changing world.

References

Global Mechanism and Réseau International d'ONG sur la Désertification (1999). 'Community Exchange and Training Programme'. Rome.

Maarleveld, Maureen (2003). 'Social Environmental Learning for Sustaining Natural Resource Management: Theory, Practice, and Facilitation'. Ph.D. Wageningen University.

Narjisse, Hamid (2000). 'Rangeland Issues and Trends in Developing Countries'. In S. Archer and A. Olafur, eds., *Rangeland Desertification*. Kluwer Academic Publishers, Dordrecht.

O'Riordan, Timothy and Susan Stoll-Kleemann, eds. (2002). *Biodiversity, Sustainability, and Human Communities: Protecting beyond the Protected*. Cambridge University Press, Cambridge.

Oettlé, Noel, Adele Arendse, B. Koelle, et al. (2002). 'Community Exchange and Training in the Suid Bokkeveld: A Pilot Project to Enhance Natural Resource Management'. Proceedings for 'Alternative Ways to Combat Desertification: Connecting Community Action with Science and Common Sense', 8–20 April, Cape Town. V. Ward, ed. Desert Research Foundation of Namibia.

Pretty, Jules and Robert Chambers (1994). 'Towards a Learning Paradigm: New Professionalism and Institutions for Sustainable Agriculture'. In I. Scoones and J. Thompson, eds., *Beyond Farmer First: Rural People's Knowledge, Agricultural Research, and Extension Practice*. Intermediate Technology Publishing, London.

United Nations Convention to Combat Desertification (1994). 'Text of the United Nations Convention to Combat Desertification'. <www.unccd.int/convention/text/convention.php> (December 2005).

United Nations General Assembly (1992). 'Report of the United Nations Conference on Environment and Development'. 3–14 June. Rio de Janeiro. <www.un.org/documents/ga/conf151/aconf15126-2.htm> (December 2005).

Chapter 12

Agriculture, Trade, and Desertification: Implications for the UNCCD

Karel Mayrand and Marc Paquin[1]

The marginalisation of developing-country farmers in the multilateral trading system is often described as a major driver of persistent worldwide poverty. The Doha Round of trade liberalisation, which was launched in 2001 under the auspices of the World Trade Organization (WTO), aims to support development and address the marginalisation of least-developed countries (LDCs) in international trade through trade liberalisation, notably in agriculture (WTO 2001, par. 3).[2] Given the stakes for developing countries, the Doha Round has been described as a 'development round' and as a major undertaking to alleviate poverty in the world. The 2002 World Summit on Sustainable Development (WSSD) also endorsed the Doha initiative, emphasising its potential for developing countries (United Nations 2002, par. 47[a]).

The Doha negotiation mandate on agriculture aims for 'substantial improvements in market access; reductions of, with a view to phasing out, all forms of export subsidies; and substantial reductions in trade-distorting domestic support' (World Trade Organization 2001, par. 13). The mandate also recognises the principle of special and differentiated treatment for developing countries and its concrete application in the outcomes of the negotiations. The Doha declaration contains a general commitment to sustainable development (par. 6) and expresses an intention to enhance the mutual supportiveness between the multilateral trade regime and multilateral environmental agreements (MEAs) (par. 31). This statement of principles, however, does not provide a mandate to analyse the potential impacts of the Doha Round on the progress of MEA objectives or to submit the negotiated texts to a sustainability assessment. There is therefore little, if any, consideration of environmental issues such as desertification in the current negotiation process.

The Doha Development Agenda rests on the assumption that agricultural trade liberalisation will raise agricultural exports and incomes in the rural developing world, thereby contributing to alleviating rural poverty. This chapter argues that such theoretical assumptions may not fully materialise in the absence of appropriate policy interventions that seek to resolve persistent structural inequities in the world commodities market and to improve agricultural productivity and rural infrastructure development in drylands. On the contrary, agricultural trade liberalisation in the context of the Doha Round could generate significant impacts on land degradation and rural livelihoods in the world's drylands.

The outcomes of the Doha Round will likely have a profound impact on agriculture and on rural livelihoods in developing countries' rural drylands, including those regions affected or threatened by desertification. These potential impacts will influence the achievement of Millennium Development Goals (MDGs) on poverty and the environment as well as the implementation of the United Nations Convention to Combat Desertification (UNCCD) in the next decade. This chapter describes and analyses the interconnections among agricultural trade liberalisation, land degradation, and poverty in rural drylands. It seeks to understand the implications of the Doha Round on UNCCD implementation, and to derive policy implications to improve the coherence among poverty alleviation, agricultural trade, and land management strategies with a specific focus on activities to combat desertification under the framework of the UNCCD.

Agricultural Trade and Desertification: Exploring the Drivers

The Food and Agriculture Organization (FAO) (2004, 6) estimates that approximately 2.5 billion people worldwide depend on agriculture for their livelihoods; Oxfam (2002b, 7) reports that 96 percent of those people live in developing countries. Trade in agriculture totalled US$552 billion in 2001, which represents 45 percent of total agricultural production (World Wildlife Fund (WWF) et al. 2004, 3). Agricultural trade liberalisation is considered by many to be a key strategy to increase rural incomes and support development in the developing world. According to Oxfam (2002a, 47–48), a 1 percent growth in agricultural exports in developing countries can increase annual per capita income by 12 percent in Southern Asia, by 4 percent in Latin America and Eastern Asia, and by up to 20 percent in sub-Saharan Africa. In the same line, Vangelis Vitalis (2004, 4) argues that agricultural trade liberalisation could reduce the number of people living in absolute poverty by 130 million, or 12 percent of the world total, with the greatest gains likely to go to sub-Saharan Africa.

Nevertheless, although agricultural trade liberalisation bears much potential for developing countries' farmers in terms of fairer competition and increased market shares resulting in higher rural incomes, it also involves socioeconomic and environmental transformations that may significantly affect long-term agricultural production, rural livelihoods, and land degradation in drylands. And it remains to be seen if developing countries, especially LDCs, will be able to capitalise on the new opportunities created by a liberalised world agricultural market, or if they will end up in a worse situation following liberalisation.

Based on a review of trade impact assessment methodologies and case studies in the agricultural sector, this chapter puts forward the proposition that agricultural trade can affect desertification by two main drivers (on methodologies, see Convention on Biological Diversity [CBD] 2003; on agricultural case studies, see, inter alia, Pomareda and Murillo 2003; United Nations Environmental Programme [UNEP] 2002b; 2002a; Werth 2002; WWF 2003). The first driver is the intensification of production and the replacement of small-scale agriculture by intensive industrial monocultures. Large-scale agriculture, through intensive use of agro-chemicals, irrigation systems, and mechanised farming techniques, can have major impacts on soil quality and dryland ecosystems. However, trade can bring new investments and technologies that can improve yields while maintaining the productive capacity of the land.

The second way in which agricultural trade may affect desertification is through its impacts on rural livelihoods, especially those of small dryland farmers. Trade liberalisation can either create opportunities or marginalise small farmers in the developing world further, depending on the policies accompanying its implementation. Small farmers tend to respond to new opportunities or income shocks by expanding land under production or intensifying the use of the land; in turn, this leads to land degradation when marginal lands are brought into production or overexploited, which is often the case in the drylands. In such situations, agricultural trade liberalisation may fuel the land degradation–poverty dynamics that are at the source of desertification.

Agricultural Trade, Poverty, and Development: Mapping the Linkages

International trade theory predicts that under perfect market conditions, countries will specialise according to their respective comparative advantages and that goods and services will be produced at a lower cost, thereby increasing welfare globally. This increased welfare will lead to higher average incomes, and to higher consumption and investment levels that will fuel economic growth. Generally speaking, higher investment rates will translate into productivity gains that will make the economy more efficient and competitive. Trade theory also predicts that the process of reallocating productive resources will produce losers and winners, and that losers should be compensated or redirected toward new productive activities. Those winners and losers will be, *inter alia*, countries or regions, economic sectors or industries, and consumers or producers.

It is generally accepted that developing countries have a strong comparative advantage in agriculture and that they will benefit most from a liberalisation of trade in that sector. In theory, developing countries will specialise in producing commodities for the world market. Sales on the world market will generate foreign currency earnings to sustain investments in agricultural productivity improvements. In addition, liberalisation can be instrumental in attracting foreign direct investment (FDI) and new technologies to improve productivity in the agricultural sector. These productivity gains and extra earnings will raise rural incomes and stimulate consumption and economic growth. In the process of increasing agricultural productivity, farm labour will be released and absorbed into the urban manufacturing sector. Economic theory predicts that under appropriate conditions, this process will allow developing countries to realise their economic potential.

Unfortunately, evidence from the last 15–20 years tends to diverge from the classic theoretical model. According to the Organisation for Economic Co-operation and Development ([OECD] 2003, 20), 'although economic theory predicts increases in welfare and economic growth following trade liberalisation, empirical research has not demonstrated a direct link between trade openness and growth'. In 'The Least Developed Countries Report 2004', the United Nations Conference on Trade and Development (UNCTAD) (2004) reached a similar conclusion based on an analysis of 51 LDCs that saw their exports grow between 1990 and 2000. It concludes that 'there is no statistically significant relationship between export growth and changes in private consumption', and that 'export growth is simply not having a strong and sustained virtuous poverty-reduction effect in most of the LDCs'.[3] Last but not least, UNCTAD concludes that 'the relationship between trade and poverty is ... asymmetrical. Although LDCs with declining exports are almost certain to have a rising incidence of poverty, increasing exports do not necessarily lead to poverty reduction' (9).

These observations apply, to various extents, to developing countries as a whole, but are particularly relevant for LDCs, many of which are affected by land degradation and desertification.[4] Indeed, agriculture represents 25 percent of the gross domestic product (GDP) of low-income countries and employs nearly 75 percent of their workforce (United Nations Development Programme [UNDP] 2003, 125). For most LDCs, earnings from agricultural exports are essential to provide foreign currencies required to import the goods necessary for economic growth and poverty reduction, including food, manufactured consumer goods, energy, raw materials, industrial goods, and machinery (UNCTAD 2004, 3). This dependence on agricultural commodities for export earnings makes LDCs highly vulnerable to resource depletion, including land degradation, and to deteriorating terms of trade for agricultural commodities (UNDP 2003, 123).

Indeed, trade liberalisation often crystallises LDCs' specialisation and dependence as exporters of commodities instead of supporting export diversification (UNCTAD 2004, 20). According to the FAO (2004), some commodities exporters in developing countries have managed to diversify their production, but the vast majority of LDCs still depend on only a few commodities for their export earnings. More than 50 developing countries depend on three or fewer commodities for between 20 and 90 percent of their export earnings. For example, cotton represents over 50 percent of Burkina Faso's export earnings, over 35 percent for Chad and approximately 30 percent for Mali (21). The FAO further notes that '43 developing countries depend on a single commodity for more than 20 percent of their total revenues from merchandise exports. Most of these countries are in sub-Saharan Africa or Latin America and the Caribbean and depend on exports of sugar, coffee, cotton lint, or bananas. Most suffer from widespread poverty' (20).

Empirical evidence therefore differs from theoretical predictions, and this is especially true in the case of LDCs, despite the fact that they specialise in the production of agricultural commodities for which they possess theoretical comparative advantages. This is due in part to imperfect market conditions, including non-liberalised agricultural markets, concentration in the food-supply chain, declining terms of trade for commodities, and lack of capacity in developing countries — especially LDCs — to seize new market opportunities. Moreover, specialisation in commodities exports has increased the vulnerability of many developing countries and LDCs to price volatility on world markets. Under these conditions, trade liberalisation often leads to major social and environmental impacts that play a role in increasing poverty and accelerating land degradation worldwide.

According to the FAO (2004), countries that have specialised in commodity exports have not benefited from such a strategy due to what it describes as a commodity crisis that is having devastating impacts on poor farmers in the developing world. This crisis is created by a combination of factors: price fluctuations, market concentration in agricultural processing and distribution leading to lower prices paid to producers, and a decline in world prices due to surplus production as a result of World Bank and International Monetary Fund (IMF) policies that encouraged the expansion of export crops to generate earnings in foreign currency. As a consequence, the terms of trade for commodities-dependant countries have deteriorated markedly in the last decade (Lines 2004). Nonetheless, it must also be noted that the intensification and mechanisation of agriculture under the impulse of the multilateral institutions have led to higher yields and lower prices for food in developing countries.

According to the World Bank, the average real price of agricultural commodities fell by 5.3 percent between 1990 and 2003 (Lines 2004). Another study found that between 1980 and 2002 the prices of twelve tropical commodities fell by 50 points to 86 percent.[5] For example, cotton prices declined by 50 percent between 1980 and 2002 with significant impacts in countries, such as Burkina Faso, that depend on this commodity for export earnings (FAO 2004, 21). The FAO estimates that if the prices of the ten most important agricultural commodities exported by developing countries had kept pace with inflation, those countries would have earned an additional US$112 billion in 2002 (21).

While economic theory predicts that lower prices will lead producers to decrease production or to shift to other crops, LDCs' dependence on a few commodities, and their lack of capacity to diversify exports, often leads them to expand production to compensate for the declining value of commodities. Such expansion can lead to the increased application of agro-chemicals, to the expansion of cultivated land, or to reduced fallows, with a potential impact on land degradation and desertification.

The current difficulties of developing countries and LDCs in raising incomes and fostering economic development through agricultural trade are not only attributable to structural inequities in the trade regime. According to UNCTAD (2004, 13), 'international trade cannot work to reduce poverty in countries where the level and efficiency of investment are not adequate to support sustained economic growth'. Indeed, current difficulties faced by developing countries are caused in large part by structural and institutional socioeconomic weaknesses and often by adverse environmental conditions, including drought and desertification. In rural drylands, these adverse conditions coincide, thus magnifying the difficulties dryland farmers face in integrating the world market.

Competitiveness on world markets is linked not only to labour costs, but also to agricultural productivity in general. This is where rural dryland communities appear to be most disadvantaged, because they do not have the financial resources, infrastructure, technology, or human capacity to increase their productivity while simultaneously facing a deteriorating resources base. This creates a context in which their insertion in the world market is extremely difficult without outside support. This shows that in addition to improving access to foreign markets for developing countries' agricultural products, measures are needed to improve their competitiveness.

Agricultural Trade Liberalisation and Rural Livelihoods

The analysis of the impacts of agricultural trade liberalisation on social structures and small-scale farming in rural areas of the developing world is often nonexistent, incomplete, or anecdotal (WWF 2004). Generally speaking, agricultural trade liberalisation influences existing dynamics in the agricultural sector, including the scale, type, and intensity of farming, land management practices, inputs and technology use, as well as other variables that may have an impact on rural dryland livelihoods. Through its effects on these variables, agricultural trade liberalisation can affect poverty and farm incomes, rural development, migratory flows and urbanisation, food security, income inequities, gender issues, health, traditional knowledge, and culture.

The impact of agricultural trade liberalisation on poverty alleviation in rural drylands is contingent on the capacity of small farmers to seize new opportunities created by the opening of markets and to raise their incomes. According to the UN Millennium Project's Task Force on Hunger, small-scale farming is both the main engine of economic growth and a central factor in reducing poverty and hunger in developing countries (UN Millennium Project 2005a, 69).

Indeed, according to John Mellor, the greatest progress in poverty alleviation has been observed in countries where small-scale farming has driven agricultural growth (Mellor 2000, cited in Department for International Development [DFID] 2002, 5). This seems to be the case in Burkina Faso where cash-crop producers have seen their incomes and food security improve. In the opposite case, where agricultural growth has been generated by large scale agriculture, such as in Brazil, the impacts on poverty reduction have been close to neutral on average (5). This difference can be explained by the fact that large-scale farmers tend to spend their income on capital-intensive, often imported goods and services while small-scale, labour-intensive agriculture generates income that is spent on locally produced goods, services, and labour, thus having a greater impact on poverty and hunger (DFID 2004b, 13).

Small farmers are often disadvantaged on world markets compared to large intensive production units due to diseconomies of scale, lack of access to credit and inputs, low productivity, and insufficient human resources, which prevent access to foreign markets. A survey of 27 case studies on the impacts of trade liberalisation on hunger in developing countries found that trade was in many instances hurting small-scale farmers by raising the price of agricultural inputs and decreasing prices of agricultural commodities at the farm gate (Madely 2000, 8).

As a consequence of trade liberalisation, a gradual shift from small-scale to large-scale, export-oriented farming has been observed in several developing countries. This process often led to the marginalisation of small farmers.[6] This trend toward large-scale agriculture is in large part the result of new market conditions created by trade liberalisation. Indeed, agricultural trade liberalisation creates new conditions in which non-traditional markets gradually replace traditional ones. These new markets are characterised by increased transaction costs, higher risk, and the need for greater capital investment to modernise production techniques (Kydd and Dorward 2001, cited in DFID 2004a, 13–14). Large-scale farming is more adapted to this new environment because larger producers can absorb transaction costs and risks and access credit more easily. This suggests that large-scale farmers are better positioned to benefit from the new market environment and that they may reap the lion's share of new market opportunities generated by trade liberalisation, thereby intensifying existing inequities.

A report published by the International Institute for Environment and Development distinguishes three categories of agricultural producers in the rural developing world: successful competitive entrepreneurs, traditional family farmers, and the marginalised poor struggling for survival (Vorley 2002). According to the report, the last two categories are the main centres of rural poverty in the world, but the first one — successful entrepreneurs — is the main beneficiary of trade liberalisation. Small farmers are often neglected or disadvantaged by the policy environment, including the trade regime.[7] This suggests that unless something is done to facilitate the integration of small farmers in the global market, trade liberalisation will not succeed in alleviating poverty in rural drylands (see, for a detailed analysis, Vorley 2002).

Similarly, a study on Central American agriculture distinguishes four types of farmers (Pomareda and Murillo 2003, 3). Two categories — agro-industrial and transnational firms — are well integrated in the international product chain. They consist of capital-intensive, large-scale, agro-industrial farms with access to foreign capital and technology. Another category consists of medium-sized producers that are integrated in the product chain through contract farming and depend heavily on a few buyers. The last category — small farmers — comprises the majority of farmers and labour force and is not integrated in the global market. Small producers are typically labour-intensive, traditional farmers concentrated on low value lands and producing for their own consumption or the local market.

The picture described by these two studies is that of a deepening gap between small- and large-scale farmers with the latter often appropriating the bulk of state support, high-quality lands, and water resources, and the former being pushed onto marginal lands, faced with a deteriorating resource base and persistent socioeconomic marginalisation. In between, medium-sized producers try to maintain their position on local and national markets. According to a UNCCD secretariat report, the incidence of poverty decreased by 8 percentage points to 42 percent between 1994 and 1998 for cash-crop farmers, but increased by 2 percentage points to 53 percent for food-crop farmers (UNCCD 2005, 27).

This trend may exacerbate inequities and lead small poor farmers to adopt unsustainable survival strategies that deteriorate their resource base, or to abandon their land and migrate to cities or developed countries (see Chapter 4). As mentioned, economic theory predicts that rural labour released by the consolidation of the agricultural sector will be absorbed in secondary and tertiary industries that will develop in urban centres. This reallocation of labour, however, assumes that job creation in the manufacturing and services sector will keep pace with population growth and rural migration, which is often not the case in developing countries. As a result, rural dryland communities often find themselves caught in unsustainable survival patterns fuelling a downward spiral of poverty and resource degradation (see Chapter 3).

Moreover, small farmers may not respond to new market conditions the way economic theory predicts. Studies on traditional corn farmers in Mexico showed that despite adverse new market conditions following the entry into force of the North American Free Trade Agreement (NAFTA), and market share losses to U.S. corn imports, traditional corn farmers in Mexico did not abandon production but instead sought to expand their cultivated area into marginal lands, forests, and biosphere reserves to compensate for lower incomes (Nadal 2000). As a result, a declining average yield was observed among poor small-scale Mexican farmers due to the cultivation of less productive, fragile lands.

Agricultural Trade Liberalisation and the Dryland Environment

The environmental impacts of trade liberalisation relate directly to pre-existing environmental conditions, including land quality, water availability, climatic conditions, ecosystem resilience, etc. Given that drylands are particularly sensitive ecosystems that are often already subject to land degradation, and that agriculture is the single most important land use in rural drylands, they have the potential to be greatly affected by trade liberalisation.

Five different effects of agricultural trade liberalisation can generate environmental impacts (adapted from Kirton et al. 1999; see also Organisation for Economic Co-operation and Development 1994). First, trade liberalisation can affect the scale of production. Following liberalisation, agricultural production can expand or contract in a given country or region following the reshuffling of comparative advantages. Both the expansion and contraction of production can have positive or negative impacts on the environment. Generally speaking, increases in production are achieved through the expansion of land under production or through the intensification of production to generate higher yields. These two strategies each raise a series of potential environmental impacts.

A second driver is the composition of agricultural production. Countries may seek to adapt to new opportunities and price signals by changing the mix of commodities they produce. They may also respond to new opportunities by shifting from food production to export crops such as cotton, rice, or soybeans, for example. Given that each crop is characterised by a different mix of environmental impacts on land, water, and agro-chemical use, changes in the production mix following trade liberalisation can have positive, negative, or neutral impacts. For example, a gradual shift toward cotton production is occurring in Burkina Faso with impacts on land degradation (see Mayrand, Paquin, and Dionne 2005). In Pakistan this shift toward cotton and rice production in the last few decades has had important impacts on scarce water resources since these cash crops are more water intensive than traditional food crops such as wheat (Alam and Naqvi 2003).

The structure of production is a third driver of environmental change often associated with trade liberalisation. As described above, trade liberalisation favours a shift from small- to large-scale agricultural production. Such a pattern is observable in Brazil's soybean sector and is associated with intensive use of agro-chemicals and mechanised cropping techniques (WWF 2003). However, small-scale production is characterised by a low rate of application of farm inputs and low mechanisation and by land expansion strategies. Both intensive and extensive agriculture can lead to land degradation if improper land management strategies are adopted.

The fourth driver of the environmental impacts of trade liberalisation is related to the technologies used in agriculture. As trade is liberalised, farmers may gain better access to agricultural technologies, including farm inputs and machinery and genetically modified crops in areas where such crops are suitable. Moreover, trade leads to higher incomes and to improved access to foreign capital to finance the adoption of new technologies. The environmental impacts of such a process will depend on the technologies introduced. For example, drip-irrigation technologies could reduce soil salinity and improve efficiency in water use in arid regions. However, access to cheaper first-generation agro-chemicals can lead to land and water contamination.

The last driver of environmental impacts associated with trade liberalisation is the regulatory framework. Trade liberalisation is often accompanied by regulatory reforms or deregulation processes designed to decrease state intervention in the economy. This has been the case in several developing countries where the state has disengaged from the farm input sector, thereby making inputs less accessible to small farmers (Mayrand, Paquin, and Dionne 2005). In some cases, the environmental impacts may have been positive when they led to reduced use of pesticides. In other cases, such policies may have led to the under-application of fertilizers and to decreasing yields and land degradation. Policies and regulations adopted in the wake of trade liberalisation can do much to prevent negative environmental impacts. For example, new land management policies can prevent land degradation resulting from a more intensive use of the land to respond to world demand.

Trade liberalisation therefore holds the potential to magnify pressures on land, water, and ecosystem resources in drylands. This may lead to an overuse of land resources and to further land degradation, especially in countries with fragile drylands (Harou 2002). The magnitude of this degradation is difficult to assess given that local conditions as well as state interventions need to be factored in. Anna Strutt (1998) has estimated that trade liberalisation in Indonesia would lead to increased land degradation in certain crops, but that the loss would equivalent to only 0.15 percent of the global welfare gains generated by trade liberalisation. There is growing evidence, however, that trade can generate significant impacts on land degradation and be one of the key drivers in the desertification process.

The conversion of agriculture from small-scale multiple cropping systems to medium- and large-scale intensive monocultures may have significant impacts on land degradation. Anil Agarwal, Sunita Narain, and Anju Sharma (1999, 166) describe this process in the following way: 'Governments and large farmers in West Africa have become critically dependant on cash crops to pay for import and taxes. In the bargain, the intense cultivation of the lands through monoculture has led to increasing desertification.... The increased demand for cash crops discouraged food production, and virtually wiped out practices such as allowing the land to fallow and recover, or crop rotations systems, which kept the land in good condition.'

Similarly, Alejandro Nadal (2000, 3) argues that the restructuring of the corn sector following NAFTA has contributed to 'accelerating soil erosion trends both through specialization and monoculture, coupled with increased use of fertilizers (as is observed in the case of the more competitive producers), and because of a more intensive use of soils, including through the extension of the agricultural frontier to marginal lands, by traditional producers'. This means that small-scale farming can also generate negative environmental impacts in the context of trade liberalisation. Trade is especially likely to exacerbate land degradation under open access land management regimes where land property rights are ill defined, allowing for uncontrolled extension on marginal lands (Demeke 2002).

While the transformation induced by trade liberalisation can produce negative environmental impacts, trade can also bring environmental benefits through improved infrastructure, the spread of new management techniques, or improved access to new and adapted technologies. Trade can also open new market opportunities for certified products, thereby improving agricultural practices. However, such positive outcomes may not materialise if proper policies and regulations are not put in place. Capacity building and various forms of support to small farmers appear especially important in this regard.

Trade and Desertification: What Role for the UNCCD?

The UNCCD can play a role in addressing the two trade-related drivers of land degradation analysed in this chapter: the expansion and intensification of export-oriented agriculture and the marginalisation of small farmers. It can do so by contributing to the improvement of land management and agricultural practices, and by actively supporting small farmers in securing their resource base, building their capacity, improving their agricultural practices, and facilitating their access to farm inputs and financial resources.

First, the UNCCD provides a framework for the development and implementation of policy interventions that can improve land management in the export crop sector to ensure that the expansion of large-scale intensive export monoculture does not affect long-term agricultural growth by leading to widespread land degradation. The UNCCD can serve as a catalyst for the early adoption of appropriate mitigation plans promoting sustainable land management practices. Such an approach would be useful in successful large-scale commodities exporting countries, such as Brazil. It would also be valuable in countries such as Pakistan where intensive use of irrigation and agro-chemicals is leading to serious land degradation issues.

Second, the convention can also provide a basis for intervention aimed at building small farmers' capacity, improving their productivity, and breaking their isolation through infrastructure development. In addition, measures can also be adopted under the UNCCD framework to reduce small farmers' vulnerability to climate variations or price or exchange rate shocks that may force them to resort to destructive survival patterns. Moreover, in the context of increased competition for scarce land and water resources, participatory resources management principles promoted under the UNCCD can secure small farmers' access to resource bases and prevent their displacement by large-scale export-oriented agriculture.

The Millennium Project also supports measures to combat desertification as key way to improve rural livelihoods and promote economic growth. In its report on *Ending Africa's Poverty Trap*, the Millennium Project argues that the structural impediments to raising agricultural productivity in Africa's rural drylands can only be overcome with an intensive investment programme in well-targeted interventions that mostly fall within the UNCCD framework (Sachs et al. 2004). These interventions include improving agricultural varieties, providing better access to fertilizers, introducing new agro-forestry techniques, developing small-scale water management systems and irrigation techniques, and investing in transport infrastructure and education. The Millennium Project also recommends 'a massive replenishment of soil nutrients for smallholder farmers on lands with nutrient-depleted soils, through free or subsidized distribution of chemical fertilizers and agroforestry, by no later than the end of 2006' (UN Millennium Project 2005b, xxi). Moreover, it recommends that measures be taken to improve dryland farmers' access to markets, including investing in storage, marketing, and agro-processing, and improving access to credit (Sachs et al. 2004).

As mentioned in an UNCCD report, however, strategies to improve land productivity and rural incomes often prioritize 'increased application of external inputs, especially chemical fertilizers and improved seed varieties' as well as 'activities to raise rural incomes focused on agricultural commercialization, regionalization of crop production and increased production of cash and export crops' (UNCCD 2005, 12). According to the report, 'these activities may not address the root causes of poor land management and declining yields, and would not necessarily benefit the poorest and most degraded regions or the poorest households (12).

The UNCCD can play a role in expanding the focus of current strategies to improve productivity and alleviate poverty by specifically promoting measures to support small farmers. According to a report published by WWF, the International

Centre for Trade and Sustainable Development (ICTSD), the World Bank, and the government of the Netherlands (2004), key strategies to alleviate poverty in the context of agricultural trade liberalisation include strengthening and raising the productivity of subsistence farming, promoting local markets, protecting the assets of the rural poor, strengthening the management capacity of the poor, and creating investment opportunities with and for the poor.

These strategies fall within UNCCD's area of intervention. Furthermore, examples of governmental policies and programmes that may directly affect land degradation and poverty in the context of liberalising agricultural markets include land tenure frameworks and tenure security; land-use planning, particularly for marginal environments; price controls for agricultural commodities; the development of commodity processing facilities; the development of markets; investment in transportation infrastructure; and investment in agricultural research and extension (UNCCD 2005, 16).

The UNCCD can be instrumental in implementing such strategies and policy measures, provided that its framework is mainstreamed into rural and agricultural development strategies (see chapters 8, 9, and 10). The development of targeted joint programmes involving economic, environmental, and agriculture departments, and designed to specifically address the impacts of trade liberalisation on rural livelihoods and land degradation would facilitate such mainstreaming by focussing on concrete actions rather than on principles or broad national strategies.

Moreover, such a concerted, targeted approach would be easier to finance, and its results more easily measurable than broad, far-reaching national programmes. Indeed, joint action programmes to prevent land degradation and alleviate poverty make economic sense in the context of poverty reduction strategy papers (PRSPs) and national export development strategies since their objective is to preserve the environmental resource base that serves as the basis for the export sector and to maximise the positive impacts of trade on poor segments of rural populations. This aligns activities to combat desertification with major economic and international developmental processes, including the Doha Round and the MDGs.

In this perspective, National Action Programmes (NAPs) could be adapted to include concrete measures that reflect the challenges and opportunities generated by the new environment created by national and international liberalisation of the agricultural sector. Article 10.2(b) of the UNCCD allows for such modification of NAPs to reflect changing circumstances and different socioeconomic conditions.[8] A first step could be to undertake national assessments of the potential impacts of trade liberalisation on desertification so as to identify areas in which intervention is most needed to avoid perverse impacts and maximise positive ones on rural livelihoods and land degradation.

Such assessments could lead to national roundtables where stakeholders — donors, UNCCD focal points, and economic, agriculture, and environment departments, *inter alia* — would define and adopt strategic, targeted interventions. These roundtables would adopt a joint action programme on trade, poverty, and desertification that would be implemented as agricultural trade is gradually liberalised.

Ideally, these action programmes could provide for the continuous monitoring of impacts and for periodic revisions designed to adapt to the unforeseen impacts of trade liberalisation.

At the multilateral level, the Committee on the Review of the Implementation of the Convention (CRIC) could hold a special session on the impacts of agricultural trade liberalisation on desertification. Such a session could help to document national impacts and facilitate the exchange of expertise and best practices with a view to enhancing UNCCD's effectiveness in the context of rapid macroeconomic and land-use changes that are often associated with trade liberalisation. Moreover, by clearly establishing and documenting the relationship between agricultural trade liberalisation, poverty, and land degradation in drylands, the case for enhanced financial support for UNCCD implementation could be strengthened in the context of the MDGs.

Conclusion

This chapter has argued that due to structural flaws in commodities markets and to small farmers' lack of capacities, trade liberalisation may not generate the broad-based, sustainable growth in agricultural production predicted by economic theory. On the contrary, there is a significant risk that agricultural trade liberalisation will lead to accelerated land degradation through the extension and intensification of agricultural production, thus deteriorating the rural resource base of developing countries. Moreover, in the absence of appropriate policies supporting small farmers and landless peasants, there is a risk that these populations will not benefit from trade liberalisation but rather be further marginalised in terms of their access to markets and resources, and that eventually they will be forced to migrate or resort to unsustainable survival strategies.

Such outcomes are not inevitable, however, and the UNCCD can provide a framework under which action can be taken to avert these risks. The stakes of agricultural trade liberalisation are high in the world's rural drylands. In the absence of appropriate policy interventions and investments, it is unlikely that trade will lead to a generalised increase in rural incomes. Rather, it could lead to the exclusion of small farmers and to the deterioration of the agricultural resource base. The development community needs to take on the challenge of drylands agricultural productivity and to support the equitable inclusion of dryland farmers into world trade. Under current conditions, the economic boom promised through trade liberalisation is likely to turn to dust for millions of rural dwellers.

Notes

1   This chapter is based on a study published by Unisféra International Centre with financial support from the Canadian International Development Agency (CIDA). The authors wish

to thank Pierre Marc Johnson, Chantal Line Carpentier, Jocelyne Néron, and Michelle Leighton, as well as CIDA, for their comments on an earlier version. For the full report, see Mayrand (2005).
2   For the purpose of this chapter, trade liberalisation is defined as the removal of barriers to the free circulation of goods and services among countries. In the agricultural sector, the WTO distinguishes three areas for negotiation: market access, domestic support, and export subsidies.
3   Indeed, the report finds that export growth led to higher private consumption in 22 countries but had an impoverishing effect in 18 countries (UNCTAD 2004, 11). In 29 countries export growth was found to have an ambiguous effect on consumption.
4   Out of 50 LDCs, 12 have more than 50 percent of their territory covered with drylands, several of which are facing moderate to severe desertification. Other LDCs such as Haiti or Bangladesh do not have significant portions of their territory composed of drylands but nevertheless face important land degradation problems.
5   These commodities are copra, coconut oil, palm oil, sugar, cocoa, coffee, tea, pepper, groundnuts, jute, cotton, and rubber (Robbins 2003, cited in Lines 2004).
6   For example, a case study on India conducted for the European Commission by Oliver Morrissey, Dirk Willem de Velde, Ian Gillson, and Steve Wiggins (2005, 18) concludes that 'past benefits of liberalisation seem to have gone more in favour of large farmers than the small ones'.
7   See Chapter 2, in which Sally-Anne Way argues that socioeconomic and political structures are often the root cause of poor communities' marginalisation. It could be added that the international trade regime can exacerbate this adverse power structure.
8   NAPs 'allow for modifications to be made in response to changing circumstances and be sufficiently flexible at the local level to cope with different socio-economic, biological and geo-physical conditions'.

References

Agarwal, Anil, Sunita Narain, and Anju Sharma, eds. (1999). *Green Politics: Global Environmental Negotiations I*. Centre for Science and Environment, New Delhi.
Alam, S.M. and M.H. Naqvi (2003). 'Pakistan Agriculture: 2003'. *Pakistan & Gulf Economist* 19–25 May. <www.pakistaneconomist.com/database1/cover/c2003-28.asp> (December 2005).
Convention on Biological Diversity (2003). 'The Impact of Trade Liberalization on Agricultural Biological Diversity: A Synthesis of Assessment Frameworks'. UNEP/CBD/COP/7/INF/15, 18 December. <www.biodiv.org/doc/meetings/cop/cop-07/information/cop-07-inf-15-en.pdf> (2005 December).
Demeke, Bayou (2002). 'Is Globalization Bad for the Environment? International Trade and Land Degradation in Developing Countries: The Case of Small Open Economy'. Paper presented at the European Economic Association convention, 22–24 August. Venice. <www.eea-esem.com/eea-esem/eea2002/prog/viewpaper.asp?pid=1575> (December 2005).
Department for International Development (United Kingdom) (2002). 'Better Livelihoods for Poor People: The Role of Agriculture'. Consultation Document. <www.dfid.gov.uk/pubs/files/agricultureconsult.pdf> (December 2005).
Department for International Development (United Kingdom) (2004a). 'Agriculture, Growth and Poverty Reduction'. Working Paper 1, October. <dfid-agriculture-consultation.nri.org/summaries/wp1.pdf> (December 2005).

Department for International Development (United Kingdom) (2004b). 'Agriculture, Hunger, and Food Security'. Working Paper 7. <dfid-agriculture-consultation.nri.org/summaries/wp7.pdf> (December 2005).

Food and Agriculture Organization (2004). 'The State of Agricultural Commodity Markets'. Rome. <ftp://ftp.fao.org/docrep/fao/007/y5419e/y5419e00.pdf> (December 2005).

Harou, Patrice A. (2002). 'What Is the Role of Markets in Altering the Sensitivity of Arid Land Systems to Perturbation?' In J.F. Reynolds and D.M. Stafford Smith, eds., *Global Desertification: Do Humans Cause Deserts?*, vol. 253–274. Dahlem University Press, Berlin.

Kirton, John J. et al. (1999). *Assessing the Environmental Effects of the North American Free Trade Agreement (NAFTA): Final Analytic Framework and Methodological Issues and Empirical Background*. Commission for Environmental Cooperation, Montreal.

Kydd, Jonathan and Andrew Dorward (2001). 'The Washington Consensus on Poor Country Agriculture: Analysis, Prescription, and Institutional Gaps'. *Development Policy Review* vol. 19, no. 4, pp. 467–478.

Lines, Thomas (2004). 'Commodities Trade, Poverty Alleviation, and Sustainable Development: The Re-emerging Debate'. Paper presented at UNCTAD XI, 15 June. Sao Paolo.

Madely, John (2000). 'Trade and Hunger: An Overview of Case studies on the Impact of Trade Liberalization on Food Security'. Forum Syd, Stockholm.

Mayrand, Karel, Marc Paquin, and Stéphanie Dionne (2005). 'From Boom to Dust? Agricultural Trade Liberalization, Poverty, and Desertification in Rural Drylands: The Role of the UNCCD'. Unisféra International Centre, <www.unisfera.org/IMG/pdf/Unisfera_-_From_Boom_to_Dust_-_Final.pdf> (December 2005).

Mellor, John (2000). 'Agricultural Growth, Rural Employment, and Poverty Reduction: Non-tradables, Public Expenditure, and Balanced Growth'. Proceedings for 'Poverty or Prosperity: Rural People in a Globalized Economy', World Bank Rural Week, 28–31 March.

Morrissey, Oliver, Dirk Willem te Velde, Ian Gillson, et al. (2005). 'Sustainability Impact Assessment of Proposed WTO Negotiations: Mid-Term Report for the Agriculture Study'. Overseas Development Institute and Institute for Development Policy and Management, Manchester University, 31 January. <www.sia-trade.org/wto/Phase3B/Reports/MTRAgricultureJan05.pdf> (December 2005).

Nadal, Alejandro (2000). 'The Environmental and Social Impacts of Economic Liberalization on Corn Production in Mexico'. Oxfam GB and World Wildlife Fund International, <www.intermonoxfam.org/cms/HTML/espanol/520/AGR_CornStudy_OGB_WWF_0301.pdf> (December 2005).

Organisation for Economic Co-operation and Development (1994). 'Methodologies for Environmental and Trade Reviews'. OCDE/GD(94)103. Paris. <www.eldis.org/static/DOC4037.htm> (December 2005).

Organisation for Economic Co-operation and Development (2003). 'Agricultural Trade and Poverty: Making Policy Analysis Count'. Paris.

Oxfam International (2002a). 'Rigged Rules and Double Standards: Trade, Globalisation, and the Fight Against Poverty'. <www.maketradefair.com/en/index.php?file=03042002121618.htm> (December 2005).

Oxfam International (2002b). 'Boxing Match in Agricultural Trade: Will WTO Negotiations Knock Out the World's Poorest Farmers?' Oxfam Briefing Paper No. 32. <www.oxfam.org.uk/what_we_do/issues/trade/bp32_agric_trade.htm> (December 2005).

Pomareda, Carlos and Carlos Murillo (2003). 'The Relationship between Trade and Sustainable Development of Agriculture in Central America'. International Institute for Sustainable Development, Winnipeg. <www.iisd.org/publications/pub.aspx?pno=565> (December 2005).

Robbins, Peter (2003). *Stolen Fruit: The Tropical Commodities Disaster*. Zed Books, London.

Sachs, Jeffrey, John W. McArthur, Guido Schmidt-Traub, et al. (2004). 'Ending Africa's Poverty Trap'. Brookings Papers on Economic Activity 1. <www.earthinstitute.columbia.edu/about/director/documents/bpea0104.pdf> (December 2005).

Strutt, Anna (1998). 'Trade Liberalization and Land Degradation in Indonesia'. ACIAR Indonesia Research Project Working Paper 98.06. <www.adelaide.edu.au/cies/papers/iwp9806.pdf> (December 2005).

United Nations (2002). 'Plan of Implementation of the World Summit on Sustainable Development'. Johannesburg. <www.un.org/esa/sustdev/documents/WSSD_POI_PD/English/WSSD_PlanImpl.pdf> (December 2005).

United Nations Conference on Trade and Development (2004). 'The Least Developed Countries Report, 2004: Linking International Trade with Poverty Reduction'. Geneva.

United Nations Convention to Combat Desertification (2005). 'Mainstreaming of National Action Programmes and Their Contribution to Overall Poverty Eradication'. ICCD/CRIC(3)/Misc.1, 2–11 May. Bonn. <www.unccd.int/cop/officialdocs/cric3/pdf/misc1eng.pdf> (December 2005).

United Nations Development Programme (2003). 'Human Development Report 2003: Millennium Development Goals — A Compact among Nations to End Human Poverty'. New York. <hdr.undp.org/reports/global/2003/> (December 2005).

United Nations Environmental Programme (2002a). 'Integrated Assessment of Trade Liberalization and Trade-Related Policies — Round II: A Country Study on the Export Crop Sector in Nigeria'. Synthesis Report. <www.unep.ch/etu/publications/Synth_Nigeria.PDF> (December 2005).

United Nations Environmental Programme (2002b). 'Integrated Assessment of Trade Liberalization and Trade-Related Policies — Round II: A Country Study on the Ecuador Banana Sector'. Synthesis Report. <www.unep.ch/etu/publications/Synth_Ecuador.PDF> (December 2005).

United Nations Millennium Project (2005a). 'Halving Hunger: It Can Be Done'. New York. <www.unmillenniumproject.org/reports/tf_hunger.htm> (December 2005).

United Nations Millennium Project (2005b). 'Investing in Development: A Practical Plan to Achieve the Millennium Development Goals'. <www.unmillenniumproject.org/reports/index.htm> (December 2005).

Vitalis, Vangelis (2004). 'Trade, Agriculture, the Environment, and Development: Reaping the Benefits of a Win-Win-Win'. Paper presented at a Strategic Dialogue on Agriculture, Trade Negotiations, Poverty, and Sustainability, 14–16 July. Windsor UK. <www.ictsd.org/dlogue/2004-07-14/Vangelis.pdf> (December 2005).

Vorley, Bill (2002). 'Sustaining Agriculture: Policy, Governance, and the Future of Family-Based Farming'. International Institute for Environment and Development, London.

Werth, Alexander (2002). 'Agri-Environment and Rural Development in the Doha Round'. International Institute for Sustainable Development, Winnipeg. <www.tradeknowledgenetwork.net/pdf/tkn_ruraldev_doha.pdf> (December 2005).

World Trade Organization (2001). 'Ministerial Declaration'. WT/MIN(01)/DEC/1, 14 November. Doha. <www.wto.org/english/thewto_e/minist_e/min01_e/mindecl_e.htm> (December 2005).

World Wildlife Fund (2003). 'Sustainability Assessment of Export-Led Growth in Soy Production: Full Report'. <www.panda.org/downloads/policy/soylongeng.pdf> (December 2005).

World Wildlife Fund, World Bank, International Centre for Trade and Sustainable Development, et al. (2004). 'From Negotiations to Global Adjustment: Preparatory Phase Summary Report'. 14 June. Paris. <www.panda.org/downloads/policy/ntoasummary.pdf> (December 2005).

Chapter 13

Conclusion:
The UNCCD at a Crossroad

Pierre Marc Johnson, Karel Mayrand, and Marc Paquin

The year 2006 marks the tenth anniversary of the entry into force of the United Nations Convention to Combat Desertification (UNCCD) and has been officially designated the international year of deserts and desertification. This anniversary provides an opportunity to reflect on the first decade of UNCCD implementation and assess what has been achieved thus far and what remains to be done to deploy this convention as a fully effective instrument of global environmental governance.

This book has described at length the successes, weaknesses, sensitivities, areas of tension, and operational difficulties confronted by this convention since its entry into force. What comes out of this picture and from the opinions expressed by most UNCCD observers is that the convention has entered a delicate transition period between the early institutionalisation and development stages and the operational stage where energies are directed toward its concrete implementation in the affected communities. As stated by the UNCCD's Executive Secretary, Arba Diallo, at the 2002 World Summit on Sustainable Development (WSSD), 'this Convention is now ready to move from its initial conceptual phase to a much awaited implementation phase, based on its National Action Programs (or NAPs), which constitute the very building blocks of its implementation' (UNCCD 2002). While this may sound like evidence, moving toward the operational phase remains an unresolved challenge, as this concluding chapter attempts to show.

The first ten years of the UNCCD's existence have been devoted to setting up convention institutions and instruments, including the secretariat and the Global Mechanism (GM), the elaboration of NAPs as well as Regional Action Programmes (RAPs) and Sub-Regional Action Programmes (SRAPs), the creation of thematic programme networks (TPN),[1] the mobilisation of partners, and the launching of new instruments under the convention system (such as the Committee for the Review of the Implementation of the Convention [CRIC] and the Committee on Science and Technology [CST]). Taken together, these developments have created some of the underlying conditions required to undertake the effective, concrete implementation of the convention.

The UNCCD now has close to universal membership with 191 ratifications (including that of the United States), a number equivalent to that of its two fellow conventions, the Convention on Biological Diversity (CBD) and the UN Framework Convention on Climate Change (UNFCCC), also born at the 1992 United Nations Conference on Environment and Development at Rio de Janeiro in 1992 (UNCCD 2005b).[2] This brings considerable legitimacy to this multilateral instrument and contributes to asserting the global significance of desertification. While universal membership is not an indicator of the strength of the parties' commitment to UNCCD implementation, it does constitute an asset for this convention. Moreover, the convention's adoption has greatly contributed to raising international attention on desertification-related issues in rural drylands. As explained in Chapter 5, the very existence of this convention is success in itself.

Another of the convention's achievements is the elaboration of 77 NAPs by affected country parties.[3] NAPs are the result of extensive consultative processes led by UNCCD national focal points. As mentioned in several of this book's chapters, the NAPs' bottom-up, participatory approach is one of the distinctive features of this convention. Although they vary in quality and are often criticised for their shopping list approach and lack of strategic direction (as described in chapters 9 and 10), in many countries NAPs are the product of effective participative processes and enjoy considerable support from national nongovernmental organisations (NGOs) (as discussed in chapters 8 and 10).

The extent of NGO participation in multilateral discussions is another distinctive feature of this convention, highlighted in Chapter 7. NGOs have played an instrumental role in the elaboration and early implementation of this convention, through the Réseau international des ONG pour la desertification (RIOD) and other networks. This also contributes to this instrument's legitimacy. Chapter 8 also shows how the UNCCD process incorporates good governance principles and is conducive to improved — although imperfect — participation and governance in rural drylands. Improved governance and community empowerment are key to reverse the marginalisation of dryland communities described in Chapter 3, which is often at the roots of poverty and desertification.

Another major achievement is the designation of the Global Environment Facility (GEF) as a financial mechanism of the UNCCD at the sixth Conference of the Parties (COP) in Havana (UNCCD 2003a, 17–18). This designation followed the opening of a new GEF focal area on land degradation, desertification, and deforestation at the second GEF assembly in Beijing in October 2002 as a means to support the implementation of the UNCCD (GEF 2003). This brings higher levels of financing to the convention, although insufficient to meet demand from affected developing countries (see Chapter 9). It also has symbolic significance given that the UNCCD was the only one of the Rio conventions that did not benefit from initial GEF support. This situation was interpreted by many as lack of willingness on the part of the donor community to support the objectives of this instrument of global governance. Although the level of funding allocated to the UNCCD under the third GEF replenishment is significantly lower than that of its fellow conventions of the CBD and the UNFCCC, this evolution indicates a growing interest among donors to address UNCCD-related issues.

From Words to Action: Removing Stumbling Blocks

If, indeed, the UNCCD is moving from a conceptual to an operational phase as stated repeatedly at COP-6 in Havana and at the third meeting of the CRIC in Bonn, several hurdles remain to be removed to translate words into action. The institutional and political dynamics of the UNCCD regime have affected its implementation since its entry into force in 1996. In many ways, the UNCCD is the product of a fragile compromise and several of the tensions that existed before the convention was signed remain perceptible today and continue to plague the intergovernmental process and the concrete implementation of the UNCCD (as described in Chapter 5).

These tensions originate both in perceptions and facts. First, the convention is perceived by many as having been an afterthought of the member countries of the Organisation for Economic Co-operation and Development (OECD) in the Rio process, and is seen largely as an African convention. This perception was reinforced by the fact that OECD countries have been adamant that they would not increase resources substantively toward developing countries under the convention. This dynamic has produced North-South as well as South-South tensions that still persist in many cases, as Najam describes.

Another set of tension arises from confusion on the focus and scope of the UNCCD. It is considered by many as a developmental convention rather than an environmental one, which leads to some confusion over the key UNCCD constituencies. It is also seen as a convention dealing with agriculture and rural development without involving agricultural departments. This contributes to making the convention less appealing to the rural development community. Also, the intersectoral and participatory nature of the UNCCD sometimes conflicts with the objective to define cost-efficient strategies and approaches to address a focussed set of issues. This has affected the concrete implementation of the convention at the country level.

These underlying and unresolved tensions have generated confrontational discussions in the context of COPs that have contributed to direct attention toward process-oriented issues rather than substantive ones. The inherent political dynamics of the UNCCD opposes, on the one hand, the G77, which calls for increased financial support and stronger commitments from OECD countries, and, on the other hand, developed country parties, which have been reluctant to step up their efforts to support this convention. While the politicisation of discussions is not unique to the UNCCD, it has often transformed its process into a political forum for North-South debates instead of an instrument of co-operative global governance.

In many instances, this dynamic has antagonised the parties to the convention and been counterproductive for its credibility and effectiveness. It has certainly contributed to discouraging OECD countries from actively supporting this convention.[4] Their unenthusiastic support for NAPs implementation has in turn contributed to criticism from G77 countries — supported by the UNCCD secretariat — which expect stronger support from their developed country counterparts and continuously remind them of their commitments under the convention (see the discussion of the secretariat's involvement in Chapter 6). This has created a self-reinforcing pattern of confrontation that has demobilised the donor countries and affected the implementation of the convention. Parties need to find a common ground on which to reinstate a climate of confidence and co-operation if the UNCCD is to move effectively in to the operational phase.

The hesitancies over adequately financing the implementation of the UNCCD are a source of tension that cannot be ignored. Indeed, the convention is not supported by a strong commitment by OECD countries to provide new and additional funding for its implementation. Consequently, the UNCCD's financial provisions were conceived as an instrument to mobilise, catalyse, and channel existing funds into activities to combat desertification under its umbrella. As shown in Chapter 9, nothing shows it has succeeded in catalysing such investments or improving co-ordination in the field.

As a matter of fact, international agencies such as the World Bank, regional banks, the United Nations Development Programme (UNDP), the United Nations Environment Programme (UNEP), the International Fund for Agricultural Development (IFAD), and the Food and Agriculture Organization (FAO), to mention just a few, as well as bilateral development agencies, had been investing in agriculture and land degradation activities in drylands decades before the UNCCD entered into force. Redirecting these funds to channel them through the UNCCD framework has consequently proven a considerable challenge for the convention, which must constantly demonstrate its usefulness as a co-ordinating mechanism.

In theory, the UNCCD's added financial value is in improving the efficiency of desertification-related official development assistance (ODA) by increasing co-ordination among donors for integrated and participatory programmes to combat desertification. In practice, this co-ordination is still lacking and agencies continue to work separately and sometimes compete with each other in the field. The UNCCD provides that donors will act as *chef de files*, or leaders of the donor community, in specific affected countries. Only a limited number of parties have taken this initiative and the *chef de file* approach remains underused.

A decade after the UNCCD's entry into force, the fact that only a small portion of investments made by bilateral and multilateral donors in agriculture, land degradation, drylands rural development, and other desertification-related activities are channelled through the UNCCD framework constitutes a persistent weakness for which all UNCCD Parties and partners bear responsibility.[5] Instead of serving as an instrument to co-ordinate and maximise the impacts of investments in drylands, the UNCCD is at risk of becoming simply an additional legal and institutional layer in sustainable development governance. Such combination of high politicisation and low operational impact would in time seriously damage the ikts credibility.

The scientific and information base of the UNCCD is also weak when compared to other instruments of environmental governance, including the UNFCCC and the CBD. In the past, successful environmental governance instruments have been established on solid scientific grounds that facilitate the measurement of progress through indicators or other quantitative or qualitative measures. This was the case with the Vienna Convention for the Protection of the Ozone Layer and the UNFCCC in particular. Such a science-based approach facilitates consensus building and may attenuate some political or ideological conflicts by focussing discussions on science-based issues. As argued in Chapter 2, uncertainties about the scientific basis of the desertification process lead to conflicting definitions of the concept, which open the door for the politicisation of discussions.

The lack of information and data on desertification was recognised in the desertification report of the Millennium Ecosystem Assessment, which states that 'without a scientifically robust and consistent baseline of desertification, identifying priorities and monitoring the consequences of actions are seriously constrained' (Millennium Ecosystems Assessment 2005, 27). This makes the marketing of the UNCCD more difficult within the donor community, which is increasingly interested in investing in projects that deliver visible, quantifiable results. Expanding the UNCCD information and knowledge base and improving its dissemination would certainly increase the visibility of the convention and raise its financial appeal by facilitating links between the funds invested and the results that are obtained.

As mentioned in Chapter 6, UN agencies and bilateral funders have been reluctant to adopt the concept of desertification in part due to the difficulty in elaborating an operational definition that would facilitate its division into easily measurable and quantifiable indicators. The inclusion of socioeconomic aspects in the definition of desertification has also contributed to the politicisation of discussions under the UNCCD. In summary, the concept of desertification itself is partly responsible for the weak information base and politicisation of the Convention, which in turn makes it more difficult to engage multilateral and bilateral agencies in the fight against poverty and land degradation in rural drylands. In fact, the definition of desertification adopted by the UNCCD is the child of sustainable development and shares with it the advantages of inclusiveness and the difficulties of finding effective operational definitions in a politicised context.

Finding the Missing Link: Mainstreaming the UNCCD in Development Governance

The UNCCD was conceived as an innovative governance instrument with a dual focus on environment and development, making it one of the few instruments of global governance with a truly sustainable development approach. In that perspective, the convention text contributes to bridging the very distinctive worlds of environmental and developmental governance. This is a significant contribution in a context where the international community is making concerted attempts at integrating the different frameworks of world governance.

In practice, however, to date the UNCCD remains almost exclusively confined to the world of environmental governance. This situation is due to the fact, well illustrated in Chapter 10, that most UNCCD focal points at the country level are located within environmental departments. The convention constituency is therefore composed in great majority of under-resourced, politically marginalised environmental ministries with very limited influence on development planning, especially in the context of institutional reforms and decentralisation, which have been central features of governance in many African countries in the last decade.

In contrast to their developing country partners, most OECD countries have established UNCCD focal points within development agencies. Most of these development agencies interact with economic departments in developing countries, few of which ever raise desertification as one of their priorities for development. This creates a triangular relationship with development agencies facing repeated demands from UNCCD focal points for increased funding in activities to combat desertification while the economic departments in developing countries rarely mention desertification in their discussions with development agencies. This disconnect between the priorities of UNCCD focal points and those officially expressed by their countries in their discussions with development agencies has proven a major stumbling block in operationalising the UNCCD's approach to sustainable development (see Chapter 9).

This problem is not faced solely by developing countries. OECD countries' development agencies also confront this divide internally when trying to incorporate UNCCD objectives into their sectoral policies on agriculture, rural development, land degradation, and food security, or when trying to operationalise them through their regional or national branches. In summary, the UNCCD remains a marginal instrument in development governance due to a mix of institutional resistance, political disconnect, and cultural differences between the worlds of environmental and developmental governance.

This issue has brought to the forefront the challenge of 'mainstreaming' the UNCCD into development policy at the country level in order to resolve this disconnect and to make full use of the convention in development planning (see chapters 9 and 10). Concretely, mainstreaming the UNCCD involves the partial or full integration of NAPs into strategic development planning, including poverty reduction strategy papers (PRSPs). This is a considerable challenge that has been described in the following words by Diallo: 'National action programmes, the core objective of the Convention, are now being linked to poverty reduction and investment strategies. But this mainstreaming trend is still more formal than operational' (UNCCD 2005a).

The difficulty in mainstreaming the UNCCD into development planning comes in part from the weaknesses of the NAPs that have been discussed elsewhere in this book (see Chapter 10). Indeed, few NAPs provide an adequate strategic framework for financing, even though they reflect the outcomes of legitimate participatory processes. Consequently, development planning institutions have been reluctant to adhere to the NAPs framework or to finance their implementation. In some cases, the GM has accompanied countries in adapting their NAPs to the needs of development agencies, but this has proven burdensome and there are doubts that this could be done at a global scale (see Chapter 9).

Moreover, affected developing countries that have invested resources and energies in elaborating NAPs now expect to see them financed and implemented. Several of them see the mainstreaming of NAPs as an additional process-oriented requirement that contributes to delaying action to fight desertification. Further delays in implementing NAPs could have a demobilising effect on affected developing countries. Nevertheless, the need remains to mainstream the UNCCD into development planning and this issue remains unresolved.

Reinvigorating the Convention of Endangered Populations

The UNCCD has been recognised by the WSSD as a key tool for poverty alleviation and has received explicit endorsement from UN secretary general Kofi Annan as one of the selected instruments that support the achievements of the UN's mission. Moreover, the clear correlation among rural poverty, hunger, and land degradation highlights the importance of this convention as one of the few instruments of sustainable development governance that address poverty and environmental degradation in the rural drylands. Despite this political visibility and support, the UNCCD has still not seen its level of political commitment or financing significantly rise in recent years. As a consequence, it remains underused in relation to its potential.

Some have argued that in order to reinvigorate the UNCCD and avoid seeing it becoming dormant, the focus should shift from the specific application of its provisions to the realisation of its objectives. This would require revisiting the convention to go back to its objective, giving it more flexibility and dismantling some of its current institutional and political rigidities (Bassett and Talafré 2003).

Efforts should also be made to demonstrate the benefits of combating desertification systematically in terms of poverty alleviation, improved food security, and environmental protection. A renewed case would have to be made to show that a dollar spent through the UNCCD framework can produce more results than one spent outside it. Such a demonstration would have a powerful effect in mainstreaming the convention into development planning. For example, the relationships between socioeconomic marginalisation, environmental degradation, and migration that are illustrated in chapters 3 and 4 show that the UNCCD approach of addressing both the environmental and socioeconomic causes of desertification can lead to better outcomes than traditional technical responses to land degradation and desertification issues. This argument has to be translated into success stories in the field that will serve to demonstrate how this instrument of international governance can make a difference.

The 2008–09 work programme of the United Nations Commission on Sustainable Development will focus on desertification, rural development, land degradation, and drought. This will provide an opportunity to assess progress made in fighting desertification and to analyse the existing institutional arrangements and legal frameworks — including the UNCCD — on which international action to combat desertification converge. The actors of environmental and developmental governance ought to seize this opportunity to define a strategic plan to co-ordinate interventions and improve the effectiveness of their actions in reversing the desertification trends worldwide. The UNCCD should clearly be at the centre of this exercise.

This book has attempted not only to analyse the strengths and weaknesses of this convention as an instrument of global sustainable development governance, but also to highlight its importance and promise for millions of rural poor in affected regions. Although desertification does not benefit from the same level of attention and visibility as biodiversity loss or climate change, it is nevertheless a global phenomenon that affects countries on all continents. As such it requires concerted, multilateral action to prevent, mitigate, and reverse its impacts, which lead to thousands of deaths every year and condemn millions to the most absolute poverty.

The UNCCD has been conceived to co-ordinate multilateral action to fight desertification and to mobilise and channel funds: it has also been adopted to give a voice to poor rural dryland communities that are persistently marginalised, deprived of their resources, and too often forced into self-destructive survival patterns in all regions of the world. This has led one observer to qualify it as the 'convention of endangered populations' (Néron 2002). The international community clearly has a responsibility to these populations and the UNCCD provides the framework to assume this responsibility collectively.

Notes

1 TPNs are networks of institutions and agencies that conduct substantive work on themes related to desertification at a regional basis. For example, Africa has six TPNs on the following themes: integrated management of international river, lake, and hydro-geological basins; promotion of agro-forestry and soil conservation; rational use of rangelands and promotion of the development of fodder crops; ecological monitoring, natural resources mapping, remote sensing, and early warning systems; promotion of new and renewable energy sources and technologies; and promotion of sustainable agricultural farming systems.
2 The U.S. is not party to the CBD but has ratified the UNFCCC.
3 As of December 2005, 34 African countries, 23 Asian countries, 18 Latin American or Caribbean countries, 3 Northern Mediterranean countries, and 4 Central and Eastern Europe countries have adopted NAPs. In addition, five SRAPs and one Regional Action Programme have been formulated. The programmes are available from the UNCCD website at <www.unccd.int> (December 2005).
4 This divide reached unprecedented proportions at COP-6, when only one minister of an OECD country — Italy, representing the European Union — attended the meetings, while eleven heads of state or government from affected developing countries participated in the controversial high-level segment (see UNCCD 2003b, 20–21, see also Chapter 6).
5 As shown in Chapter 8, most financing under the umbrella of the convention has been directed toward process-oriented activities, while substantive investments in rural drylands continue to be channelled mostly outside of the UNCCD system. As a result, NAP-related projects remain largely unfunded.

References

Bassett, Charles and Joana Talafré (2003). 'Implementing the UNCCD: Towards a Recipe for Success'. *Review of European Community and International Environmental Law* vol. 12, no. 2, pp. 133–139.

Global Environmental Facility (2003). 'Operational Program on Sustainable Land Management'. Operational Program No. 15. <www.gefweb.org/Operational_Policies/ Operational_Programs/OP_15_English.pdf> (December 2005).

Millennium Ecosystems Assessment (2005). 'Ecosystems and Human Well-Being: Desertification Synthesis'. World Resources Institute, Washington DC. <www.maweb.org// proxy/document.355.aspx> (December 2005).

Néron, Jocelyne (2002). 'La convention des «populations» menacées: Le Sommet saura-t-il redonner l'élan nécessaire au règlement du problème de la désertification?' *Le Devoir*, 24–25 August.

United Nations Convention to Combat Desertification (2002). 'Statement by Mr. Arba Diallo, Executive Secretary of the UNCCD'. World Summit on Sustainable Development, 30 August. Johannesburg. <www.un.org/events/wssd/statements/unccdE.htm> (December 2005).

United Nations Convention to Combat Desertification (2003a). 'Report of the Conference of the Parties on Its Sixth Session, Held in Havana from 25 August to 5 September 2003: Part Two — Action Taken by the Conference of the Parties at Its Sixth Session'. ICCD/ COP(6)/11/Add.1, 7 November. <www.unccd.int/cop/officialdocs/cop6/pdf/11add1eng.pdf> (December 2005).

United Nations Convention to Combat Desertification (2003b). 'Report of the Conference of the Parties on Its Sixth Session, Held in Havana from 25 August to 5 September 2003: Part One — Proceedings'. ICCD/COP(6)/11, 3 November 2003. <www.unccd.int/cop/ officialdocs/cop6/pdf/11eng.pdf> (December 2005).

United Nations Convention to Combat Desertification (2005a). 'UN Maps New Ways to Mainstream Desertification'. 12 May. Bonn. <www.unccd.int/publicinfo/pressrel/ showpressrel.php?pr=press12_05_05> (December 2005).

United Nations Convention to Combat Desertification (2005b). 'Status of Ratification and Entry into Force'. <www.unccd.int/convention/ratif/doeif.php> (December 2005).

Appendix

Appendix

United Nations Convention to Combat Desertification in those Countries Experiencing Serious Drought and/or Desertification, Particularly in Africa

The Parties to this Convention,

Affirming that human beings in affected or threatened areas are at the centre of concerns to combat desertification and mitigate the effects of drought,

Reflecting the urgent concern of the international community, including States and international organizations, about the adverse impacts of desertification and drought,

Aware that arid, semi-arid and dry sub-humid areas together account for a significant proportion of the Earth's land area and are the habitat and source of livelihood for a large segment of its population,

Acknowledging that desertification and drought are problems of global dimension in that they affect all regions of the world and that joint action of the international community is needed to combat desertification and/or mitigate the effects of drought,

Noting the high concentration of developing countries, notably the least developed countries, among those experiencing serious drought and/or desertification, and the particularly tragic consequences of these phenomena in Africa,

Noting also that desertification is caused by complex interactions among physical, biological, political, social, cultural and economic factors,

Considering the impact of trade and relevant aspects of international economic relations on the ability of affected countries to combat desertification adequately,

Conscious that sustainable economic growth, social development and poverty eradication are priorities of affected developing countries, particularly in Africa, and are essential to meeting sustainability objectives,

Mindful that desertification and drought affect sustainable development through their interrelationships with important social problems such as poverty, poor health and nutrition, lack of food security, and those arising from migration, displacement of persons and demographic dynamics,

Appreciating the significance of the past efforts and experience of States and international organizations in combating desertification and mitigating the effects of drought, particularly in implementing the Plan of Action to Combat Desertification which was adopted at the United Nations Conference on Desertification in 1977,

Realizing that, despite efforts in the past, progress in combating desertification and mitigating the effects of drought has not met expectations and that a new and more effective approach is needed at all levels within the framework of sustainable development,

Recognizing the validity and relevance of decisions adopted at the United Nations Conference on Environment and Development, particularly of Agenda 21 and its chapter 12, which provide a basis for combating desertification,

Reaffirming in this light the commitments of developed countries as contained in paragraph 13 of chapter 33 of Agenda 21,

Recalling General Assembly resolution 47/188, particularly the priority in it prescribed for Africa, and all other relevant United Nations resolutions, decisions and programmes on desertification and drought, as well as relevant declarations by African countries and those from other regions,

Reaffirming the Rio Declaration on Environment and Development which states, in its Principle 2, that States have, in accordance with the Charter of the United Nations and the principles of international law, the sovereign right to exploit their own resources pursuant to their own environmental and developmental policies, and the responsibility to ensure that activities within their jurisdiction or control do not cause damage to the environment of other States or of areas beyond the limits of national jurisdiction,

Recognizing that national Governments play a critical role in combating desertification and mitigating the effects of drought and that progress in that respect depends on local implementation of action programmes in affected areas,

Recognizing also the importance and necessity of international cooperation and partnership in combating desertification and mitigating the effects of drought,

Recognizing further the importance of the provision to affected developing countries, particularly in Africa, of effective means, *inter alia* substantial financial resources, including new and additional funding, and access to technology, without which it will be difficult for them to implement fully their commitments under this Convention,

Expressing concern over the impact of desertification and drought on affected countries in Central Asia and the Transcaucasus,

Stressing the important role played by women in regions affected by desertification and/or drought, particularly in rural areas of developing countries, and the importance of ensuring the full participation of both men and women at all levels in programmes to combat desertification and mitigate the effects of drought,

Emphasizing the special role of non-governmental organizations and other major groups in programmes to combat desertification and mitigate the effects of drought,

Bearing in mind the relationship between desertification and other environmental problems of global dimension facing the international and national communities,

Appendix

Bearing also in mind the contribution that combating desertification can make to achieving the objectives of the United Nations Framework Convention on Climate Change, the Convention on Biological Diversity and other related environmental conventions,

Believing that strategies to combat desertification and mitigate the effects of drought will be most effective if they are based on sound systematic observation and rigorous scientific knowledge and if they are continuously re-evaluated,

Recognizing the urgent need to improve the effectiveness and coordination of international cooperation to facilitate the implementation of national plans and priorities,

Determined to take appropriate action in combating desertification and mitigating the effects of drought for the benefit of present and future generations,

Have agreed as follows:

Part I: Introduction

Article 1. Use of Terms

For the purposes of this Convention:

(a) "desertification" means land degradation in arid, semi-arid and dry sub-humid areas resulting from various factors, including climatic variations and human activities;

(b) "combating desertification" includes activities which are part of the integrated development of land in arid, semi-arid and dry sub-humid areas for sustainable development which are aimed at:

    (i) prevention and/or reduction of land degradation;

    (ii) rehabilitation of partly degraded land; and

    (iii) reclamation of desertified land;

(c) "drought" means the naturally occurring phenomenon that exists when precipitation has been significantly below normal recorded levels, causing serious hydrological imbalances that adversely affect land resource production systems;

(d) "mitigating the effects of drought" means activities related to the prediction of drought and intended to reduce the vulnerability of society and natural systems to drought as it relates to combating desertification;

(e) "land" means the terrestrial bio-productive system that comprises soil, vegetation, other biota, and the ecological and hydrological processes that operate within the system;

(f) "land degradation" means reduction or loss, in arid, semi-arid and dry sub-humid areas, of the biological or economic productivity and complexity of rainfed cropland, irrigated cropland, or range, pasture, forest and woodlands resulting from land uses or from a process or combination of processes, including processes arising from human activities and habitation patterns, such as:

    (i) soil erosion caused by wind and/or water;

(ii) deterioration of the physical, chemical and biological or economic properties of soil; and

(iii) long-term loss of natural vegetation;

(g) "arid, semi-arid and dry sub-humid areas" means areas, other than polar and sub-polar regions, in which the ratio of annual precipitation to potential evapotranspiration falls within the range from 0.05 to 0.65;

(h) "affected areas" means arid, semi-arid and/or dry sub-humid areas affected or threatened by desertification;

(i) "affected countries" means countries whose lands include, in whole or in part, affected areas;

(j) "regional economic integration organization" means an organization constituted by sovereign States of a given region which has competence in respect of matters governed by this Convention and has been duly authorized, in accordance with its internal procedures, to sign, ratify, accept, approve or accede to this Convention;

(k) "developed country Parties" means developed country Parties and regional economic integration organizations constituted by developed countries.

Article 2. Objective

1. The objective of this Convention is to combat desertification and mitigate the effects of drought in countries experiencing serious drought and/or desertification, particularly in Africa, through effective action at all levels, supported by international cooperation and partnership arrangements, in the framework of an integrated approach which is consistent with Agenda 21, with a view to contributing to the achievement of sustainable development in affected areas.

2. Achieving this objective will involve long-term integrated strategies that focus simultaneously, in affected areas, on improved productivity of land, and the rehabilitation, conservation and sustainable management of land and water resources, leading to improved living conditions, in particular at the community level.

Article 3. Principles

In order to achieve the objective of this Convention and to implement its provisions, the Parties shall be guided, *inter alia*, by the following:

(a) the Parties should ensure that decisions on the design and implementation of programmes to combat desertification and/or mitigate the effects of drought are taken with the participation of populations and local communities and that an enabling environment is created at higher levels to facilitate action at national and local levels;

(b) the Parties should, in a spirit of international solidarity and partnership, improve cooperation and coordination at subregional, regional and international levels, and better focus financial, human, organizational and technical resources where they are needed;

(c) the Parties should develop, in a spirit of partnership, cooperation among all levels of government, communities, non-governmental organizations and landholders to establish a better understanding of the nature and value of land and scarce water resources in affected areas and to work towards their sustainable use; and

(d) the Parties should take into full consideration the special needs and circumstances of affected developing country Parties, particularly the least developed among them.

Part II: General Provisions

Article 4. General Obligations

1. The Parties shall implement their obligations under this Convention, individually or jointly, either through existing or prospective bilateral and multilateral arrangements or a combination thereof, as appropriate, emphasizing the need to coordinate efforts and develop a coherent long-term strategy at all levels.

2. In pursuing the objective of this Convention, the Parties shall:

(a) adopt an integrated approach addressing the physical, biological and socio-economic aspects of the processes of desertification and drought;

(b) give due attention, within the relevant international and regional bodies, to the situation of affected developing country Parties with regard to international trade, marketing arrangements and debt with a view to establishing an enabling international economic environment conducive to the promotion of sustainable development;

(c) integrate strategies for poverty eradication into efforts to combat desertification and mitigate the effects of drought;

(d) promote cooperation among affected country Parties in the fields of environmental protection and the conservation of land and water resources, as they relate to desertification and drought;

(e) strengthen subregional, regional and international cooperation;

(f) cooperate within relevant intergovernmental organizations;

(g) determine institutional mechanisms, if appropriate, keeping in mind the need to avoid duplication; and

(h) promote the use of existing bilateral and multilateral financial mechanisms and arrangements that mobilize and channel substantial financial resources to affected developing country Parties in combating desertification and mitigating the effects of drought.

3. Affected developing country Parties are eligible for assistance in the implementation of the Convention.

Article 5. Obligations of Affected Country Parties

In addition to their obligations pursuant to article 4, affected country Parties undertake to:

(a) give due priority to combating desertification and mitigating the effects of drought, and allocate adequate resources in accordance with their circumstances and capabilities;

(b) establish strategies and priorities, within the framework of sustainable development plans and/or policies, to combat desertification and mitigate the effects of drought;

(c) address the underlying causes of desertification and pay special attention to the socio-economic factors contributing to desertification processes;

(d) promote awareness and facilitate the participation of local populations, particularly women and youth, with the support of non-governmental organizations, in efforts to combat desertification and mitigate the effects of drought; and

(e) provide an enabling environment by strengthening, as appropriate, relevant existing legislation and, where they do not exist, enacting new laws and establishing long-term policies and action programmes.

Article 6. Obligations of Developed Country Parties

In addition to their general obligations pursuant to article 4, developed country Parties undertake to:

(a) actively support, as agreed, individually or jointly, the efforts of affected developing country Parties, particularly those in Africa, and the least developed countries, to combat desertification and mitigate the effects of drought;

(b) provide substantial financial resources and other forms of support to assist affected developing country Parties, particularly those in Africa, effectively to develop and implement their own long-term plans and strategies to combat desertification and mitigate the effects of drought;

(c) promote the mobilization of new and additional funding pursuant to article 20, paragraph 2 (b);

(d) encourage the mobilization of funding from the private sector and other non-governmental sources; and

(e) promote and facilitate access by affected country Parties, particularly affected developing country Parties, to appropriate technology, knowledge and know-how.

Article 7. Priority for Africa

In implementing this Convention, the Parties shall give priority to affected African country Parties, in the light of the particular situation prevailing in that region, while not neglecting affected developing country Parties in other regions.

Article 8. Relationship with Other Conventions

1. The Parties shall encourage the coordination of activities carried out under this Convention and, if they are Parties to them, under other relevant international

agreements, particularly the United Nations Framework Convention on Climate Change and the Convention on Biological Diversity, in order to derive maximum benefit from activities under each agreement while avoiding duplication of effort. The Parties shall encourage the conduct of joint programmes, particularly in the fields of research, training, systematic observation and information collection and exchange, to the extent that such activities may contribute to achieving the objectives of the agreements concerned.

2. The provisions of this Convention shall not affect the rights and obligations of any Party deriving from a bilateral, regional or international agreement into which it has entered prior to the entry into force of this Convention for it.

Part III: Action Programmes, Scientific and Technical Cooperation, and Supporting Measures

Section 1: Action Programmes

Article 9. Basic Approach

1. In carrying out their obligations pursuant to article 5, affected developing country Parties and any other affected country Party in the framework of its regional implementation annex or, otherwise, that has notified the Permanent Secretariat in writing of its intention to prepare a national action programme, shall, as appropriate, prepare, make public and implement national action programmes, utilizing and building, to the extent possible, on existing relevant successful plans and programmes, and subregional and regional action programmes, as the central element of the strategy to combat desertification and mitigate the effects of drought. Such programmes shall be updated through a continuing participatory process on the basis of lessons from field action, as well as the results of research. The preparation of national action programmes shall be closely interlinked with other efforts to formulate national policies for sustainable development.

2. In the provision by developed country Parties of different forms of assistance under the terms of article 6, priority shall be given to supporting, as agreed, national, subregional and regional action programmes of affected developing country Parties, particularly those in Africa, either directly or through relevant multilateral organizations or both.

3. The Parties shall encourage organs, funds and programmes of the United Nations system and other relevant intergovernmental organizations, academic institutions, the scientific community and non-governmental organizations in a position to cooperate, in accordance with their mandates and capabilities, to support the elaboration, implementation and follow-up of action programmes.

Article 10. National Action Programmes

1. The purpose of national action programmes is to identify the factors contributing to desertification and practical measures necessary to combat desertification and mitigate the effects of drought.

2. National action programmes shall specify the respective roles of government, local communities and land users and the resources available and needed. They shall, *inter alia*:

(a) incorporate long-term strategies to combat desertification and mitigate the effects of drought, emphasize implementation and be integrated with national policies for sustainable development;

(b) allow for modifications to be made in response to changing circumstances and be sufficiently flexible at the local level to cope with different socio-economic, biological and geo-physical conditions;

(c) give particular attention to the implementation of preventive measures for lands that are not yet degraded or which are only slightly degraded;

(d) enhance national climatological, meteorological and hydrological capabilities and the means to provide for drought early warning;

(e) promote policies and strengthen institutional frameworks which develop cooperation and coordination, in a spirit of partnership, between the donor community, governments at all levels, local populations and community groups, and facilitate access by local populations to appropriate information and technology;

(f) provide for effective participation at the local, national and regional levels of non-governmental organizations and local populations, both women and men, particularly resource users, including farmers and pastoralists and their representative organizations, in policy planning, decision-making, and implementation and review of national action programmes; and

(g) require regular review of, and progress reports on, their implementation.

3. National action programmes may include, *inter alia*, some or all of the following measures to prepare for and mitigate the effects of drought:

(a) establishment and/or strengthening, as appropriate, of early warning systems, including local and national facilities and joint systems at the subregional and regional levels, and mechanisms for assisting environmentally displaced persons;

(b) strengthening of drought preparedness and management, including drought contingency plans at the local, national, subregional and regional levels, which take into consideration seasonal to interannual climate predictions;

(c) establishment and/or strengthening, as appropriate, of food security systems, including storage and marketing facilities, particularly in rural areas;

(d) establishment of alternative livelihood projects that could provide incomes in drought prone areas; and

(e) development of sustainable irrigation programmes for both crops and livestock.

4. Taking into account the circumstances and requirements specific to each affected country Party, national action programmes include, as appropriate,

inter alia, measures in some or all of the following priority fields as they relate to combating desertification and mitigating the effects of drought in affected areas and to their populations: promotion of alternative livelihoods and improvement of national economic environments with a view to strengthening programmes aimed at the eradication of poverty and at ensuring food security; demographic dynamics; sustainable management of natural resources; sustainable agricultural practices; development and efficient use of various energy sources; institutional and legal frameworks; strengthening of capabilities for assessment and systematic observation, including hydrological and meteorological services, and capacity building, education and public awareness.

Article 11. Subregional and Regional Action Programmes

Affected country Parties shall consult and cooperate to prepare, as appropriate, in accordance with relevant regional implementation annexes, subregional and/or regional action programmes to harmonize, complement and increase the efficiency of national programmes. The provisions of article 10 shall apply *mutatis mutandis* to subregional and regional programmes. Such cooperation may include agreed joint programmes for the sustainable management of transboundary natural resources, scientific and technical cooperation, and strengthening of relevant institutions.

Article 12. International Cooperation

Affected country Parties, in collaboration with other Parties and the international community, should cooperate to ensure the promotion of an enabling international environment in the implementation of the Convention. Such cooperation should also cover fields of technology transfer as well as scientific research and development, information collection and dissemination and financial resources.

Article 13. Support for the Elaboration and Implementation of Action Programmes

1. Measures to support action programmes pursuant to article 9 include, *inter alia*:

(a) financial cooperation to provide predictability for action programmes, allowing for necessary long-term planning;

(b) elaboration and use of cooperation mechanisms which better enable support at the local level, including action through non-governmental organizations, in order to promote the replicability of successful pilot programme activities where relevant;

(c) increased flexibility in project design, funding and implementation in keeping with the experimental, iterative approach indicated for participatory action at the local community level; and

(d) as appropriate, administrative and budgetary procedures that increase the efficiency of cooperation and of support programmes.

2. In providing such support to affected developing country Parties, priority shall be given to African country Parties and to least developed country Parties.

Article 14. Coordination in the Elaboration and Implementation of Action Programmes

1. The Parties shall work closely together, directly and through relevant intergovernmental organizations, in the elaboration and implementation of action programmes.
2. The Parties shall develop operational mechanisms, particularly at the national and field levels, to ensure the fullest possible coordination among developed country Parties, developing country Parties and relevant intergovernmental and non-governmental organizations, in order to avoid duplication, harmonize interventions and approaches, and maximize the impact of assistance. In affected developing country Parties, priority will be given to coordinating activities related to international cooperation in order to maximize the efficient use of resources, to ensure responsive assistance, and to facilitate the implementation of national action programmes and priorities under this Convention.

Article 15. Regional Implementation Annexes

Elements for incorporation in action programmes shall be selected and adapted to the socio-economic, geographical and climatic factors applicable to affected country Parties or regions, as well as to their level of development. Guidelines for the preparation of action programmes and their exact focus and content for particular subregions and regions are set out in the regional implementation annexes.

Section 2: Scientific and Technical Cooperation

Article 16. Information Collection, Analysis and Exchange

The Parties agree, according to their respective capabilities, to integrate and coordinate the collection, analysis and exchange of relevant short term and long term data and information to ensure systematic observation of land degradation in affected areas and to understand better and assess the processes and effects of drought and desertification. This would help accomplish, *inter alia*, early warning and advance planning for periods of adverse climatic variation in a form suited for practical application by users at all levels, including especially local populations. To this end, they shall, as appropriate:
    (a) facilitate and strengthen the functioning of the global network of institutions and facilities for the collection, analysis and exchange of information, as well as for systematic observation at all levels, which shall, *inter alia*:
        (i) aim to use compatible standards and systems;
        (ii) encompass relevant data and stations, including in remote areas;
        (iii) use and disseminate modern technology for data collection, transmission and assessment on land degradation; and
        (iv) link national, subregional and regional data and information centres more closely with global information sources;

(b) ensure that the collection, analysis and exchange of information address the needs of local communities and those of decision makers, with a view to resolving specific problems, and that local communities are involved in these activities;

(c) support and further develop bilateral and multilateral programmes and projects aimed at defining, conducting, assessing and financing the collection, analysis and exchange of data and information, including, *inter alia*, integrated sets of physical, biological, social and economic indicators;

(d) make full use of the expertise of competent intergovernmental and non-governmental organizations, particularly to disseminate relevant information and experiences among target groups in different regions;

(e) give full weight to the collection, analysis and exchange of socio-economic data, and their integration with physical and biological data;

(f) exchange and make fully, openly and promptly available information from all publicly available sources relevant to combating desertification and mitigating the effects of drought; and

(g) subject to their respective national legislation and/or policies, exchange information on local and traditional knowledge, ensuring adequate protection for it and providing appropriate return from the benefits derived from it, on an equitable basis and on mutually agreed terms, to the local populations concerned.

Article 17. Research and Development

1. The Parties undertake, according to their respective capabilities, to promote technical and scientific cooperation in the fields of combating desertification and mitigating the effects of drought through appropriate national, subregional, regional and international institutions. To this end, they shall support research activities that:

(a) contribute to increased knowledge of the processes leading to desertification and drought and the impact of, and distinction between, causal factors, both natural and human, with a view to combating desertification and mitigating the effects of drought, and achieving improved productivity as well as sustainable use and management of resources;

(b) respond to well defined objectives, address the specific needs of local populations and lead to the identification and implementation of solutions that improve the living standards of people in affected areas;

(c) protect, integrate, enhance and validate traditional and local knowledge, know-how and practices, ensuring, subject to their respective national legislation and/or policies, that the owners of that knowledge will directly benefit on an equitable basis and on mutually agreed terms from any commercial utilization of it or from any technological development derived from that knowledge;

(d) develop and strengthen national, subregional and regional research capabilities in affected developing country Parties, particularly in Africa, including the development of local skills and the strengthening of appropriate capacities, especially in countries with a weak research base, giving particular attention to multidisciplinary and participative socio-economic research;

(e) take into account, where relevant, the relationship between poverty, migration caused by environmental factors, and desertification;

(f) promote the conduct of joint research programmes between national, subregional, regional and international research organizations, in both the public and private sectors, for the development of improved, affordable and accessible technologies for sustainable development through effective participation of local populations and communities; and

(g) enhance the availability of water resources in affected areas, by means of, *inter alia*, cloud-seeding.

2. Research priorities for particular regions and subregions, reflecting different local conditions, should be included in action programmes. The Conference of the Parties shall review research priorities periodically on the advice of the Committee on Science and Technology.

Article 18. Transfer, Acquisition, Adaptation, and Development of Technology

1. The Parties undertake, as mutually agreed and in accordance with their respective national legislation and/or policies, to promote, finance and/or facilitate the financing of the transfer, acquisition, adaptation and development of environmentally sound, economically viable and socially acceptable technologies relevant to combating desertification and/or mitigating the effects of drought, with a view to contributing to the achievement of sustainable development in affected areas. Such cooperation shall be conducted bilaterally or multilaterally, as appropriate, making full use of the expertise of intergovernmental and non-governmental organizations. The Parties shall, in particular:

(a) fully utilize relevant existing national, subregional, regional and international information systems and clearing-houses for the dissemination of information on available technologies, their sources, their environmental risks and the broad terms under which they may be acquired;

(b) facilitate access, in particular by affected developing country Parties, on favourable terms, including on concessional and preferential terms, as mutually agreed, taking into account the need to protect intellectual property rights, to technologies most suitable to practical application for specific needs of local populations, paying special attention to the social, cultural, economic and environmental impact of such technology;

(c) facilitate technology cooperation among affected country Parties through financial assistance or other appropriate means;

(d) extend technology cooperation with affected developing country Parties, including, where relevant, joint ventures, especially to sectors which foster alternative livelihoods; and

(e) take appropriate measures to create domestic market conditions and incentives, fiscal or otherwise, conducive to the development, transfer, acquisition and adaptation of suitable technology, knowledge, know-how and practices, including measures to ensure adequate and effective protection of intellectual property rights.

2. The Parties shall, according to their respective capabilities, and subject to their respective national legislation and/or policies, protect, promote and use in particular relevant traditional and local technology, knowledge, know-how and practices and, to that end, they undertake to:

(a) make inventories of such technology, knowledge, know-how and practices and their potential uses with the participation of local populations, and disseminate such information, where appropriate, in cooperation with relevant intergovernmental and non-governmental organizations;

(b) ensure that such technology, knowledge, know-how and practices are adequately protected and that local populations benefit directly, on an equitable basis and as mutually agreed, from any commercial utilization of them or from any technological development derived therefrom;

(c) encourage and actively support the improvement and dissemination of such technology, knowledge, know-how and practices or of the development of new technology based on them; and

(d) facilitate, as appropriate, the adaptation of such technology, knowledge, know-how and practices to wide use and integrate them with modern technology, as appropriate.

Section 3: Supporting Measures

Article 19. Capacity Building, Education, and Public Awareness

1. The Parties recognize the significance of capacity building — that is to say, institution building, training and development of relevant local and national capacities — in efforts to combat desertification and mitigate the effects of drought. They shall promote, as appropriate, capacity-building:

(a) through the full participation at all levels of local people, particularly at the local level, especially women and youth, with the cooperation of non-governmental and local organizations;

(b) by strengthening training and research capacity at the national level in the field of desertification and drought;

(c) by establishing and/or strengthening support and extension services to disseminate relevant technology methods and techniques more effectively, and by training field agents and members of rural organizations in participatory approaches for the conservation and sustainable use of natural resources;

(d) by fostering the use and dissemination of the knowledge, know-how and practices of local people in technical cooperation programmes, wherever possible;

(e) by adapting, where necessary, relevant environmentally sound technology and traditional methods of agriculture and pastoralism to modern socio-economic conditions;

(f) by providing appropriate training and technology in the use of alternative energy sources, particularly renewable energy resources, aimed particularly at reducing dependence on wood for fuel;

(g) through cooperation, as mutually agreed, to strengthen the capacity of affected developing country Parties to develop and implement programmes in the field of collection, analysis and exchange of information pursuant to article 16;

(h) through innovative ways of promoting alternative livelihoods, including training in new skills;

(i) by training of decision makers, managers, and personnel who are responsible for the collection and analysis of data for the dissemination and use of early warning information on drought conditions and for food production;

(j) through more effective operation of existing national institutions and legal frameworks and, where necessary, creation of new ones, along with strengthening of strategic planning and management; and

(k) by means of exchange visitor programmes to enhance capacity building in affected country Parties through a long-term, interactive process of learning and study.

2. Affected developing country Parties shall conduct, in cooperation with other Parties and competent intergovernmental and non-governmental organizations, as appropriate, an interdisciplinary review of available capacity and facilities at the local and national levels, and the potential for strengthening them.

3. The Parties shall cooperate with each other and through competent intergovernmental organizations, as well as with non-governmental organizations, in undertaking and supporting public awareness and educational programmes in both affected and, where relevant, unaffected country Parties to promote understanding of the causes and effects of desertification and drought and of the importance of meeting the objective of this Convention. To that end, they shall:

(a) organize awareness campaigns for the general public;

(b) promote, on a permanent basis, access by the public to relevant information, and wide public participation in education and awareness activities;

(c) encourage the establishment of associations that contribute to public awareness;

(d) develop and exchange educational and public awareness material, where possible in local languages, exchange and second experts to train personnel of affected developing country Parties in carrying out relevant education and awareness programmes, and fully utilize relevant educational material available in competent international bodies;

(e) assess educational needs in affected areas, elaborate appropriate school curricula and expand, as needed, educational and adult literacy programmes and opportunities for all, in particular for girls and women, on the identification, conservation and sustainable use and management of the natural resources of affected areas; and

(f) develop interdisciplinary participatory programmes integrating desertification and drought awareness into educational systems and in non-formal, adult, distance and practical educational programmes.

4. The Conference of the Parties shall establish and/or strengthen networks of regional education and training centres to combat desertification and mitigate the effects of drought. These networks shall be coordinated by an institution created or

designated for that purpose, in order to train scientific, technical and management personnel and to strengthen existing institutions responsible for education and training in affected country Parties, where appropriate, with a view to harmonizing programmes and to organizing exchanges of experience among them. These networks shall cooperate closely with relevant intergovernmental and non-governmental organizations to avoid duplication of effort.

Article 20. Financial Resources

1. Given the central importance of financing to the achievement of the objective of the Convention, the Parties, taking into account their capabilities, shall make every effort to ensure that adequate financial resources are available for programmes to combat desertification and mitigate the effects of drought.

2. In this connection, developed country Parties, while giving priority to affected African country Parties without neglecting affected developing country Parties in other regions, in accordance with article 7, undertake to:

(a) mobilize substantial financial resources, including grants and concessional loans, in order to support the implementation of programmes to combat desertification and mitigate the effects of drought;

(b) promote the mobilization of adequate, timely and predictable financial resources, including new and additional funding from the Global Environment Facility of the agreed incremental costs of those activities concerning desertification that relate to its four focal areas, in conformity with the relevant provisions of the Instrument establishing the Global Environment Facility;

(c) facilitate through international cooperation the transfer of technology, knowledge and know-how; and

(d) explore, in cooperation with affected developing country Parties, innovative methods and incentives for mobilizing and channelling resources, including those of foundations, non-governmental organizations and other private sector entities, particularly debt swaps and other innovative means which increase financing by reducing the external debt burden of affected developing country Parties, particularly those in Africa.

3. Affected developing country Parties, taking into account their capabilities, undertake to mobilize adequate financial resources for the implementation of their national action programmes.

4. In mobilizing financial resources, the Parties shall seek full use and continued qualitative improvement of all national, bilateral and multilateral funding sources and mechanisms, using consortia, joint programmes and parallel financing, and shall seek to involve private sector funding sources and mechanisms, including those of non-governmental organizations. To this end, the Parties shall fully utilize the operational mechanisms developed pursuant to article 14.

5. In order to mobilize the financial resources necessary for affected developing country Parties to combat desertification and mitigate the effects of drought, the Parties shall:

(a) rationalize and strengthen the management of resources already allocated for combating desertification and mitigating the effects of drought by using them more effectively and efficiently, assessing their successes and shortcomings, removing hindrances to their effective use and, where necessary, reorienting programmes in light of the integrated long-term approach adopted pursuant to this Convention;

(b) give due priority and attention within the governing bodies of multilateral financial institutions, facilities and funds, including regional development banks and funds, to supporting affected developing country Parties, particularly those in Africa, in activities which advance implementation of the Convention, notably action programmes they undertake in the framework of regional implementation annexes; and

(c) examine ways in which regional and subregional cooperation can be strengthened to support efforts undertaken at the national level.

6. Other Parties are encouraged to provide, on a voluntary basis, knowledge, know-how and techniques related to desertification and/or financial resources to affected developing country Parties.

7. The full implementation by affected developing country Parties, particularly those in Africa, of their obligations under the Convention will be greatly assisted by the fulfilment by developed country Parties of their obligations under the Convention, including in particular those regarding financial resources and transfer of technology. In fulfilling their obligations, developed country Parties should take fully into account that economic and social development and poverty eradication are the first priorities of affected developing country Parties, particularly those in Africa.

Article 21. Financial Mechanisms

1. The Conference of the Parties shall promote the availability of financial mechanisms and shall encourage such mechanisms to seek to maximize the availability of funding for affected developing country Parties, particularly those in Africa, to implement the Convention. To this end, the Conference of the Parties shall consider for adoption *inter alia* approaches and policies that:

(a) facilitate the provision of necessary funding at the national, subregional, regional and global levels for activities pursuant to relevant provisions of the Convention;

(b) promote multiple-source funding approaches, mechanisms and arrangements and their assessment, consistent with article 20;

(c) provide on a regular basis, to interested Parties and relevant intergovernmental and non-governmental organizations, information on available sources of funds and on funding patterns in order to facilitate coordination among them;

(d) facilitate the establishment, as appropriate, of mechanisms, such as national desertification funds, including those involving the participation of non-governmental organizations, to channel financial resources rapidly and efficiently to the local level in affected developing country Parties; and

(e) strengthen existing funds and financial mechanisms at the subregional and regional levels, particularly in Africa, to support more effectively the implementation of the Convention.

2. The Conference of the Parties shall also encourage the provision, through various mechanisms within the United Nations system and through multilateral financial institutions, of support at the national, subregional and regional levels to activities that enable developing country Parties to meet their obligations under the Convention.

3. Affected developing country Parties shall utilize, and where necessary, establish and/or strengthen, national coordinating mechanisms, integrated in national development programmes, that would ensure the efficient use of all available financial resources. They shall also utilize participatory processes involving non-governmental organizations, local groups and the private sector, in raising funds, in elaborating as well as implementing programmes and in assuring access to funding by groups at the local level. These actions can be enhanced by improved coordination and flexible programming on the part of those providing assistance.

4. In order to increase the effectiveness and efficiency of existing financial mechanisms, a Global Mechanism to promote actions leading to the mobilization and channelling of substantial financial resources, including for the transfer of technology, on a grant basis, and/or on concessional or other terms, to affected developing country Parties, is hereby established. This Global Mechanism shall function under the authority and guidance of the Conference of the Parties and be accountable to it.

5. The Conference of the Parties shall identify, at its first ordinary session, an organization to house the Global Mechanism. The Conference of the Parties and the organization it has identified shall agree upon modalities for this Global Mechanism to ensure *inter alia* that such Mechanism:

(a) identifies and draws up an inventory of relevant bilateral and multilateral cooperation programmes that are available to implement the Convention;

(b) provides advice, on request, to Parties on innovative methods of financing and sources of financial assistance and on improving the coordination of cooperation activities at the national level;

(c) provides interested Parties and relevant intergovernmental and non-governmental organizations with information on available sources of funds and on funding patterns in order to facilitate coordination among them; and

(d) reports to the Conference of the Parties, beginning at its second ordinary session, on its activities.

6. The Conference of the Parties shall, at its first session, make appropriate arrangements with the organization it has identified to house the Global Mechanism for the administrative operations of such Mechanism, drawing to the extent possible on existing budgetary and human resources.

7. The Conference of the Parties shall, at its third ordinary session, review the policies, operational modalities and activities of the Global Mechanism accountable to it pursuant to paragraph 4, taking into account the provisions of article 7. On the basis of this review, it shall consider and take appropriate action.

Part IV: Institutions

Article 22. Conference of the Parties

1. A Conference of the Parties is hereby established.
2. The Conference of the Parties is the supreme body of the Convention. It shall make, within its mandate, the decisions necessary to promote its effective implementation. In particular, it shall:

(a) regularly review the implementation of the Convention and the functioning of its institutional arrangements in the light of the experience gained at the national, subregional, regional and international levels and on the basis of the evolution of scientific and technological knowledge;

(b) promote and facilitate the exchange of information on measures adopted by the Parties, and determine the form and timetable for transmitting the information to be submitted pursuant to article 26, review the reports and make recommendations on them;

(c) establish such subsidiary bodies as are deemed necessary for the implementation of the Convention;

(d) review reports submitted by its subsidiary bodies and provide guidance to them;

(e) agree upon and adopt, by consensus, rules of procedure and financial rules for itself and any subsidiary bodies;

(f) adopt amendments to the Convention pursuant to articles 30 and 31;

(g) approve a programme and budget for its activities, including those of its subsidiary bodies, and undertake necessary arrangements for their financing;

(h) as appropriate, seek the cooperation of, and utilize the services of and information provided by, competent bodies or agencies, whether national or international, intergovernmental or non-governmental;

(i) promote and strengthen the relationship with other relevant conventions while avoiding duplication of effort; and

(j) exercise such other functions as may be necessary for the achievement of the objective of the Convention.

3. The Conference of the Parties shall, at its first session, adopt its own rules of procedure, by consensus, which shall include decision-making procedures for matters not already covered by decision-making procedures stipulated in the Convention. Such procedures may include specified majorities required for the adoption of particular decisions.

4. The first session of the Conference of the Parties shall be convened by the interim secretariat referred to in article 35 and shall take place not later than one year after the date of entry into force of the Convention. Unless otherwise decided by the Conference of the Parties, the second, third and fourth ordinary sessions shall be held yearly, and thereafter, ordinary sessions shall be held every two years.

5. Extraordinary sessions of the Conference of the Parties shall be held at such other times as may be decided either by the Conference of the Parties in ordinary

session or at the written request of any Party, provided that, within three months of the request being communicated to the Parties by the Permanent Secretariat, it is supported by at least one third of the Parties.

6. At each ordinary session, the Conference of the Parties shall elect a Bureau. The structure and functions of the Bureau shall be determined in the rules of procedure. In appointing the Bureau, due regard shall be paid to the need to ensure equitable geographical distribution and adequate representation of affected country Parties, particularly those in Africa.

7. The United Nations, its specialized agencies and any State member thereof or observers thereto not Party to the Convention, may be represented at sessions of the Conference of the Parties as observers. Any body or agency, whether national or international, governmental or non-governmental, which is qualified in matters covered by the Convention, and which has informed the Permanent Secretariat of its wish to be represented at a session of the Conference of the Parties as an observer, may be so admitted unless at least one third of the Parties present object. The admission and participation of observers shall be subject to the rules of procedure adopted by the Conference of the Parties.

8. The Conference of the Parties may request competent national and international organizations which have relevant expertise to provide it with information relevant to article 16, paragraph (g), article 17, paragraph 1 (c) and article 18, paragraph 2(b).

Article 23. Permanent Secretariat

1. A Permanent Secretariat is hereby established.

2. The functions of the Permanent Secretariat shall be:

(a) to make arrangements for sessions of the Conference of the Parties and its subsidiary bodies established under the Convention and to provide them with services as required;

(b) to compile and transmit reports submitted to it;

(c) to facilitate assistance to affected developing country Parties, on request, particularly those in Africa, in the compilation and communication of information required under the Convention;

(d) to coordinate its activities with the secretariats of other relevant international bodies and conventions;

(e) to enter, under the guidance of the Conference of the Parties, into such administrative and contractual arrangements as may be required for the effective discharge of its functions;

(f) to prepare reports on the execution of its functions under this Convention and present them to the Conference of the Parties; and

(g) to perform such other secretariat functions as may be determined by the Conference of the Parties.

3. The Conference of the Parties, at its first session, shall designate a Permanent Secretariat and make arrangements for its functioning.

Article 24. Committee on Science and Technology

1. A Committee on Science and Technology is hereby established as a subsidiary body of the Conference of the Parties to provide it with information and advice on scientific and technological matters relating to combating desertification and mitigating the effects of drought. The Committee shall meet in conjunction with the ordinary sessions of the Conference of the Parties and shall be multidisciplinary and open to the participation of all Parties. It shall be composed of government representatives competent in the relevant fields of expertise. The Conference of the Parties shall decide, at its first session, on the terms of reference of the Committee.

2. The Conference of the Parties shall establish and maintain a roster of independent experts with expertise and experience in the relevant fields. The roster shall be based on nominations received in writing from the Parties, taking into account the need for a multidisciplinary approach and broad geographical representation.

3. The Conference of the Parties may, as necessary, appoint ad hoc panels to provide it, through the Committee, with information and advice on specific issues regarding the state of the art in fields of science and technology relevant to combating desertification and mitigating the effects of drought. These panels shall be composed of experts whose names are taken from the roster, taking into account the need for a multidisciplinary approach and broad geographical representation. These experts shall have scientific backgrounds and field experience and shall be appointed by the Conference of the Parties on the recommendation of the Committee. The Conference of the Parties shall decide on the terms of reference and the modalities of work of these panels.

Article 25. Networking of Institutions, Agencies, and Bodies

1. The Committee on Science and Technology shall, under the supervision of the Conference of the Parties, make provision for the undertaking of a survey and evaluation of the relevant existing networks, institutions, agencies and bodies willing to become units of a network. Such a network shall support the implementation of the Convention.

2. On the basis of the results of the survey and evaluation referred to in paragraph 1, the Committee on Science and Technology shall make recommendations to the Conference of the Parties on ways and means to facilitate and strengthen networking of the units at the local, national and other levels, with a view to ensuring that the thematic needs set out in articles 16 to 19 are addressed.

3. Taking into account these recommendations, the Conference of the Parties shall:

(a) identify those national, subregional, regional and international units that are most appropriate for networking, and recommend operational procedures, and a time frame, for them; and

(b) identify the units best suited to facilitating and strengthening such networking at all levels.

Part V: Procedures

Article 26. Communication of Information

1. Each Party shall communicate to the Conference of the Parties for consideration at its ordinary sessions, through the Permanent Secretariat, reports on the measures which it has taken for the implementation of the Convention. The Conference of the Parties shall determine the timetable for submission and the format of such reports.

2. Affected country Parties shall provide a description of the strategies established pursuant to article 5 and of any relevant information on their implementation.

3. Affected country Parties which implement action programmes pursuant to articles 9 to 15 shall provide a detailed description of the programmes and of their implementation.

4. Any group of affected country Parties may make a joint communication on measures taken at the subregional and/or regional levels in the framework of action programmes.

5. Developed country Parties shall report on measures taken to assist in the preparation and implementation of action programmes, including information on the financial resources they have provided, or are providing, under the Convention.

6. Information communicated pursuant to paragraphs 1 to 4 shall be transmitted by the Permanent Secretariat as soon as possible to the Conference of the Parties and to any relevant subsidiary body.

7. The Conference of the Parties shall facilitate the provision to affected developing countries, particularly those in Africa, on request, of technical and financial support in compiling and communicating information in accordance with this article, as well as identifying the technical and financial needs associated with action programmes.

Article 27. Measures to Resolve Questions on Implementation

The Conference of the Parties shall consider and adopt procedures and institutional mechanisms for the resolution of questions that may arise with regard to the implementation of the Convention.

Article 28. Settlement of Disputes

1. Parties shall settle any dispute between them concerning the interpretation or application of the Convention through negotiation or other peaceful means of their own choice.

2. When ratifying, accepting, approving, or acceding to the Convention, or at any time thereafter, a Party which is not a regional economic integration organization may declare in a written instrument submitted to the Depositary that, in respect of any dispute concerning the interpretation or application of the Convention, it recognizes one or both of the following means of dispute settlement as compulsory in relation to any Party accepting the same obligation:

(a) arbitration in accordance with procedures adopted by the Conference of the Parties in an annex as soon as practicable;

(b) submission of the dispute to the International Court of Justice.

3. A Party which is a regional economic integration organization may make a declaration with like effect in relation to arbitration in accordance with the procedure referred to in paragraph 2 (a).

4. A declaration made pursuant to paragraph 2 shall remain in force until it expires in accordance with its terms or until three months after written notice of its revocation has been deposited with the Depositary.

5. The expiry of a declaration, a notice of revocation or a new declaration shall not in any way affect proceedings pending before an arbitral tribunal or the International Court of Justice unless the Parties to the dispute otherwise agree.

6. If the Parties to a dispute have not accepted the same or any procedure pursuant to paragraph 2 and if they have not been able to settle their dispute within twelve months following notification by one Party to another that a dispute exists between them, the dispute shall be submitted to conciliation at the request of any Party to the dispute, in accordance with procedures adopted by the Conference of the Parties in an annex as soon as practicable.

Article 29. Status of Annexes

1. Annexes form an integral part of the Convention and, unless expressly provided otherwise, a reference to the Convention also constitutes a reference to its annexes.

2. The Parties shall interpret the provisions of the annexes in a manner that is in conformity with their rights and obligations under the articles of this Convention.

Article 30. Amendments to the Convention

1. Any Party may propose amendments to the Convention.

2. Amendments to the Convention shall be adopted at an ordinary session of the Conference of the Parties. The text of any proposed amendment shall be communicated to the Parties by the Permanent Secretariat at least six months before the meeting at which it is proposed for adoption. The Permanent Secretariat shall also communicate proposed amendments to the signatories to the Convention.

3. The Parties shall make every effort to reach agreement on any proposed amendment to the Convention by consensus. If all efforts at consensus have been exhausted and no agreement reached, the amendment shall, as a last resort, be adopted by a two-thirds majority vote of the Parties present and voting at the meeting. The adopted amendment shall be communicated by the Permanent Secretariat to the Depositary, who shall circulate it to all Parties for their ratification, acceptance, approval or accession.

4. Instruments of ratification, acceptance, approval or accession in respect of an amendment shall be deposited with the Depositary. An amendment adopted

pursuant to paragraph 3 shall enter into force for those Parties having accepted it on the ninetieth day after the date of receipt by the Depositary of an instrument of ratification, acceptance, approval or accession by at least two thirds of the Parties to the Convention which were Parties at the time of the adoption of the amendment.

5. The amendment shall enter into force for any other Party on the ninetieth day after the date on which that Party deposits with the Depositary its instrument of ratification, acceptance or approval of, or accession to the said amendment.

6. For the purposes of this article and article 31, "Parties present and voting" means Parties present and casting an affirmative or negative vote.

Article 31. Adoption and Amendment of Annexes

1. Any additional annex to the Convention and any amendment to an annex shall be proposed and adopted in accordance with the procedure for amendment of the Convention set forth in article 30, provided that, in adopting an additional regional implementation annex or amendment to any regional implementation annex, the majority provided for in that article shall include a two-thirds majority vote of the Parties of the region concerned present and voting. The adoption or amendment of an annex shall be communicated by the Depositary to all Parties.

2. An annex, other than an additional regional implementation annex, or an amendment to an annex, other than an amendment to any regional implementation annex, that has been adopted in accordance with paragraph 1, shall enter into force for all Parties to the Convention six months after the date of communication by the Depositary to such Parties of the adoption of such annex or amendment, except for those Parties that have notified the Depositary in writing within that period of their non-acceptance of such annex or amendment. Such annex or amendment shall enter into force for Parties which withdraw their notification of non-acceptance on the ninetieth day after the date on which withdrawal of such notification has been received by the Depositary.

3. An additional regional implementation annex or amendment to any regional implementation annex that has been adopted in accordance with paragraph 1, shall enter into force for all Parties to the Convention six months after the date of the communication by the Depositary to such Parties of the adoption of such annex or amendment, except with respect to:

(a) any Party that has notified the Depositary in writing, within such six month period, of its non-acceptance of that additional regional implementation annex or of the amendment to the regional implementation annex, in which case such annex or amendment shall enter into force for Parties which withdraw their notification of non-acceptance on the ninetieth day after the date on which withdrawal of such notification has been received by the Depositary; and

(b) any Party that has made a declaration with respect to additional regional implementation annexes or amendments to regional implementation annexes in accordance with article 34, paragraph 4, in which case any such annex or amendment shall enter into force for such a Party on the ninetieth day after the date of deposit with the Depositary of its instrument of ratification, acceptance, approval or accession with respect to such annex or amendment.

4. If the adoption of an annex or an amendment to an annex involves an amendment to the Convention, that annex or amendment to an annex shall not enter into force until such time as the amendment to the Convention enters into force.

Article 32. Right to Vote

1. Except as provided for in paragraph 2, each Party to the Convention shall have one vote.

2. Regional economic integration organizations, in matters within their competence, shall exercise their right to vote with a number of votes equal to the number of their member States that are Parties to the Convention. Such an organization shall not exercise its right to vote if any of its member States exercises its right, and vice versa.

Part VI: Final Provisions

Article 33. Signature

This Convention shall be opened for signature at Paris, on 14-15 October 1994, by States Members of the United Nations or any of its specialized agencies or that are Parties to the Statute of the International Court of Justice and by regional economic integration organizations. It shall remain open for signature, thereafter, at the United Nations Headquarters in New York until 13 October 1995.

Article 34. Ratification, Acceptance, Approval, and Accession

1. The Convention shall be subject to ratification, acceptance, approval or accession by States and by regional economic integration organizations. It shall be open for accession from the day after the date on which the Convention is closed for signature. Instruments of ratification, acceptance, approval or accession shall be deposited with the Depositary.

2. Any regional economic integration organization which becomes a Party to the Convention without any of its member States being a Party to the Convention shall be bound by all the obligations under the Convention. Where one or more member States of such an organization are also Party to the Convention, the organization and its member States shall decide on their respective responsibilities for the performance of their obligations under the Convention. In such cases, the organization and the member States shall not be entitled to exercise rights under the Convention concurrently.

3. In their instruments of ratification, acceptance, approval or accession, regional economic integration organizations shall declare the extent of their competence with respect to the matters governed by the Convention. They shall also promptly inform the Depositary, who shall in turn inform the Parties, of any substantial modification in the extent of their competence.

4. In its instrument of ratification, acceptance, approval or accession, any Party may declare that, with respect to it, any additional regional implementation annex or any amendment to any regional implementation annex shall enter into force only upon the deposit of its instrument of ratification, acceptance, approval or accession with respect thereto.

Article 35. Interim Arrangements

The secretariat functions referred to in article 23 will be carried out on an interim basis by the secretariat established by the General Assembly of the United Nations in its resolution 47/188 of 22 December 1992, until the completion of the first session of the Conference of the Parties.

Article 36. Entry into Force

1. The Convention shall enter into force on the ninetieth day after the date of deposit of the fiftieth instrument of ratification, acceptance, approval or accession.

2. For each State or regional economic integration organization ratifying, accepting, approving or acceding to the Convention after the deposit of the fiftieth instrument of ratification, acceptance, approval or accession, the Convention shall enter into force on the ninetieth day after the date of deposit by such State or regional economic integration organization of its instrument of ratification, acceptance, approval or accession.

3. For the purposes of paragraphs 1 and 2, any instrument deposited by a regional economic integration organization shall not be counted as additional to those deposited by States members of the organization.

Article 37. Reservations

No reservations may be made to this Convention.

Article 38. Withdrawal

1. At any time after three years from the date on which the Convention has entered into force for a Party, that Party may withdraw from the Convention by giving written notification to the Depositary.

2. Any such withdrawal shall take effect upon expiry of one year from the date of receipt by the Depositary of the notification of withdrawal, or on such later date as may be specified in the notification of withdrawal.

Article 39. Depositary

The Secretary-General of the United Nations shall be the Depositary of the Convention.

Article 40. Authentic Texts

The original of the present Convention, of which the Arabic, Chinese, English, French, Russian and Spanish texts are equally authentic, shall be deposited with the Secretary-General of the United Nations.

IN WITNESS WHEREOF the undersigned, being duly authorized to that effect, have signed the present Convention.

DONE AT Paris, this 17th day of June one thousand nine hundred and ninety-four.

Annex I — Regional Implementation Annex for Africa

Article 1. Scope

This Annex applies to Africa, in relation to each Party and in conformity with the Convention, in particular its article 7, for the purpose of combating desertification and/or mitigating the effects of drought in its arid, semi-arid and dry sub-humid areas.

Article 2. Purpose

The purpose of this Annex, at the national, subregional and regional levels in Africa and in the light of its particular conditions, is to:

(a) identify measures and arrangements, including the nature and processes of assistance provided by developed country Parties, in accordance with the relevant provisions of the Convention;

(b) provide for the efficient and practical implementation of the Convention to address conditions specific to Africa; and

(c) promote processes and activities relating to combating desertification and/or mitigating the effects of drought within the arid, semi-arid and dry sub-humid areas of Africa.

Article 3. Particular Conditions of the African region

In carrying out their obligations under the Convention, the Parties shall, in the implementation of this Annex, adopt a basic approach that takes into consideration the following particular conditions of Africa:

(a) the high proportion of arid, semi-arid and dry sub-humid areas;

(b) the substantial number of countries and populations adversely affected by desertification and by the frequent recurrence of severe drought;

(c) the large number of affected countries that are landlocked;

(d) the widespread poverty prevalent in most affected countries, the large number of least developed countries among them, and their need for significant amounts of external assistance, in the form of grants and loans on concessional terms, to pursue their development objectives;

(e) the difficult socio-economic conditions, exacerbated by deteriorating and fluctuating terms of trade, external indebtedness and political instability, which induce internal, regional and international migrations;

(f) the heavy reliance of populations on natural resources for subsistence which, compounded by the effects of demographic trends and factors, a weak technological base and unsustainable production practices, contributes to serious resource degradation;

(g) the insufficient institutional and legal frameworks, the weak infrastructural base and the insufficient scientific, technical and educational capacity, leading to substantial capacity building requirements; and

(h) the central role of actions to combat desertification and/or mitigate the effects of drought in the national development priorities of affected African countries.

Article 4. Commitments and Obligations of African Country Parties

1. In accordance with their respective capabilities, African country Parties undertake to:

(a) adopt the combating of desertification and/or the mitigation of the effects of drought as a central strategy in their efforts to eradicate poverty;

(b) promote regional cooperation and integration, in a spirit of solidarity and partnership based on mutual interest, in programmes and activities to combat desertification and/or mitigate the effects of drought;

(c) rationalize and strengthen existing institutions concerned with desertification and drought and involve other existing institutions, as appropriate, in order to make them more effective and to ensure more efficient use of resources;

(d) promote the exchange of information on appropriate technology, knowledge, know-how and practices between and among them; and

(e) develop contingency plans for mitigating the effects of drought in areas degraded by desertification and/or drought.

2. Pursuant to the general and specific obligations set out in articles 4 and 5 of the Convention, affected African country Parties shall aim to:

(a) make appropriate financial allocations from their national budgets consistent with national conditions and capabilities and reflecting the new priority Africa has accorded to the phenomenon of desertification and/or drought;

(b) sustain and strengthen reforms currently in progress toward greater decentralization and resource tenure as well as reinforce participation of local populations and communities; and

(c) identify and mobilize new and additional national financial resources, and expand, as a matter of priority, existing national capabilities and facilities to mobilize domestic financial resources.

Article 5. Commitments and Obligations of Developed Country Parties

1. In fulfilling their obligations pursuant to articles 4, 6 and 7 of the Convention, developed country Parties shall give priority to affected African country Parties and, in this context, shall:

(a) assist them to combat desertification and/or mitigate the effects of drought by, *inter alia*, providing and/or facilitating access to financial and/or other resources, and promoting, financing and/or facilitating the financing of the transfer, adaptation and access to appropriate environmental technologies and know-how, as mutually agreed and in accordance with national policies, taking into account their adoption of poverty eradication as a central strategy;

(b) continue to allocate significant resources and/or increase resources to combat desertification and/or mitigate the effects of drought; and

(c) assist them in strengthening capacities to enable them to improve their institutional frameworks, as well as their scientific and technical capabilities, information collection and analysis, and research and development for the purpose of combating desertification and/or mitigating the effects of drought.

2. Other country Parties may provide, on a voluntary basis, technology, knowledge and know-how relating to desertification and/or financial resources, to affected African country Parties. The transfer of such knowledge, know-how and techniques is facilitated by international cooperation.

Article 6. Strategic Planning Framework for Sustainable Development

1. National action programmes shall be a central and integral part of a broader process of formulating national policies for the sustainable development of affected African country Parties.

2. A consultative and participatory process involving appropriate levels of government, local populations, communities and non-governmental organizations shall be undertaken to provide guidance on a strategy with flexible planning to allow maximum participation from local populations and communities. As appropriate, bilateral and multilateral assistance agencies may be involved in this process at the request of an affected African country Party.

Article 7. Timetable for Preparation of Action Programmes

Pending entry into force of this Convention, the African country Parties, in cooperation with other members of the international community, as appropriate, shall, to the extent possible, provisionally apply those provisions of the Convention relating to the preparation of national, subregional and regional action programmes.

Article 8. Content of National Action Programmes

1. Consistent with article 10 of the Convention, the overall strategy of national action programmes shall emphasize integrated local development programmes for affected areas, based on participatory mechanisms and on integration of strategies for poverty eradication into efforts to combat desertification and mitigate the effects of drought. The programmes shall aim at strengthening the capacity of local authorities and ensuring the active involvement of local populations, communities and groups, with emphasis on education and training, mobilization of non-governmental organizations with proven expertise and strengthening of decentralized governmental structures.

2. National action programmes shall, as appropriate, include the following general features:

(a) the use, in developing and implementing national action programmes, of past experiences in combating desertification and/or mitigating the effects of drought, taking into account social, economic and ecological conditions;

(b) the identification of factors contributing to desertification and/or drought and the resources and capacities available and required, and the setting up of appropriate policies and institutional and other responses and measures necessary to combat those phenomena and/or mitigate their effects; and

(c) the increase in participation of local populations and communities, including women, farmers and pastoralists, and delegation to them of more responsibility for management.

3. National action programmes shall also, as appropriate, include the following:

(a) measures to improve the economic environment with a view to eradicating poverty:

(i) increasing incomes and employment opportunities, especially for the poorest members of the community, by:

- developing markets for farm and livestock products;
- creating financial instruments suited to local needs;
- encouraging diversification in agriculture and the setting-up of agricultural enterprises; and
- developing economic activities of a para-agricultural or non-agricultural type;

(ii) improving the long-term prospects of rural economies by the creation of:

- incentives for productive investment and access to the means of production; and
- price and tax policies and commercial practices that promote growth;

(iii) defining and applying population and migration policies to reduce population pressure on land; and

(iv) promoting the use of drought resistant crops and the application of integrated dryland farming systems for food security purposes;

(b) measures to conserve natural resources:

(i) ensuring integrated and sustainable management of natural resources, including:

- agricultural land and pastoral land;
- vegetation cover and wildlife;
- forests;
- water resources; and
- biological diversity;

(ii) training with regard to, and strengthening, public awareness and environmental education campaigns and disseminating knowledge of techniques relating to the sustainable management of natural resources; and

(iii) ensuring the development and efficient use of diverse energy sources, the promotion of alternative sources of energy, particularly solar energy, wind energy and bio-gas, and specific arrangements for the transfer, acquisition and adaptation of relevant technology to alleviate the pressure on fragile natural resources;

(c) measures to improve institutional organization:

(i) defining the roles and responsibilities of central government and local authorities within the framework of a land use planning policy;

(ii) encouraging a policy of active decentralization, devolving responsibility for management and decision-making to local authorities, and encouraging initiatives and the assumption of responsibility by local communities and the establishment of local structures; and

(iii) adjusting, as appropriate, the institutional and regulatory framework of natural resource management to provide security of land tenure for local populations;

(d) measures to improve knowledge of desertification:

(i) promoting research and the collection, processing and exchange of information on the scientific, technical and socio-economic aspects of desertification;

(ii) improving national capabilities in research and in the collection, processing, exchange and analysis of information so as to increase understanding and to translate the results of the analysis into operational terms; and

(iii) encouraging the medium and long term study of:
- socio-economic and cultural trends in affected areas;
- qualitative and quantitative trends in natural resources; and
- the interaction between climate and desertification; and

(e) measures to monitor and assess the effects of drought:

(i) developing strategies to evaluate the impacts of natural climate variability on regional drought and desertification and/or to utilize predictions of climate variability on seasonal to interannual time scales in efforts to mitigate the effects of drought;

(ii) improving early warning and response capacity, efficiently managing emergency relief and food aid, and improving food stocking and distribution systems, cattle protection schemes and public works and alternative livelihoods for drought prone areas; and

(iii) monitoring and assessing ecological degradation to provide reliable and timely information on the process and dynamics of resource degradation in order to facilitate better policy formulations and responses.

Article 9. Preparation of National Action Programmes and Implementation and Evaluation Indicators

Each affected African country Party shall designate an appropriate national coordinating body to function as a catalyst in the preparation, implementation and evaluation of its national action programme. This coordinating body shall, in the light of article 3 and as appropriate:

(a) undertake an identification and review of actions, beginning with a locally driven consultation process, involving local populations and communities and with the cooperation of local administrative authorities, developed country Parties and intergovernmental and non-governmental organizations, on the basis of initial consultations of those concerned at the national level;

(b) identify and analyze the constraints, needs and gaps affecting development and sustainable land use and recommend practical measures to avoid duplication by making full use of relevant ongoing efforts and promote implementation of results;

(c) facilitate, design and formulate project activities based on interactive, flexible approaches in order to ensure active participation of the population in affected areas, to minimize the negative impact of such activities, and to identify and prioritize requirements for financial assistance and technical cooperation;

(d) establish pertinent, quantifiable and readily verifiable indicators to ensure the assessment and evaluation of national action programmes, which encompass actions in the short, medium and long terms, and of the implementation of such programmes; and

(e) prepare progress reports on the implementation of the national action programmes.

Article 10. Organizational Framework of Subregional Action Programmes

1. Pursuant to article 4 of the Convention, African country Parties shall cooperate in the preparation and implementation of subregional action programmes for central, eastern, northern, southern and western Africa and, in that regard, may delegate the following responsibilities to relevant subregional intergovernmental organizations:

(a) acting as focal points for preparatory activities and coordinating the implementation of the subregional action programmes;

(b) assisting in the preparation and implementation of national action programmes;

(c) facilitating the exchange of information, experience and know-how as well as providing advice on the review of national legislation; and

(d) any other responsibilities relating to the implementation of subregional action programmes.

2. Specialized subregional institutions may provide support, upon request, and/or be entrusted with the responsibility to coordinate activities in their respective fields of competence.

Article 11. Content and Preparation of Subregional Action Programmes

Subregional action programmes shall focus on issues that are better addressed at the subregional level. They shall establish, where necessary, mechanisms for the management of shared natural resources. Such mechanisms shall effectively handle transboundary problems associated with desertification and/or drought and shall provide support for the harmonious implementation of national action programmes. Priority areas for subregional action programmes shall, as appropriate, focus on:

(a) joint programmes for the sustainable management of transboundary natural resources through bilateral and multilateral mechanisms, as appropriate;

(b) coordination of programmes to develop alternative energy sources;

(c) cooperation in the management and control of pests as well as of plant and animal diseases;

(d) capacity building, education and public awareness activities that are better carried out or supported at the subregional level;

(e) scientific and technical cooperation, particularly in the climatological, meteorological and hydrological fields, including networking for data collection and assessment, information sharing and project monitoring, and coordination and prioritization of research and development activities;

(f) early warning systems and joint planning for mitigating the effects of drought, including measures to address the problems resulting from environmentally induced migrations;

(g) exploration of ways of sharing experiences, particularly regarding participation of local populations and communities, and creation of an enabling environment for improved land use management and for use of appropriate technologies;

(h) strengthening of the capacity of subregional organizations to coordinate and provide technical services, as well as establishment, reorientation and strengthening of subregional centres and institutions; and

(i) development of policies in fields, such as trade, which have impact upon affected areas and populations, including policies for the coordination of regional marketing regimes and for common infrastructure.

Article 12. Organizational Framework of the Regional Action Programme

1. Pursuant to article 11 of the Convention, African country Parties shall jointly determine the procedures for preparing and implementing the regional action programme.

2. The Parties may provide appropriate support to relevant African regional institutions and organizations to enable them to assist African country Parties to fulfil their responsibilities under the Convention.

Article 13. Content of the Regional Action Programme

The regional action programme includes measures relating to combating desertification and/or mitigating the effects of drought in the following priority areas, as appropriate:

(a) development of regional cooperation and coordination of sub-regional action programmes for building regional consensus on key policy areas, including through regular consultations of sub-regional organizations;

(b) promotion of capacity building in activities which are better implemented at the regional level;

(c) the seeking of solutions with the international community to global economic and social issues that have an impact on affected areas taking into account article 4, paragraph 2 (b) of the Convention;

(d) promotion among the affected country Parties of Africa and its subregions, as well as with other affected regions, of exchange of information and appropriate techniques, technical know-how and relevant experience; promotion of scientific and technological cooperation particularly in the fields of climatology, meteorology, hydrology, water resource development and alternative energy sources; coordination of sub-regional and regional research activities; and identification of regional priorities for research and development;

(e) coordination of networks for systematic observation and assessment and information exchange, as well as their integration into world wide networks; and

(f) coordination of and reinforcement of sub-regional and regional early warning systems and drought contingency plans.

Article 14. Financial Resources

1. Pursuant to article 20 of the Convention and article 4, paragraph 2, affected African country Parties shall endeavour to provide a macroeconomic framework conducive to the mobilization of financial resources and shall develop policies and establish procedures to channel resources more effectively to local development programmes, including through non-governmental organizations, as appropriate.

2. Pursuant to article 21, paragraphs 4 and 5 of the Convention, the Parties agree to establish an inventory of sources of funding at the national, subregional, regional and international levels to ensure the rational use of existing resources and to identify gaps in resource allocation, to facilitate implementation of the action programmes. The inventory shall be regularly reviewed and updated.

3. Consistent with article 7 of the Convention, the developed country Parties shall continue to allocate significant resources and/or increased resources as well as other forms of assistance to affected African country Parties on the basis of partnership agreements and arrangements referred to in article 18, giving, *inter alia*, due attention to matters related to debt, international trade and marketing arrangements in accordance with article 4, paragraph 2 (b) of the Convention.

Article 15. Financial Mechanisms

1. Consistent with article 7 of the Convention underscoring the priority to affected African country Parties and considering the particular situation prevailing in this region, the Parties shall pay special attention to the implementation in Africa of the provisions of article 21, paragraph 1 (d) and (e) of the Convention, notably by:

(a) facilitating the establishment of mechanisms, such as national desertification funds, to channel financial resources to the local level; and

(b) strengthening existing funds and financial mechanisms at the subregional and regional levels.

2. Consistent with articles 20 and 21 of the Convention, the Parties which are also members of the governing bodies of relevant regional and subregional financial institutions, including the African Development Bank and the African Development Fund, shall promote efforts to give due priority and attention to the activities of those institutions that advance the implementation of this Annex.

3. The Parties shall streamline, to the extent possible, procedures for channelling funds to affected African country Parties.

Article 16. Technical Assistance and Cooperation

The Parties undertake, in accordance with their respective capabilities, to rationalize technical assistance to, and cooperation with, African country Parties with a view to increasing project and programme effectiveness by, *inter alia*:

(a) limiting the costs of support measures and backstopping, especially overhead costs; in any case, such costs shall only represent an appropriately low percentage of the total cost of the project so as to maximize project efficiency;

(b) giving preference to the utilization of competent national experts or, where necessary, competent experts from within the subregion and/or region, in project design, preparation and implementation, and to the building of local expertise where it does not exist; and

(c) effectively managing and coordinating, as well as efficiently utilizing, technical assistance to be provided.

Article 17. Transfer, Acquisition, Adaptation, and Access to Environmentally Sound Technology

In implementing article 18 of the Convention relating to transfer, acquisition, adaptation and development of technology, the Parties undertake to give priority to African country Parties and, as necessary, to develop with them new models of partnership and cooperation with a view to strengthening capacity building in the fields of scientific research and development and information collection and dissemination to enable them to implement their strategies to combat desertification and mitigate the effects of drought.

Article 18. Coordination and Partnership Agreements

1. African country Parties shall coordinate the preparation, negotiation and implementation of national, subregional and regional action programmes. They may involve, as appropriate, other Parties and relevant intergovernmental and non-governmental organizations in this process.

2. The objectives of such coordination shall be to ensure that financial and technical cooperation is consistent with the Convention and to provide the necessary continuity in the use and administration of resources.

3. African country Parties shall organize consultative processes at the national, subregional and regional levels. These consultative processes may:

(a) serve as a forum to negotiate and conclude partnership agreements based on national, subregional and regional action programmes; and

(b) specify the contribution of African country Parties and other members of the consultative groups to the programmes and identify priorities and agreements on implementation and evaluation indicators, as well as funding arrangements for implementation.

4. The Permanent Secretariat may, at the request of African country Parties, pursuant to article 23 of the Convention, facilitate the convocation of such consultative processes by:

(a) providing advice on the organization of effective consultative arrangements, drawing on experiences from other such arrangements;

(b) providing information to relevant bilateral and multilateral agencies concerning consultative meetings or processes, and encouraging their active involvement; and

(c) providing other information that may be relevant in establishing or improving consultative arrangements.

5. The subregional and regional coordinating bodies shall, *inter alia*:

(a) recommend appropriate adjustments to partnership agreements;

(b) monitor, assess and report on the implementation of the agreed subregional and regional programmes; and

(c) aim to ensure efficient communication and cooperation among African country Parties.

6. Participation in the consultative groups shall, as appropriate, be open to Governments, interested groups and donors, relevant organs, funds and programmes of the United Nations system, relevant subregional and regional organizations, and representatives of relevant non-governmental organizations. Participants of each consultative group shall determine the modalities of its management and operation.

7. Pursuant to article 14 of the Convention, developed country Parties are encouraged to develop, on their own initiative, an informal process of consultation and coordination among themselves, at the national, subregional and regional levels, and, at the request of an affected African country Party or of an appropriate subregional or regional organization, to participate in a national, subregional or regional consultative process that would evaluate and respond to assistance needs in order to facilitate implementation.

Article 19. Follow-up Arrangements

Follow-up of this Annex shall be carried out by African country Parties in accordance with the Convention as follows:

(a) at the national level, by a mechanism the composition of which should be determined by each affected African country Party and which shall include representatives of local communities and shall function under the supervision of the national coordinating body referred to in article 9;

(b) at the subregional level, by a multidisciplinary scientific and technical consultative committee, the composition and modalities of operation of which shall be determined by the African country Parties of the subregion concerned; and

(c) at the regional level, by mechanisms defined in accordance with the relevant provisions of the Treaty establishing the African Economic Community, and by an African Scientific and Technical Advisory Committee.

Annex II — Regional Implementation Annex for Asia

Article 1. Purpose

The purpose of this Annex is to provide guidelines and arrangements for the effective implementation of the Convention in the affected country Parties of the Asian region in the light of its particular conditions.

Article 2. Particular Conditions of the Asian Region

In carrying out their obligations under the Convention, the Parties shall, as appropriate, take into consideration the following particular conditions which apply in varying degrees to the affected country Parties of the region:

(a) the high proportion of areas in their territories affected by, or vulnerable to, desertification and drought and the broad diversity of these areas with regard to climate, topography, land use and socio-economic systems;

(b) the heavy pressure on natural resources for livelihoods;

(c) the existence of production systems, directly related to widespread poverty, leading to land degradation and to pressure on scarce water resources;

(d) the significant impact of conditions in the world economy and social problems such as poverty, poor health and nutrition, lack of food security, migration, displaced persons and demographic dynamics;

(e) their expanding, but still insufficient, capacity and institutional frameworks to deal with national desertification and drought problems; and

(f) their need for international cooperation to pursue sustainable development objectives relating to combating desertification and mitigating the effects of drought.

Article 3. Framework for National Action Programmes

1. National action programmes shall be an integral part of broader national policies for sustainable development of the affected country Parties of the region.

2. The affected country Parties shall, as appropriate, develop national action programmes pursuant to articles 9 to 11 of the Convention, paying special attention to article 10, paragraph 2 (f). As appropriate, bilateral and multilateral cooperation agencies may be involved in this process at the request of the affected country Party concerned.

Article 4. National Action Programmes

1. In preparing and implementing national action programmes, the affected country Parties of the region, consistent with their respective circumstances and policies, may, *inter alia*, as appropriate:

(a) designate appropriate bodies responsible for the preparation, coordination and implementation of their action programmes;

(b) involve affected populations, including local communities, in the elaboration, coordination and implementation of their action programmes through a locally driven consultative process, with the cooperation of local authorities and relevant national and non-governmental organizations;

(c) survey the state of the environment in affected areas to assess the causes and consequences of desertification and to determine priority areas for action;

(d) evaluate, with the participation of affected populations, past and current programmes for combating desertification and mitigating the effects of drought, in order to design a strategy and elaborate activities in their action programmes;

(e) prepare technical and financial programmes based on the information derived from the activities in subparagraphs (a) to (d);

(f) develop and utilize procedures and benchmarks for evaluating implementation of their action programmes;

(g) promote the integrated management of drainage basins, the conservation of soil resources, and the enhancement and efficient use of water resources;

(h) strengthen and/or establish information, evaluation and follow up and early warning systems in regions prone to desertification and drought, taking account of climatological, meteorological, hydrological, biological and other relevant factors; and

(i) formulate in a spirit of partnership, where international cooperation, including financial and technical resources, is involved, appropriate arrangements supporting their action programmes.

2. Consistent with article 10 of the Convention, the overall strategy of national action programmes shall emphasize integrated local development programmes for affected areas, based on participatory mechanisms and on the integration of strategies for poverty eradication into efforts to combat desertification and mitigate the effects of drought. Sectoral measures in the action programmes shall be grouped in priority fields which take account of the broad diversity of affected areas in the region referred to in article 2 (a).

Article 5. Subregional and Joint Action Programmes

1. Pursuant to article 11 of the Convention, affected country Parties in Asia may mutually agree to consult and cooperate with other Parties, as appropriate, to prepare and implement subregional or joint action programmes, as appropriate, in order to complement, and increase effectiveness in the implementation of, national action programmes. In either case, the relevant Parties may jointly agree to entrust subregional, including bilateral or national organizations, or specialized institutions, with responsibilities relating to the preparation, coordination and implementation of programmes. Such organizations or institutions may also act as focal points for the promotion and coordination of actions pursuant to articles 16 to 18 of the Convention.

2. In preparing and implementing subregional or joint action programmes, the affected country Parties of the region shall, *inter alia*, as appropriate:

(a) identify, in cooperation with national institutions, priorities relating to combating desertification and mitigating the effects of drought which can better be met by such programmes, as well as relevant activities which could be effectively carried out through them;

(b) evaluate the operational capacities and activities of relevant regional, subregional and national institutions;

(c) assess existing programmes relating to desertification and drought among all or some parties of the region or subregion and their relationship with national action programmes; and

(d) formulate in a spirit of partnership, where international cooperation, including financial and technical resources, is involved, appropriate bilateral and/or multilateral arrangements supporting the programmes.

3. Subregional or joint action programmes may include agreed joint programmes for the sustainable management of transboundary natural resources relating to desertification, priorities for coordination and other activities in the fields of capacity building, scientific and technical cooperation, particularly drought early warning systems and information sharing, and means of strengthening the relevant subregional and other organizations or institutions.

Article 6. Regional Activities

Regional activities for the enhancement of subregional or joint action programmes may include, *inter alia*, measures to strengthen institutions and mechanisms for coordination and cooperation at the national, subregional and regional levels, and to promote the implementation of articles 16 to 19 of the Convention. These activities may also include:

(a) promoting and strengthening technical cooperation networks;

(b) preparing inventories of technologies, knowledge, know-how and practices, as well as traditional and local technologies and know-how, and promoting their dissemination and use;

(c) evaluating the requirements for technology transfer and promoting the adaptation and use of such technologies; and

(d) encouraging public awareness programmes and promoting capacity building at all levels, strengthening training, research and development and building systems for human resource development.

Article 7. Financial Resources and Mechanisms

1. The Parties shall, in view of the importance of combating desertification and mitigating the effects of drought in the Asian region, promote the mobilization of substantial financial resources and the availability of financial mechanisms, pursuant to articles 20 and 21 of the Convention.

2. In conformity with the Convention and on the basis of the coordinating mechanism provided for in article 8 and in accordance with their national development policies, affected country Parties of the region shall, individually or jointly:

(a) adopt measures to rationalize and strengthen mechanisms to supply funds through public and private investment with a view to achieving specific results in action to combat desertification and mitigate the effects of drought;

(b) identify international cooperation requirements in support of national efforts, particularly financial, technical and technological; and

(c) promote the participation of bilateral and/or multilateral financial cooperation institutions with a view to ensuring implementation of the Convention.

3. The Parties shall streamline, to the extent possible, procedures for channelling funds to affected country Parties in the region.

Article 8. Cooperation and Coordination Mechanisms

1. Affected country Parties, through the appropriate bodies designated pursuant to article 4, paragraph 1 (a), and other Parties in the region, may, as appropriate, set up a mechanism for, *inter alia*, the following purposes:

(a) exchange of information, experience, knowledge and know-how;

(b) cooperation and coordination of actions, including bilateral and multilateral arrangements, at the subregional and regional levels;

(c) promotion of scientific, technical, technological and financial cooperation pursuant to articles 5 to 7;

(d) identification of external cooperation requirements; and

(e) follow-up and evaluation of the implementation of action programmes.

2. Affected country Parties, through the appropriate bodies designated pursuant to article 4, paragraph 1 (a), and other Parties in the region, may also, as appropriate, consult and coordinate as regards the national, subregional and joint action programmes. They may involve, as appropriate, other Parties and relevant intergovernmental and non-governmental organizations in this process. Such coordination shall, *inter alia*, seek to secure agreement on opportunities for international cooperation in accordance with articles 20 and 21 of the Convention, enhance technical cooperation and channel resources so that they are used effectively.

3. Affected country Parties of the region shall hold periodic coordination meetings, and the Permanent Secretariat may, at their request, pursuant to article 23 of the Convention, facilitate the convocation of such coordination meetings by:

(a) providing advice on the organization of effective coordination arrangements, drawing on experience from other such arrangements;

(b) providing information to relevant bilateral and multilateral agencies concerning coordination meetings, and encouraging their active involvement; and

(c) providing other information that may be relevant in establishing or improving coordination processes.

Annex III — Regional Implementation Annex for Latin America and the Caribbean

Article 1. Purpose

The purpose of this Annex is to provide general guidelines for the implementation of the Convention in the Latin American and Caribbean region, in light of its particular conditions.

Article 2. Particular Conditions of the Latin American and Caribbean Region

The Parties shall, in accordance with the provisions of the Convention, take into consideration the following particular conditions of the region:

(a) the existence of broad expanses which are vulnerable and have been severely affected by desertification and/or drought and in which diverse characteristics may be observed, depending on the area in which they occur; this cumulative and intensifying process has negative social, cultural, economic and environmental effects which are all the more serious in that the region contains one of the largest resources of biological diversity in the world;

(b) the frequent use of unsustainable development practices in affected areas as a result of complex interactions among physical, biological, political, social, cultural and economic factors, including international economic factors such as external indebtedness, deteriorating terms of trade and trade practices which affect markets for agricultural, fishery and forestry products; and

(c) a sharp drop in the productivity of ecosystems being the main consequence of desertification and drought, taking the form of a decline in agricultural, livestock and forestry yields and a loss of biological diversity; from the social point of view, the results are impoverishment, migration, internal population movements, and the deterioration of the quality of life; the region will therefore have to adopt an integrated approach to problems of desertification and drought by promoting sustainable development models that are in keeping with the environmental, economic and social situation in each country.

Article 3. Action Programmes

1. In conformity with the Convention, in particular its articles 9 to 11, and in accordance with their national development policies, affected country Parties of the region shall, as appropriate, prepare and implement national action programmes to combat desertification and mitigate the effects of drought as an integral part of their national policies for sustainable development. Subregional and regional programmes may be prepared and implemented in accordance with the requirements of the region.

2. In the preparation of their national action programmes, affected country Parties of the region shall pay particular attention to article 10, paragraph 2 (f) of the Convention.

Article 4. Content of National Action Programmes

In the light of their respective situations, the affected country Parties of the region may take account, *inter alia*, of the following thematic issues in developing their national strategies for action to combat desertification and/or mitigate the effects of drought, pursuant to article 5 of the Convention:

(a) increasing capacities, education and public awareness, technical, scientific and technological cooperation and financial resources and mechanisms;

(b) eradicating poverty and improving the quality of human life;

(c) achieving food security and sustainable development and management of agricultural, livestock-rearing, forestry and multipurpose activities;

(d) sustainable management of natural resources, especially the rational management of drainage basins;

(e) sustainable management of natural resources in high-altitude areas;

(f) rational management and conservation of soil resources and exploitation and efficient use of water resources;

(g) formulation and application of emergency plans to mitigate the effects of drought;

(h) strengthening and/or establishing information, evaluation and follow-up and early warning systems in areas prone to desertification and drought, taking account of climatological, meteorological, hydrological, biological, soil, economic and social factors;

(i) developing, managing and efficiently using diverse sources of energy, including the promotion of alternative sources;

(j) conservation and sustainable use of biodiversity in accordance with the provisions of the Convention on Biological Diversity;

(k) consideration of demographic aspects related to desertification and drought; and

(l) establishing or strengthening institutional and legal frameworks permitting application of the Convention and aimed, *inter alia*, at decentralizing administrative structures and functions relating to desertification and drought, with the participation of affected communities and society in general.

Article 5. Technical, Scientific, and Technological Cooperation

In conformity with the Convention, in particular its articles 16 to 18, and on the basis of the coordinating mechanism provided for in article 7, affected country Parties of the region shall, individually or jointly:

(a) promote the strengthening of technical cooperation networks and national, subregional and regional information systems, as well as their integration, as appropriate, in worldwide sources of information;

(b) prepare an inventory of available technologies and know-how and promote their dissemination and use;

(c) promote the use of traditional technology, knowledge, know-how and practices pursuant to article 18, paragraph 2 (b), of the Convention;

(d) identify transfer of technology requirements; and

(e) promote the development, adaptation, adoption and transfer of relevant existing and new environmentally sound technologies.

Article 6. Financial Resources and Mechanisms

In conformity with the Convention, in particular its articles 20 and 21, on the basis of the coordinating mechanism provided for in article 7 and in accordance with their national development policies, affected country Parties of the region shall, individually or jointly:

(a) adopt measures to rationalize and strengthen mechanisms to supply funds through public and private investment with a view to achieving specific results in action to combat desertification and mitigate the effects of drought;

(b) identify international cooperation requirements in support of national efforts; and

(c) promote the participation of bilateral and/or multilateral financial cooperation institutions with a view to ensuring implementation of the Convention.

Article 7. Institutional Framework

1. In order to give effect to this Annex, affected country Parties of the region shall:

(a) establish and/or strengthen national focal points to coordinate action to combat desertification and/or mitigate the effects of drought; and

(b) set up a mechanism to coordinate the national focal points for the following purposes:

(i) exchanges of information and experience;

(ii) coordination of activities at the subregional and regional levels;

(iii) promotion of technical, scientific, technological and financial cooperation;

(iv) identification of external cooperation requirements; and

(v) follow-up and evaluation of the implementation of action programmes.

2. Affected country Parties of the region shall hold periodic coordination meetings and the Permanent Secretariat may, at their request, pursuant to article 23 of the Convention, facilitate the convocation of such coordination meetings, by:

(a) providing advice on the organization of effective coordination arrangements, drawing on experience from other such arrangements;

(b) providing information to relevant bilateral and multilateral agencies concerning coordination meetings, and encouraging their active involvement; and

(c) providing other information that may be relevant in establishing or improving coordination processes.

Annex IV — Regional Implementation Annex for the Northern Mediterranean

Article 1. Purpose

The purpose of this Annex is to provide guidelines and arrangements necessary for the effective implementation of the Convention in affected country Parties of the northern Mediterranean region in the light of its particular conditions.

Article 2. Particular Conditions of the Northern Mediterranean Region

The particular conditions of the northern Mediterranean region referred to in article 1 include:

(a) semi-arid climatic conditions affecting large areas, seasonal droughts, very high rainfall variability and sudden and high-intensity rainfall;

(b) poor and highly erodible soils, prone to develop surface crusts;

(c) uneven relief with steep slopes and very diversified landscapes;

(d) extensive forest coverage losses due to frequent wildfires;

(e) crisis conditions in traditional agriculture with associated land abandonment and deterioration of soil and water conservation structures;

(f) unsustainable exploitation of water resources leading to serious environmental damage, including chemical pollution, salinization and exhaustion of aquifers; and

(g) concentration of economic activity in coastal areas as a result of urban growth, industrial activities, tourism and irrigated agriculture.

Article 3. Strategic Planning Framework for Sustainable Development

1. National action programmes shall be a central and integral part of the strategic planning framework for sustainable development of the affected country Parties of the northern Mediterranean.

2. A consultative and participatory process, involving appropriate levels of government, local communities and non-governmental organizations, shall be undertaken to provide guidance on a strategy with flexible planning to allow maximum local participation, pursuant to article 10, paragraph 2 (f) of the Convention.

Article 4. Obligation to Prepare National Action Programmes and Timetable

Affected country Parties of the northern Mediterranean region shall prepare national action programmes and, as appropriate, subregional, regional or joint action programmes. The preparation of such programmes shall be finalized as soon as practicable.

Article 5. Preparation and Implementation of National Action Programmes

In preparing and implementing national action programmes pursuant to articles 9 and 10 of the Convention, each affected country Party of the region shall, as appropriate:

(a) designate appropriate bodies responsible for the preparation, coordination and implementation of its programme;

(b) involve affected populations, including local communities, in the elaboration, coordination and implementation of the programme through a locally driven consultative process, with the cooperation of local authorities and relevant non-governmental organizations;

(c) survey the state of the environment in affected areas to assess the causes and consequences of desertification and to determine priority areas for action;

(d) evaluate, with the participation of affected populations, past and current programmes in order to design a strategy and elaborate activities in the action programme;

(e) prepare technical and financial programmes based on the information gained through the activities in subparagraphs (a) to (d); and

(f) develop and utilize procedures and benchmarks for monitoring and evaluating the implementation of the programme.

Article 6. Content of National Action Programmes

Affected country Parties of the region may include, in their national action programmes, measures relating to:

(a) legislative, institutional and administrative areas;

(b) land use patterns, management of water resources, soil conservation, forestry, agricultural activities and pasture and range management;

(c) management and conservation of wildlife and other forms of biological diversity;

(d) protection against forest fires;

(e) promotion of alternative livelihoods; and

(f) research, training and public awareness.

Article 7. Subregional, Regional, and Joint Action Programmes

1. Affected country Parties of the region may, in accordance with article 11 of the Convention, prepare and implement subregional and/or regional action programmes in order to complement and increase the efficiency of national action programmes. Two or more affected country Parties of the region, may similarly agree to prepare a joint action programme between or among them.

2. The provisions of articles 5 and 6 shall apply *mutatis mutandis* to the preparation and implementation of subregional, regional and joint action programmes. In

addition, such programmes may include the conduct of research and development activities concerning selected ecosystems in affected areas.

3. In preparing and implementing subregional, regional or joint action programmes, affected country Parties of the region shall, as appropriate:

(a) identify, in cooperation with national institutions, national objectives relating to desertification which can better be met by such programmes and relevant activities which could be effectively carried out through them;

(b) evaluate the operational capacities and activities of relevant regional, subregional and national institutions; and

(c) assess existing programmes relating to desertification among Parties of the region and their relationship with national action programmes.

Article 8. Coordination of Subregional, Regional, and Joint Action Programmes

Affected country Parties preparing a subregional, regional or joint action programme may establish a coordination committee composed of representatives of each affected country Party concerned to review progress in combating desertification, harmonize national action programmes, make recommendations at the various stages of preparation and implementation of the subregional, regional or joint action programme, and act as a focal point for the promotion and coordination of technical cooperation pursuant to articles 16 to 19 of the Convention.

Article 9. Non-eligibility for Financial Assistance

In implementing national, subregional, regional and joint action programmes, affected developed country Parties of the region are not eligible to receive financial assistance under this Convention.

Article 10. Coordination with Other Subregions and Regions

Subregional, regional and joint action programmes in the northern Mediterranean region may be prepared and implemented in collaboration with those of other subregions or regions, particularly with those of the subregion of northern Africa.

Annex V — Regional Implementation Annex for Central and Eastern Europe

Article 1. Purpose

The purpose of this Annex is to provide guidelines and arrangements for the effective implementation of te Convention in affected country Parties of the Central and Eastern European region, in the light of its particular conditions.

Article 2. Particular Conditions of the Central and Eastern European Region

The particular conditions of the Central and Eastern European region referred to in article 1, which apply in varying degrees to the affected country Parties of the region, include:

(a) specific problems and challenges related to the current process of economic transition, including macroeconomic and financial problems and the need for strengthening the social and political framework for economic and market reforms;

(b) the variety of forms of land degradation in the different ecosystems of the region, including the effects of drought and the risks of desertification in regions prone to soil erosion caused by water and wind;

(c) crisis conditions in agriculture due, *inter alia*, to depletion of arable land, problems related to inappropriate irrigation systems and gradual deterioration of soil and water conservation structures;

(d) unsustainable exploitation of water resources leading to serious environmental damage, including chemical pollution, salinisation and exhaustion of aquifers;

(e) forest coverage losses due to climatic factors, consequences of air pollution and frequent wildfires;

(f) the use of unsustainable development practices in affected areas as a result of complex interactions among physical, biological, political, social and economic factors;

(g) the risks of growing economic hardships and deteriorating social conditions in areas affected by land degradation, desertification and drought;

(h) the need to review research objectives and the policy and legislative framework for the sustainable management of natural resources; and

(i) the opening up of the region to wider international cooperation and the pursuit of broad objectives of sustainable development.

Article 3. Action Programmes

1. National action programmes shall be an integral part of the policy framework for sustainable development and address in an appropriate manner the various forms of land degradation, desertification and drought affecting the Parties of the region.

2. A consultative and participatory process, involving appropriate levels of government, local communities and non-governmental organizations, shall be undertaken to provide guidance on a strategy with flexible planning to allow maximum

local participation, pursuant to article 10, paragraph 2(f), of the Convention. As appropriate, bilateral and multilateral cooperation agencies may be involved in this process at the request of the affected country Party concerned.

Article 4. Preparation and Implementation of National Action Programmes

In preparing and implementing national action programmes pursuant to articles 9 and 10 of the Convention, each affected country Party of the region shall, as appropriate:

(a) designate appropriate bodies responsible for the preparation, coordination and implementation of its programme;

(b) involve affected populations, including local communities, in the elaboration, coordination and implementation of the programme through a locally driven consultative process, with the cooperation of local authorities and relevant non-governmental organizations;

(c) survey the state of the environment in affected areas to assess the causes and consequences of desertification and to determine priority areas for action;

(d) evaluate, with the participation of affected populations, past and current programmes in order to design a strategy and elaborate actions in the action programme;

(e) prepare technical and financial programmes based on the information gained through the activities in subparagraphs (a) to (d); and

(f) develop and utilize procedures and benchmarks for monitoring and evaluating the implementation of the programme.

Article 5. Subregional, Regional, and Joint Action Programmes

1. Affected country Parties of the region, in accordance with articles 11 and 12 of the Convention, may prepare and implement subregional and/or regional action programmes in order to complement and increase the effectiveness and efficiency of national action programmes. Two or more affected country Parties of the region may similarly agree to prepare a joint action programme between or among them.

2. Such programmes may be prepared and implemented in collaboration with other Parties or regions. The objective of such collaboration would be to secure an enabling international environment and to facilitate financial and/or technical support or other forms of assistance to address more effectively desertification and drought issues at different levels.

3. The provisions of articles 3 and 4 shall apply, mutatis mutandis, to the preparation and implementation of subregional, regional and joint action programmes. In addition, such programmes may include the conduct of research and development activities concerning selected ecosystems in affected areas.

4. In preparing and implementing subregional, regional or joint action programmes, affected country Parties of the region shall, as appropriate:

(a) identify, in cooperation with national institutions, national objectives relating to desertification which can better be met by such programmes, and relevant activities, which could be effectively carried out through them;

(b) evaluate the operational capacities and activities of relevant regional, subregional and national institutions;

(c) assess existing programmes relating to desertification among Parties of the region and their relationship with national action programmes; and

(d) consider action for the coordination of subregional, regional and joint action programmes, including, as appropriate, the establishment of coordination committees composed of representatives of each affected country Party concerned to review progress in combating desertification, harmonize national action programmes, make recommendations at the various stages of preparation and implementation of the subregional, regional or joint action programmes, and act as focal points for the promotion and coordination of technical cooperation pursuant to articles 16 to 19 of the Convention.

Article 6. Technical, Scientific, and Technological Cooperation

In conformity with the objective and principles of the Convention, Parties of the region shall, individually or jointly:

(a) promote the strengthening of scientific and technical cooperation networks, of monitoring indicators and of information systems at all levels, as well as their integration, as appropriate, in worldwide systems of information; and

(b) promote the development, adaptation and transfer of relevant existing and new environmentally sound technologies within and outside the region.

Article 7. Financial Resources and Mechanisms

In conformity with the objective and principles of the Convention, affected country Parties of the region shall, individually or jointly:

(a) adopt measures to rationalize and strengthen mechanisms to supply funds through public and private investment with a view to achieving concrete results in action to combat land degradation and desertification and mitigate the effects of drought;

(b) identify international cooperation requirements in support of national efforts, thereby creating, in particular, an enabling environment for investments and encouraging active investment policies and an integrated approach to effectively combating desertification, including early identification of the problems caused by this process;

(c) seek the participation of bilateral and/or multilateral partners and financial cooperation institutions with a view to ensuring implementation of the Convention, including programme activities which take into account the specific needs of affected country Parties of the region; and

(d) assess the possible impact of article 2(a) on the implementation of articles 6, 13 and 20 and other related provisions of the Convention.

Article 8. Institutional Framework

1. In order to give effect to this Annex, Parties of the region shall:

(a) establish and/or strengthen national focal points to coordinate action to combat desertification and/or mitigate the effects of drought; and

(b) consider mechanisms to strengthen regional cooperation, as appropriate.

2. The Permanent Secretariat may, at the request of Parties of the region and pursuant to article 23 of the Convention, facilitate the convocation of coordination meetings in the region by:

(a) providing advice on the organization of effective coordination arrangements, drawing on experience from other such arrangements; and

(b) providing other information that may be relevant in establishing or improving coordination processes.

Bibliography

Adamo, Susana B. (2003). 'Vulnerable People in Fragile Lands: Migration and Desertification in the Drylands of Argentina — The Case of the Department of Jáchal'. Ph.D. thesis. University of Texas at Austin.

Adams, William M. (1991). *Green Development: Environment and Sustainability in the Third World*. Routledge, London.

African Union (2005). 'Draft Strategic Framework for Policy on Migration in Africa'. Third Ordinary Session of the African Union Labour and Social Affairs Commission, 18–23 April. Johannesburg.

Agarwal, Anil, Sunita Narain, and Anju Sharma, eds. (1999). *Green Politics: Global Environmental Negotiations I*. Centre for Science and Environment, New Delhi.

Ahmed, Nessim (2000). 'Economic, Social, and Cultural Causes and Consequences of Drought and Desertification'. International Fund for Agriculture Technical Advisory Division. Rome. <www.unccd.int/knowledge/INCDinfoSeg/partiii.php#para1> (December 2005).

Alam, S.M. and M.H. Naqvi (2003). 'Pakistan Agriculture: 2003'. *Pakistan & Gulf Economist* 19–25 May. <www.pakistaneconomist.com/database1/cover/c2003-28.asp> (December 2005).

Ambler, John (1999). 'Attacking Poverty While Improving the Environment: Towards Win-Win Policy Options'. Poverty and Environment Initiative of the United Nations Development Programme and the European Commission. <www.undp.org/pei/pdfs/PEIPhase1SummaryPaper2.pdf> (December 2005).

Andresen, Steinar and Jon Birger Skjaerseth (1999). 'Can International Environmental Secretariats Promote Effective Co-operation?' Paper presented at the United Nations University's International Conference on Synergies and Co-ordination between Multilateral Environmental Agreements, 14–16 July. Tokyo. <www.geic.or.jp/interlinkages/docs/Andresen.PDF> (December 2005).

Arab Republic of Egypt (2004). 'National Report for Combating Desertification'. Report prepared for the Third Session of the Committee for the Review of the Implementation, 2–11 May. Desert Research Center, Cairo. <www.unccd.int/cop/reports/africa/national/2004/egypt-eng.pdf> (December 2005).

Arbeitsgruppe Desertifikation (1998). 'The Combat against Desertification in Global Partnership: A Challenge for Civil Society'. Offprint.

Barbier, Edward B. (1998). 'The Economics of Land Degradation and Rural Poverty Linkages in Africa'. United Nations University Institute for Natural Resources in Africa Annual Lectures. <www.inra.unu.edu/documents/annual_lectures1998.htm> (December 2005).

Barlow, Mathew, Heidi Cullen, and Bradfield Lyon (2002). 'Drought in Central and Southwest Asia: La Nina, the Warm Pool, and Indian Ocean Precipitation'. *Journal of Climate* vol. 15, no. 7, pp. 697–700.

Barnett, Michael N. and Martha Finnemore (1999). 'The Politics, Power, and Pathologies of International Organizations'. *International Organization* vol. 53, no. 4, pp. 699–732.

Barnett, Michael N. and Martha Finnemore (2004). *Rules for the World: International Organizations in Global Politics*. Cornell University Press, Ithaca.

Barry, Roger G. and Richard J. Chorley (1998). *Atmosphere, Weather, and Climate*. 7th ed. Routledge, London.

Bassett, Charles and Joana Talafré (2003). 'Implementing the UNCCD: Towards a Recipe for Success'. *Review of European Community and International Environmental Law* vol. 12, no. 2, pp. 133–139.

Bauer, Steffen (2006, in press). 'Does Bureaucracy Really Matter? The Authority of Intergovernmental Treaty Secretariats in Global Environmental Politics'. *Global Environmental Politics* vol. 6, no. 1.

Bauer, Steffen (forthcoming). 'The Ozone Secretariat: Administering the Vienna Convention and the Montreal Protocol on Substances That Deplete the Ozone Layer'. In F. Biermann and B. Siebenhüner, eds., *Managers of Global Change: Effectiveness and Learning of International Environmental Organizations*.

Bauer, Steffen (forthcoming). 'The Secretariat of the United Nations Convention to Combat Desertification'. In F. Biermann and B. Siebenhüner, eds., *Managers of Global Change: Effectiveness and Learning of International Environmental Organizations*,

Beaumont, Peter (1993). *Drylands: Environmental Management and Development*. Routledge, London .

Betsill, Michele M. and Elisabeth Corell (2001a). 'NGO Influence in International Environmental Negotiations: A Framework for Analysis'. *Global Environmental Politics* vol. 1, no. 4, pp. 65–85.

Betsill, Michele M. and Elisabeth Corell (2001b). 'A Comparative Look at NGO Influence in International Environmental Negotiations: Desertification and Climate Change'. *Global Environmental Politics* vol. 1, no. 4, pp. 86–107.

Biermann, Frank and Steffen Bauer (2004). 'Assessing the Effectiveness of Intergovernmental Organisations in International Environmental Politics'. *Global Environmental Change* vol. 14, no. 2, pp. 189–193.

Biermann, Frank and Steffen Bauer (2005). 'Managers of Global Governance: Assessing and Explaining the Influence of International Bureaucracies'. Global Governance Working Paper No. 15. Amsterdam, Berlin, Oldenburg, Potsdam. <glogov.org > working papers> (December 2005).

Biermann, Frank and Steffen Bauer, eds. (2005). *A World Environmental Organization: Solution or Threat to Effective International Environmental Governance?* Ashgate, Aldershot.

Bilsborrow, R.E. (1992). 'Rural Poverty, Migration, and Environment in Developing Countries: Three Case Studies'. Country Economics Department Paper No. 1017. World Bank, Washington DC.

Boserup, Ester (1965). *The Conditions of Agricultural Growth: The Economics of Agrarian Change under Population Pressure*. Aldine Publishing, Chicago.

Breitmeier, Helmut and Volker Rittberger (1998). 'Environmental NGOs in an Emerging Global Society'. Tübinger Arbeitspapiere zur Internationalen Politik und Friedensforschung No. 32. <www.uni-tuebingen.de/uni/spi/taps/tap32.htm> (December 2005).

Briske, David D., Samuel D. Fuhlendorf, and Fred E. Smeins (2003). 'Vegetation Dynamics on Rangelands: A Critique of the Current Paradigm'. *Journal of Applied Ecology* vol. 40, no. 4, pp. 601–614.

Broad, Robin (1994). 'The Poor and the Environment: Friends or Foes?' *World Development* vol. 22, no. 6, pp. 811–822.

Brown, David, Kathrin Schreckenberg, Gill Shepherd, et al. (2003). 'Good Governance: What Can We Learn from the Forest Sector?' Forest Policy and Environment Programme, Overseas Development Institute. <www.odifpeg.org.uk/activities/environmental_governance/IN1/good%20gov_group_paper.pdf> (December 2005).

Bruins, Hendrick J. and Pedro R. Berliner (1998). 'Bioclimatic Aridity, Climatic Variability, Drought, and Desertification: Definitions and Management Options'. In H.J. Bruins and H. Lithwick, eds., *The Arid Frontier: Interactive Management of Environment and Development*, pp. 97–116. Kluwer Academic Publishers, Dordrecht.

Burkina Faso (2004). 'Troisième rapport national du Burkina Faso sur la mise en oeuvre de la Convention des Nations Unies sur la lutte contre la désertification'. Report prepared for the Third Session of the Committee for the Review of the Implementation, 2–11 May. Ouagadougou. <www.unccd.int/cop/reports/africa/national/2004/burkina_faso-fre.pdf> (December 2005).

Burns, W.C. (1995). 'The International Convention to Combat Desertification: Drawing a Line in the Sand?' *Michigan Journal of International Law* vol. 16, no. 3, pp. 831–882.

Burundi (2004). 'Troisième rapport national sur la mise en oeuvre de la Convention des Nations Unies sur la lutte contre la désertification'. Report prepared for the Third Session of the Committee for the Review of the Implementation, 2–11 May. Ministere de l'Aménagement du territoire, de l'Environnement et du Tourisme, Bujumbura. <www.unccd.int/cop/reports/africa/national/2004/burundi-fre.pdf> (December 2005).

Busch, Per-Olof (forthcoming). 'Making a Living in a Strait-Jacket: The Secretariat to the United Nations Framework Convention on Climate Change'. In F. Biermann and B. Siebenhüner, eds., *Managers of Global Change: Effectiveness and Learning of International Environmental Organizations*,

Campbell, D.J. (1984). 'Response to Drought among Farmers and Herders in Southern Kajiado District, Kenya'. *Human Ecology* vol. 12, no. 1, pp. 35–64.

Canada (2005). 'Canada's International Policy Statement: A Role of Pride and influence in the World — Development'. April. <www.acdi-cida.gc.ca/IPS> (December 2005).

Canadian International Development Agency (1998). 'Update from Dakar: Second Session of the Conference of the Parties (COP2) to the United Nations Convention to Cmbat Desertification'. <www.acdi-cida.gc.ca/cida_ind.nsf/0/a7ba07ff629071 c885256761006030e3?OpenDocument> (December 2005).

Castles, Stephen (2001). 'Environmental Change and Forced Migration'. Paper prepared for the Westmorland General Meeting 'Preparing for Peace' Initiative, 6 December. <www.preparingforpeace.org/castles__environmental_change_and_ forced_migration.htm> (December 2005).

Charney, Julie, Peter H. Stone, and William J. Quirk (1975). 'Drought in the Sahara: A Biogeophysical Feedback Mechanism'. *Science* vol. 187, no. 4175, pp. 434–435.

Chasek, Pamela S. (1997). 'The Convention to Combat Desertification: Lessons Learned for Sustainable Development'. *Journal of Environment and Development* vol. 6, no. 2, pp. 147–169.

Chasek, Pamela S. and Elisabeth Corell (2002). 'Addressing Desertification at the International Level: The Institutional System'. In J.F. Reynolds and D.M. Stafford Smith, eds., *Global Desertification: Do Humans Cause Deserts*. Dahlem University Press, Berlin.

Clark, John D. (2003). *Worlds Apart: Civil Society and the Battle for Ethical Globalization*. Kumarian Press, Bloomfield CT.

Clarke, John Innes and Daniel Noin (1997). *Population and Environment in Arid Regions*. Parthenon Publishing, New York.

Cleaver, Kevin M. and Götz A. Schreiber (1994). *Reversing the Spiral: The Population, Agriculture, and Environment Nexus in Sub-Saharan Africa*. World Bank, Washington DC.

Clements, Frederic E. (1916). *Plant Succession: An Analysis of the Development of Vegetation*. Carnegie Institution of Washington, Washington DC.

Comores (2004). 'Rapport national sur les mesures prises en vue de lutter contre la désertification dans le cadre de la Convention des Nations Unies sur la lutte contre la désertification'. Report prepared for the Third Session of the Committee for the Review of the Implementation, 2–11 May. <www.unccd.int/cop/reports/ africa/national/2004/comoros-fre.pdf> (December 2005).

Conseil National pour l'Environnement et le Développement Durable (2003). 'Revue des projets et programmes en cours d'exécution en vue de l'évaluation des ressources mobilisées pouvant entrer dans le cadre de la mise en oeuvre du Programme d'Action National de lutte contre la désertification au Burkina Faso'. Ouagadougou.

Conseil National pour l'Environnement et le Développement Durable (2003). 'Funding Orientations in Desertification Control'. Ouagadougou.

Convention on Biological Diversity (2003). 'The Impact of Trade Liberalization on Agricultural Biological Diversity: A Synthesis of Assessment Frameworks'. UNEP/CBD/COP/7/INF/15, 18 December. <www.biodiv.org/doc/meetings/cop/ cop-07/information/cop-07-inf-15-en.pdf> (2005 December).

Corell, Elisabeth (1998). 'North-South Financial Tensions: Desertification after UNGASS'. *Environmental Politics* vol. 7, no. 1, pp. 222–226.

Corell, Elisabeth (1999). 'The Negotiable Desert: Expert Knowledge in the Negotiations of the Convention to Combat Desertification'. Ph.D. thesis. Linköping University, Linköping, Sweden.

Corell, Elisabeth and Michele M. Betsill (2001). 'A Comparative Look at NGO Influence in Intenrational Environmental Negotiations: Desertification and Climate Change'. *Global Environmental Politics* vol. 1, no. 4, pp. 86–107.

Corell, Elisabeth (2003). 'Dryland Degradation — Africa's Main Environmental Challenge: International Activities from the 1970s to the 1990s and the Future of the United Nations Convention to Combat Desertification'. In B. Chaytor and K. Gray, eds., *International Environmental Law and Policy in Africa*. Kluwer Academic Publishers, Dordrecht.

Cornet, Antoine (2002). 'La désertification à la croisée du développement et de l'environnement: un problème qui nous concerne'. In R. Barbault, A. Cornet, J. Jouzel et al., eds., *Johannesburg: Sommet Mondial du Développement Durable 2002 — Quels enjoux? Quelle contribution des scientifiques?*, pp. 93–129. Ministère des Affaires étrangères, Paris. <www.adpf.asso.fr/adpf-publi/folio/johannesburg/pdf/joburg-desert-4.pdf> (December 2005).

Danish, K.W. (1995). 'International Environmental Law and the "Bottom-Up" Approach: A Review of the Desertification Convention'. *Indiana Journal of Global Legal Studies* vol. 3, no. 1, pp. 133–176.

Darkoh, M.B.K. (1998). 'The Nature, Causes, and Consequences of Desertification in the Drylands of Africa'. *Land Degradation and Devopment* vol. 9, no. 1, pp. 1–20.

de Haas, Hein (1998). 'Socio-economic Transformations and Oasis Agriculture in Southern Morocco'. In P.M. Blaikie and L. de Haan, eds., *Looking at Maps in the Dark: Directions for Geographical Research in Land Management and Sustainable Development in Rural and Urban Environments of the Third World*. Royal Dutch Geographical Society and Faculty of Environmental Sciences, University of Amsterdam, Utrecht and Amsterdam.

de Haas, Hein (2003). 'Migration and Development in Southern Morocco: The Disparate Socio-economic Impacts of Out-Miration on the Todgha Oasis Valley'. Ph.D. thesis thesis. University of Amsterdam.

De Waal, Alexander (1989). *Famine that Kills: Darfur, Sudan, 1984–1985*. Clarendon Press, Oxford.

Demeke, Bayou (2002). 'Is Globalization Bad for the Environment? International Trade and Land Degradation in Developing Countries: The Case of Small Open Economy'. Paper presented at the European Economic Association convention, 22–24 August. Venice. <www.eea-esem.com/eea-esem/eea2002/prog/viewpaper.asp?pid=1575> (December 2005).

Department for International Development (United Kingdom) (2002). 'Better Livelihoods for Poor People: The Role of Agriculture'. Consultation Document. <www.dfid.gov.uk/pubs/files/agricultureconsult.pdf> (December 2005).

Department for International Development (United Kingdom) (2003). 'Agriculture and Poverty Reduction: Unlocking the Potential'. DFID Policy Paper, December. <www.dfid.gov.uk/pubs/files/agripovertyreduction.pdf> (December 2005).

Department for International Development (United Kingdom) (2004). 'Agriculture, Hunger, and Food Security'. Working Paper 7. <dfid-agriculture-consultation.nri.org/summaries/wp7.pdf> (December 2005).

Department for International Development (United Kingdom) (2004). 'Agriculture, Growth and Poverty Reduction'. Working Paper 1, October. <dfid-agriculture-consultation.nri.org/summaries/wp1.pdf> (December 2005).
Department for International Development (United Kingdom) (2004). 'Making Agricultural Markets Work for the Poor'. Working Paper 2, September. <dfid-agriculture-consultation.nri.org/summaries/dfidwp2.pdf> (December 2005).
'Desertificao do Nordeste e Tema de Debate'. (1991). *Jornal de Brasil*, 1 April.
Development Co-operation Directorate (2005). 'The DAC Journal: Development Co-operation Report 2004'. Organisation for Economic Co-operation and Development. Paris.
Devereux, Stephen (2000). 'Food Insecurity in Ethiopia'. Discussion paper prepared for the Department for International Development. <www.ids.ac.uk/ids/pvty/pdf-files/FoodSecEthiopia4.pdf> (December 2005).
Dia, I. (1992). 'Les migrations comme stratégie des unités de production rurale: Une étude de cas au Sénégal'. In A. Blokland and F. van der Staay, eds., *Sustainable Development in Semi-Arid Sub-Saharan Africa*. Ministry of Foreign Affairs, Netherlands, The Hague.
Diatta, Marie Angelique and Ndiaga Mbow (1999). 'Releasing the Development Potential of Return Migrants: The Case of Senegal'. *International Migration* vol. 37, no. 1, pp. 243–266.
Dobie, Philip (2001). 'Poverty and the Drylands'. United Nations Development Programme. <www.undp.org/seed/unso/gdp/docs/Poverty-and-the-Drylands.doc> (December 2005).
Dregne, Harold E. (1983). *Desertification of Arid Lands*. Harwood Academic Publishers, Chur, Switzerland.
Duraiappah, Anantha K. (1998). 'Poverty and Environmental Degradation: A Review and Analysis of the Nexus'. *World Development* vol. 26, no. 12, pp. 2169–2179.
Dyksteruis (1949). 'Condition and Management of Range Land Based on Quantitative Ecology'. *Journal of Range Management* vol. 2, pp. 104–115.
Eklundh, Lars and Lennart Olsson (2003). 'Vegetation Index Trends for the African Sahel 1982–1999'. *Geophysical Research Letters* vol. 30, no. 8, pp. 13-11–13-14.
El-Hinnawi, Essam (1985). 'Environmental Refugees'. United Nations Environmental Programme, Nairobi.
Ellis, Frank (2000). *Rural Livelihoods and Diversity in Developing Countries*. Oxford University Press, Oxford.
Environment and Development Action in the Third World (1999). 'Proposal for the Organization of a RIOD General Meeting'. ENDA TM. <www.enda.sn/energie/desertif/gm.htm> (December 2005).
Environmental Liaison Centre International (1994). 'Réseau International d'ONG sur la Désertification — RIOD: General Framework for Operation'. NGO Planning Meeting. Nairobi.
Environmental Liaison Centre International (1994). 'The NGO Action Plan on Desertification: Facilitating the Voice of the Grassroots'. Réseau international des ONG pour la desertification.

Eritrea (2002). 'The National Action Programme for Eritrea to Combat Desertficiation and Mitigate the Effects of Drought (NAP)'. Ministry of Agriculture, Asmara. <www.unccd.int/actionprogrammes/africa/national/2002/eritrea-eng.pdf> (December 2005).
Escher, Anton (1994). 'Migrant Network: An Answer to Contain Desertification; A Case Study of Southern Syria (Gabal al-Arab)'. Paper presented at the International Symposium on Desertification and Migrations, 9–11 February. Almeria, Spain.
Evers, Yvette D. (1996). *The Social Dimensions of Desertification: Annotated Bibliography and Literature Review*. United Nations Environment Programme, Nairobi.
Ezra, Markos (2001). 'Ecological Degradation, Rural Poverty, and Migration in Ethiopia: A Contextual Analysis'. Policy Research Division Working Paper No. 149. Population Council, New York. <www.popcouncil.org/publications/wp/prd/149.html> (December 2005).
Fairhead, James and Melissa Leach (1996). *Misreading the African Landscape: Society and Ecology in the Forest-Savanna Mosaic*. Cambridge University Press, Cambridge.
Farer, Tom (1995). 'How the International System copes with Involuntary Migration: Norms, Institutions, and State Practice'. In M.S. Teitelbaum and M. Weiner, eds., *Threatened Peoples, Threatened Borders: World Migration and U.S. Policy*. W.W. Norton, New York.
Federal Democratic Republic of Ethiopia (1999). 'National Report on the Implementation of the United Nations Convention to Combat Desertification'. Environmental Protection Authority, Addis Ababa. <www.unccd.int/cop/reports/africa/national/1999/ethiopia-eng.pdf> (December 2005).
Federal Democratic Republic of Ethiopia (2004). 'The 3rd National Report on the Implementation of the UNCCD/NAP in Ethiopia'. Report prepared for the Third Session of the Committee for the Review of the Implementation, 2–11 May. Environmental Protection Authority, Addis Ababa. <www.unccd.int/cop/reports/africa/national/2004/ethiopia-eng.pdf> (December 2005).
Findley, Sally (1994). 'Does Drought Increase Migration? A Study of Migration from Rural Mali during the 1983–1985 Drought'. *International Migration Review* vol. 28, no. 3, pp. 539–553.
Finland (2002). 'Report by Finland'. Report prepared for the First Session of the Committee for the Review of the Implementation, 11–22 November. Helsinki. <www.unccd.int/cop/reports/developed/2002/finland-eng.pdf> (December 2005).
Finland (2004). 'National Report by Finland for the Committee for the Review of Implementation of the United Nations Convention to Combat Desertification'. Report prepared for the Third Session of the Committee for the Review of the Implementation, 2–11 May. Ministry for Foreign Affairs, Helsinki. <www.unccd.int/cop/reports/africa/developed/2004/finland-eng.pdf> (December 2005).
Folland, C., J. Owen, M.N. Ward, et al. (1991). 'Prediction of Seasonal Rainfall in the Sahel Region Using Empirical and Dynamical Methods'. *Journal of Forecasting* vol. 10, no. 1-2, pp. 21–56.

Food and Agriculture Organization (2000). 'State of Food Insecurity in the World'. Rome. <www.fao.org/sof/sofi/index_en.htm> (December 2005).

Food and Agriculture Organization (2000). 'Assessment of the World Food Security Situation'. 26th Session of the Committee on World Food Security, 18–21 September. Rome. <www.fao.org/docrep/meeting/x7921e.htm> (December 2005).

Food and Agriculture Organization (2003). 'Voluntary Guidelines to Support the Progressive Realization of the Right to Adequate Food in the Context of National Food Security'. 27–29 October. Rome. <www.fao.org/DOCREP/MEETING/007/J0492E.HTM> (December 2005).

Food and Agriculture Organization (2004). 'The State of Agricultural Commodity Markets'. Rome. <ftp://ftp.fao.org/docrep/fao/007/y5419e/y5419e00.pdf> (December 2005).

Fornos, Werner (1993). 'Population Distribution and Migration'. Proceedings for United Nations Expert Group Meeting on Population Distribution and Migration, 18–22 January 1993, Santa Cruz, Bolivia.

Forsyth, Tim, Melissa Leach, and Ian Scoones (1998). 'Poverty and Environment: Priorities for Research and Policy'. Prepared for the United Nations Developoment Programme and European Commission, September. <www.eldis.org/fulltext/povenv.pdf> (December 2005).

Gadgil, Madhav (undated). 'Desertification'. <ces.iisc.ernet.in/hpg/cesmg/2-6.html> (December 2005).

Georgia (2002). 'Second National Report of Georgia on the Implementation of the United Nations Convention to Combat Desertification'. Report prepared for the First Session of the Committee for the Review of the Implementation, 11–22 November. Ministry of Environment and Natural Resources Protection, Tblisi. <www.unccd.int/cop/reports/centraleu/national/2002/georgia-eng.pdf> (December 2005).

Glantz, Michael H. and Nicolai Orlovsky (1983). 'Desertification: A Review of the Concept'. *Desertification Control Bulletin* vol. 9, pp. 15–22.

Glantz, Michael H. (1987). 'Drought and Economic Development in Sub-Saharan Africa'. In M.H. Glantz, ed., *Drought and Hunger in Africa: Denying Famine a Future*. Cambridge University Press, Cambridge.

Glazovsky, Nikita and Alexander Shestakov (1994). 'Environmental Migration Caused by Desertification in Central Asia and Russia'. Paper presented at the International Symposium on Desertification and Migrations, 9–11 February. Almeria, Spain. <geographytsu.freehomepage.com/CIS%20env%20migration.htm> (December 2005).

Gliese, Jürgen (2000). 'COP 4 NGO Opening Statement'. 11 December 2000. <www.forumue.de > Positionspapiere> (December 2005).

Global Environmental Facility (2002). 'Focusing on the Global Environment: The First Decade of the GEF'. Second Overall Performance Study (OPS2). <www.gefweb.org/1Full_Report-FINAL-2-26-02.pdf> (December 2005).

Global Environmental Facility (2003). 'Operational Program on Sustainable Land Management'. Operational Program No. 15. <www.gefweb.org/Operational_ Policies/Operational_Programs/OP_15_English.pdf> (December 2005).
Global Mechanism and Réseau International d'ONG sur la Désertification (1999). 'Community Exchange and Training Programme'. Rome.
Global Mechanism (2005). 'SADC RIOD Community Exchange and Training Programme'. <www.gm-unccd.org/FIELD/Multi/GM/FR_SADC.htm>(December 2005).
Goria, Alessandra (1998). 'Desertification and Migration in the Mediterranean: An Analytical Framework'. Fondazione Eni Enrico Mattei, Milan.
Gosline, Anna (2005). 'Early Warnings of Niger Famine Disregarded'. *New Scientist* 25 July. <www.newscientist.com/article.ns?id=dn7725> (December 2005).
Griffin, Keith, ed. (2003). *Poverty Reduction in Mongolia*. Asia Pacific Press, Canberra.
GTZ (2005). 'Fact Sheet Desertification: Africa'. Convention Project to Combat Desertification. <www2.gtz.de/desert/download/basicfactsheets/factsheet_africa.pdf> (December 2005).
Haas, Peter M., Robert O. Keohane, and Marc A. Levy, eds. (1993). *Institutions for the Earth: Sources of Effective International Environmental Protection*. MIT Press, Cambridge MA.
Hagman, Gunnar (1985). *Prevention Better than Cure: Report on Human and Environmental Disasters in the Third World*. 3rd ed. Red Cross, Stockholm.
Hardin, Garrett (1968). 'Tragedy of the Commons'. *Science* vol. 162, no. 3859, pp. 1243–1248.
Harou, Patrice A. (2002). 'What Is the Role of Markets in Altering the Sensitivity of Arid Land Systems to Perturbation?' In J.F. Reynolds and D.M. Stafford Smith, eds., *Global Desertification: Do Humans Cause Deserts?*, vol. 253–274. Dahlem University Press, Berlin.
Hashemite Kingdom of Jordan (2002). 'The Hashemite Kingdom of Jordan National Report on the UNCCD Impelementation'. Report prepared for the First Session of the Committee for the Review of the Implementation, 11–22 November. General Corporation for the Environment Protection, Amman. <www.unccd.int/cop/reports/asia/national/2002/jordan-eng.pdf> (December 2005).
Hazarika, Sanjoy (1993). 'Bangladesh and Assam: Land Pressures, Migration, and Ethnic Conflict'. Occasional Paper No. 3. Project on Environmental Change and Acute Conflict.
Held, David, Anthony G. McGrew, David Goldblatt, et al. (1999). *Global Transformations:Politics, Economics, and Culture*. Polity Press, Oxford.
Hill, Tony (2004). 'Three Generations of UN-Civil Society Relations: A Quick Sketch'. UN Non-Governmental Liaison Service. <www.globalpolicy.org/ngos/ngo-un/gen/2004/0404generation.htm> (December 2005).
Hoben, A. (1996). 'The Cultural Construction of Environment Policy: Paradigms and Politics in Ethiopia'. In M. Leach and R. Mearns, eds., *The Lie of the Land:*

Challenging Received Wisdom on the African Environment. International African Institute and Heinemann, Oxford.

Holling, C.S. (1973). 'Resilience and Stability of Ecological Systems'. *Annual Review of Ecology and Systematics* vol. 4, pp. 1–23.

Holmén, Hans and Magnus Jirström (1997). 'Strengthening NGO Networking for Sustainable Development in Drylands'. Report to the United Nations Development Programme and the United Nations Office for Project Services.

Homer-Dixon, Thomas, Jeffrey H. Boutwell, and George Rathjens (1993). 'Environmental Change and Violent Conflict'. *Scientific American* vol. 268, no. 2, pp. 4, 38–45.

Hugo, Graeme (1995). 'Environmental Concerns and International Migration'. *International Migration Review* vol. 30, no. 1, pp. 105–142.

Hulme, Mike and Mick Kelly (1993). 'Exploring the Links between Desertification and Climate Change'. *Environment* vol. 35, no. 6, pp. 6–11, 39–45.

Hulme, Mike (2001). 'Climatic Perspectives on Sahelian Desiccation: 1973–1998'. *Global Environmental Change* vol. 11, pp. 19–29.

Hulme, Mike, Ruth Doherty, Todd Ngara, et al. (2001). 'African Climate Change: 1900–2100'. *Climate Research* vol. 17, no. 2, pp. 145–168.

Illius, A.W. and T.G. O'Connor (1999). 'On the Relevance of Nonequilibrium Concepts to Arid and Semiarid Grazing Systems'. *Ecological Applications* vol. 9, no. 3, pp. 798–813.

Inter-Agency Standing Committee (1994). 'Working Paper on the Definition of Complex Emergency'. Geneva.

International Federation of Red Cross and Red Crescent Societies (2003). 'World Disasters Report: Focus on Ethics in Aid'. Geneva. <www.ifrc.org/publicat/wdr2003> (December 2005).

International Fund for Agricultural Development (2001). 'Rural Poverty Report 2001: The Challenge of Ending Rural Poverty'. <www.ifad.org/poverty> (December 2005).

International Fund for Agricultural Development (2002). 'Theme Paper for the Twenty-Fifth Session of the Governing Council: Financing Development — The Rural Dimension'. GC 25/L.11, 19–20 February. Rome. <www.ifad.org/events/gc/25/e/GC-25-L-11.pdf> (December 2005).

International Institute for Environment and Development (1995). 'The Desertification Convention: The Strategic Agenda for the EU'. EC Aid and Sustainable Development Briefing Paper No. 4. <europa.eu.int/comm/development/body/theme/environment/env_theme/land_resources_desertification/documents/04.htm> (December 2005).

International Institute for Sustainable Development (1993). 'A Brief History of the INCD'. *Earth Negotiations Bulletin* vol. 4, no. 45. <www.iisd.ca/vol04/0445000e.html> (December 2005).

International Institute for Sustainable Development (1993). 'Summary of the First Session of the INCD: 24 May–3 June 1993'. *Earth Negotiations Bulletin* vol. 4, no. 11. <www.iisd.ca/vol04/0411000e.html> (December 2005).

International Institute for Sustainable Development (1993). 'Summary of the Second Session of the INC for the International Convention to Combat Desertification: 13–24 September 1993'. *Earth Negotiations Bulletin* vol. 4, no. 22. <www.iisd.ca/vol04/0422000e.html> (December 2005).

International Institute for Sustainable Development (1993). 'Elaboration of an International Convention to Combat Desertification'. *Earth Negotiations Bulletin* vol. 4, no. 8. <www.iisd.ca/vol04/0408001e.html> (December 2005).

International Institute for Sustainable Development (1994). 'Summary of the Third Session of the INC for the Elaboration of an International Convention to Combat Desertification: 17–28 January 1994'. *Earth Negotiations Bulletin* vol. 4, no. 34. <www.iisd.ca/vol04/0434000e.html> (December 2005).

International Institute for Sustainable Development (1994). 'Summary of the Sixth Session of the INC for the Elaboration of an International Convention to Combat Desertification: 9–18 January 1995'. *Earth Negotiations Bulletin* vol. 4, no. 65. <www.iisd.ca/vol04/0465000e.html> (December 2005).

International Institute for Sustainable Development (1994). 'Summary of the Fifth Session of the INC for the Elaboration of an International Convention to Combat Desertification: 6–17 June 1994'. *Earth Negotiations Bulletin* vol. 4, no. 55. <www.iisd.ca/vol04/0455000e.html> (December 2005).

International Institute for Sustainable Development (1994). 'Summary of the Fourth Session of the INC for the Elaboration of an International Convention to Combat Desertification: 21–31 March 1994'. *Earth Negotiations Bulletin* vol. 4, no. 44. <www.iisd.ca/vol04/0444000e.html> (December 2005).

International Institute for Sustainable Development (1995). 'Summary of the Seventh Session of the INC for the Convention to Combat Desertification: 7–17 August 1995'. *Earth Negotiations Bulletin* vol. 4, no. 75. <www.iisd.ca/vol04/0475000e.html> (December 2005).

International Institute for Sustainable Development (1996). 'Summary of the Ninth Session of the INC for the Convention to Combat Desertification: 3–13 September 1996'. *Earth Negotiations Bulletin* vol. 4, no. 95. <www.iisd.ca/vol04/0495000e.html> (December 2005).

International Institute for Sustainable Development (1996). 'Summary of the Eighth Session of the INC for the Convention to Combat Desertification: 5–15 February'. *Earth Negotiations Bulletin* vol. 4, no. 86. <www.iisd.ca/vol04/0486000e.html> (December 2005).

International Institute for Sustainable Development (1997). 'Summary of the Resumed Tenth Session of the INC for the Convention to Combat Desertification: 18–22 August 1997'. *Earth Negotiations Bulletin* vol. 4, no. 106. <www.iisd.ca/vol04/enb04106e.html> (December 2005).

International Institute for Sustainable Development (1997). 'Tenth Session of the INC for the Convention to Combat Desertification: 6–17 January 1997'. *Earth Negotiations Bulletin* vol. 4, no. 105. <www.iisd.ca/vol04/04105000e.html> (December 2005).

International Institute for Sustainable Development (1998). 'CCD COP-2 Highlights'. *Earth Negotiations Bulletin* vol. 4, no. 121. <www.iisd.ca/vol04/enb04121e.html> (December 2005).

International Institute for Sustainable Development (2002). 'Summary of the First Session of the Committee for the Review of the Implementation of the Convention to Combat Desertification: 11–22 November 2002'. *Earth Negotiations Bulletin* vol. 4, no. 162. <www.iisd.ca/vol04/enb04162e.html> (December 2005).

International Institute for Sustainable Development (2003). 'Summary of the Sixth Conference of the Parties to the Convention to Combat Desertification: 25 August–6 September 2003'. *Earth Negotiations Bulletin* vol. 4, no. 173. <www.iisd.ca/vol04/enb04173e.html> (December 2005).

International Institute for Sustainable Development (2005). 'Summary of the Third Session of the Committee for the Review of hte Implementation of the Convention to Combat Desertification: 2–11 May 2005'. *Earth Negotiations Bulletin* vol. 4, no. 175. <www.iisd.ca/vol04/enb04175e.html> (December 2005).

International Organization for Migration and Refugee Policy Group (1992). 'Migration and the Environment'. Geneva.

International Organization for Migration (2000). 'IOM Migration Policy Framework for Sub-Saharan Africa'. MC/INF/244, 17 November. Geneva. <www.iom.int/en/PDF_Files/other/policysubsaharan.pdf> (December 2005).

International Organization for Migration (2005). 'World Migration 2005: Costs and Benefits of International Migration'. Geneva.

International Symposium on Desertification and Migrations (1991). 'The Almeria Statement on Desertification and Migration'. 9–11 February. Almeria, Spain. <www.unccd.int/regional/northmed/meetings/others/1994AlmeriaSpain.pdf> (December 2005).

Jacobson, Jodi (1988). 'Environmental Refugees: A Yardstick of Habitability'. World Watch Paper No. 86. World Watch Institute, Washington DC.

Jodha, N.S. (1986). 'Common Property Resources and Rural Poor in Dry Regions of India'. *Economic and Political Weekly* vol. 21, no. 27, pp. 169–181.

Kassas, Mohammed (1995). 'Negotiations for the International Convention on Desertification'. *International Environmental Affairs* vol. 7, no. 2, pp. 176–186.

Kates, Robert W. (1991). 'Hunger, Poverty, and the Human Environment'. DS-9. Center for Advanced Study of International Development.

Katyal, Jagdish C. and Paul L.G. Vlek (2000). 'Desertification: Concept, Causes, and Amelioration'. ZEF Discussion Papers on Development Policy No. 33. Center for Development Research (ZEF), Bonn.

Keane, David (2004). 'The Environmental Causes and Consequences Migration: A Search for Meaning of "Environmental Refugees"'. *Georgetown International Environmental Law Review* vol. 16, no. 2, pp. 209–214.

Keane, John (2003). *Global Civil Society?* Cambridge University Press, Cambridge.

Kimball, B.A., J.R. Mauney, F.S. Nakayama, et al. (1993). 'Effects of Increasing Atmospheric CO_2 on Vegetation'. *Plant Ecology* vol. 104-105, no. 1, pp. 65–75.

Kingdom of Belgium (2002). '2002 Report to the United Nations Convention to Combat Desertification (UNCCD)'. Report prepared for the First Session of the Committee for the Review of the Implementation, 11–22 November. Ministry of Foreign Affairs, Brussels. <www.unccd.int/cop/reports/developed/2002/belgium-eng.pdf> (December 2005).

Kirton, John J. et al. (1999). *Assessing the Environmental Effects of the North American Free Trade Agreement (NAFTA): Final Analytic Framework and Methodological Issues and Empirical Background*. Commission for Environmental Cooperation, Montreal.

Knabe, Friederike (2003). 'Evaluation of the RIOD Facilitation Unit during Solidarité Canada Sahel's Mandate December 2000–August 2003'. Solidarité Canada Sahel.

Knabe, Friederike (2004). 'Final Report on the FU-RIOD/SCS Transition Mandate'. Solidarité Canada Sahel.

Knabe, Friederike and Lene Poulsen (2004). 'Promoting Equality between Men and Women through Multilateral Environmental Agreements: How the UN Convention to Combat Desertification and Drought Has Promoted the Role of Women in Decision-Making'. Paper prepared for IUCN, 18 May. <www.iucn.org/themes/cem/documents/drylands/unccdgender.pdf> (December 2005).

Knerr, Béatrice (2004). 'Desertification and Human Migration'. In D. Werner, ed., *Biological Resources and Migration*, pp. 317–338. Springer, Berlin.

Köppen, W. (1931). *Die Klimate der Erde*. Walter de Gruyter, Berlin.

Kydd, Jonathan and Andrew Dorward (2001). 'The Washington Consensus on Poor Country Agriculture: Analysis, Prescription, and Institutional Gaps'. *Development Policy Review* vol. 19, no. 4, pp. 467–478.

Le Houerou, H.N. (1996). 'Climate Change, Drought, and Desertification'. *Journal of Arid Environments* vol. 34, no. 2, pp. 133–185.

Le Houerou, H.N. (2002). 'Man-Made Deserts: Desertization Processes and Threats'. *Arid Land Research and Management* vol. 16, pp. 1–36.

Leach, Melissa and Robin Mearns (1992). 'Poverty and Environment in Developing Countries: An Overview Study'. Final Report to the Economic and Social Research Council and Overseas Development Administration. Institute of Development Studies, Sussex.

Ledgar, Richard (1997). 'NGO Report on the Convention to Combat Desertification for Rio + 5'. Prepared on behalf of the RIOD Steering Committee, Réseau International d'ONG sur la Désertification, January. <www.ecouncil.ac.cr/rio/focus/report/english/riod.htm> (December 2005).

Leighton, Michelle (1997). 'Environmental Degradation and Migration: The U.S./Mexico Case Study'. December. Natural Heritage Institute.

Leighton, Michelle (1998). 'Environmental Degradation and Migration: The U.S.-Mexico Case Study'. *Environmental Change and Security Project Report* no. 4, pp. 61–67.

Leonard, H. Jeffrey (1989). *Environment and the Poor: Development Strategies for a Common Agenda*. Transaction Books, New Brunswick NJ.

Lin, Lin, Virginia McKenzie, Jenifer Piesse, et al. (2001). 'Agricultural Productivity and Poverty in Developing Countries'. Extension to DFID Report No. 7946. Department for International Development (United Kingdom), London.

Lines, Thomas (2004). 'Commodities Trade, Poverty Alleviation, and Sustainable Development: The Re-emerging Debate'. Paper presented at UNCTAD XI, 15 June. Sao Paolo.

Lohrmann, Reinhard (1994). 'The Need for Enhanced International Cooperation in Addressing Environmental Migration Issues'. Paper presented at the International Symposium on Desertification and Migrations, 9–11 February. Almeria, Spain.

Lohrmann, Reinhard (1996). 'Environmentally Induced Population Displacements and Environmental Impacts from Mass Migrations: Conference Report'. *International Migration* vol. 34, no. 2, pp. 335–339.

Long, Marybeth (2000). 'Grains of Truth: Science and the Evolution of International Desertification Policymaking'. Ph.D. thesis thesis. Massachussetts Institute of Technology, Cambridge MA.

Maarleveld, Maureen (2003). 'Social Environmental Learning for Sustaining Natural Resource Management: Theory, Practice, and Facilitation'. Ph.D. thesis. Wageningen University.

Madely, John (2000). 'Trade and Hunger: An Overview of Case studies on the Impact of Trade Liberalization on Food Security'. Forum Syd, Stockholm.

Mainguet, Monique (1991). *Desertification: Natural Background and Human Mismanagement*. Springer, Berlin.

Mainguet, Monique (1999). *Aridity: Droughts and Human Development* T. Reimer, trans. Springer, Berlin.

Maloney, Clarence (1991). 'Environmental Displacement of Population in India'. Field Staff Reports No. 14. UFSI Inc. and National Heritage Institute.

Mayrand, Karel, Marc Paquin, and Stéphanie Dionne (2005). 'From Boom to Dust? Agricultural Trade Liberalization, Poverty, and Desertification in Rural Drylands: The Role of the UNCCD'. Unisféra International Centre. <www.unisfera.org/IMG/pdf/Unisfera_-_From_Boom_to_Dust_-_Final.pdf> (December 2005).

McCulloch, Anna Knox, Suresh Babu, and Peter Hazell, eds. (1999). *Strategies for Poverty Alleviation and Sustainable Resource Management in the Fragile Lands of Sub-Saharan Africa*. International Food Policy Research Institute, Washington DC.

Meigs, Peveril (1953). 'World Distribution of Arid and Semi-Arid Homoclimates'. UNESCO, Paris.

Mellor, John (2000). 'Agricultural Growth, Rural Employment, and Poverty Reduction: Non-tradables, Public Expenditure, and Balanced Growth'. Proceedings for 'Poverty or Prosperity: Rural People in a Globalized Economy', World Bank Rural Week, 28–31 March.

Mensching, H. (1990). *Desertifikation: ein weltweites Problem der økologischen Verwüstung in den Trockengebieten der Erde*. Winnenschaftliche Buchgesellschaft, Darmstadt.

Middleton, Nick and David S.G. Thomas (1992). *World Atlas of Desertification*. Edward Arnold, London.

Miles, Edward L. (2002). *Environmental Regime Effectiveness: Confronting Theory with Evidence*. MIT Press, Cambridge MA.

Millennium Ecosystems Assessment (2005). 'Ecosystems and Human Well-Being: Desertification Synthesis'. World Resources Institute, Washington DC. <www.maweb.org//proxy/document.355.aspx> (December 2005).

Mitchell, Ronald B. (2002). 'International Environment'. In W. Carlsnaes, T. Risse-Kappen and B.A. Simmons, eds., *Handbook of International Relations*. Sage, London.

Morrissey, Oliver, Dirk Willem te Velde, Ian Gillson, et al. (2005). 'Sustainability Impact Assessment of Proposed WTO Negotiations: Mid-Term Report for the Agriculture Study'. Overseas Development Institute and Institute for Development Policy and Management, Manchester University, 31 January. <www.sia-trade.org/wto/Phase3B/Reports/MTRAgricultureJan05.pdf> (December 2005).

Mortimore, Michael J. (1988). *Adapting to Drought: Farmers, Famines, and Desertification in West Africa*. Cambridge University Press, Cambridge.

Mossige, Anne, Yonis Berkele, and Sina Maiga (2001). 'Participation of Civil Society in the National Action Programs of the United Nation's Convention to Combat Desertification: Synthesis of an Assessment in Ethiopia and Mali'. DCG Report 16A. Drylands Coordination Group. <www.drylands-group.org/Articles/305.html> (December 2005).

Movik, Synne, Sileshi Dejene, and Gry Synnevåg (2003). 'Poverty and Environmental Degradation in the Drylands: An Overview of Problems'. Noragric Working Paper No. 29. Agricultural University of Norway. <www.eldis.org/static/DOC13294.htm> (December 2005).

Müller, Peter M. (1993). 'Tragfähigkeitsveränderung durch Bevölkerungsverlust'. *Geographische Rundschau* vol. 45, no. 3, pp. 173–179.

Myers, Norman (1991). *Population, Resources, and the Environment: The Critical Challenges*. United Nations Population Fund, New York.

Myers, Norman (1997). 'Environmental Refugees'. *Population and Environment* vol. 19, no. 2, pp. 167–182.

Myneni, Ranga B., Sietse O. Los, and Compton J. Tucker (1996). 'Satellite-Based Identification of Linked Vegetation Index and Sea Surface Temperature Anomaly Areas from 1982 to 1990 for Africa, Australia, and South America'. *Geophysical Research Letters* vol. 23, pp. 729–732.

Nadal, Alejandro (2000). 'The Environmental and Social Impacts of Economic Liberalization on Corn Production in Mexico'. Oxfam GB and World Wildlife Fund International. <www.intermonoxfam.org/cms/HTML/espanol/520/AGR_CornStudy_OGB_WWF_0301.pdf> (December 2005).

Najam, Adil (1995). 'An Environmental Negotiation Strategy for the South'. *International Environmental Affairs* vol. 7, no. 3, pp. 249–287.

Najam, Adil (2004). 'Dynamics of the Southern Collective: Developing Countries in desertification Negotiations'. *Global Environmental Politics* vol. 4, no. 3, pp. 128–154.

Namibia (2004). 'Namibia's Third National Report on the Implementation of the United Nations Convention to Combat Desertification'. Report prepared for the Third Session of the Committee for the Review of the Implementation, 2–11 May. Desert Research Foundation of Namibia, Windhoek. <www.unccd.int/cop/reports/africa/national/2004/namibia-eng.pdf> (December 2005).

Narjisse, Hamid (2000). 'Rangeland Issues and Trends in Developing Countries'. In S. Archer and A. Olafur, eds., *Rangeland Desertification*. Kluwer Academic Publishers, Dordrecht.

Natural Heritage Institute (1996). 'North-South NGO Forum on Desertification in the American Hemsiphere'. 10–12 November. San Francisco.

Nelson, Mike (1997). 'Report of the Study on CGIAR Research Priorities for Marginal Lands'. TAC Working Document, March. Technical Advisory Committee Secretariat, Food and Agriculture Organization.

Néron, Jocelyne (2002). 'La convention des «populations» menacées: Le Sommet saura-t-il redonner l'élan nécessaire au règlement du problème de la désertification?' *Le Devoir*, 24–25 August.

Netherlands (2004). 'Netherlands Communication to the CRIC on the Convention to Combat Desertification (CCD): Policy and Programmes in Support of the African Region'. Report prepared for the Third Session of the Committee for the Review of the Implementation, 2–11 May. Ministry for Foreign Affairs, The Hague. <www.unccd.int/cop/reports/africa/developed/2004/netherlands-eng.pdf> (December 2005).

New Zealand (2002). 'Report prepared for the First Session of the Committee for the Review of the Implementation, 11–22 November'. Report prepared for the First Session of the Committee for the Review of the Implementation, 11–22 November. Wellington. <www.unccd.int/cop/reports/developed/2002/new_zealand-eng.pdf> (December 2005).

Nicholson, Sharon E. (2001). 'Climatic and Environmental Change in Africa during the Last Two Centuries'. *Climate Research* vol. 17, pp. 123–144.

Nicholson, Sharon E., Compton J. Tucker, and M.B. Ba (1998). 'Desertification, Drought, and Surface Vegetation: An Example from the West African Sahel'. *Bulletin of the American Meteorological Society* vol. 79, no. 5, pp. 815–829.

Niemeijer, David and Valentina Mazzucato (2002). 'Soil Degradation in the West African Sahel'. *Environment* vol. 44, no. 2, pp. 20–31.

Noy-Meir, Imanuel (1973). 'Desert Ecosystems: Environment and Producers'. *Annual Review of Ecology and Systematics* vol. 4, pp. 51–58.

Oberthür, Sebastian, Matthias Buck, Sebastian Müller, et al. (2002). 'Participation of Non-Governmental Organisations in International Environmental Governance: Legal Basis and Practice Experience'. Ecologic. <www.ecologic.de/download/projekte/18501899/1890/-report_ngos_en.pdf> (December 2005).

Oettlé, Noel, Adele Arendse, B. Koelle, et al. (2002). 'Community Exchange and Training in the Suid Bokkeveld: A Pilot Project to Enhance Natural Resource Management'. Proceedings for 'Alternative Ways to Combat Desertification: Connecting Community Action with Science and Common Sense', 8–20 April, Cape Town. V. Ward, ed. Desert Research Foundation of Namibia.

Office of the United Nations High Commissioner for Human Rights (2005). 'Special Rapporteur of the Commission on Human Rights on the Right to Food'. <www.ohchr.org/english/issues/food/index.htm> (December 2005).

O'Riordan, Timothy and Susan Stoll-Kleemann, eds. (2002). *Biodiversity, Sustainability, and Human Communities: Protecting beyond the Protected*. Cambridge University Press, Cambridge.

Organisation for Economic Co-operation and Development (1994). 'Methodologies for Environmental and Trade Reviews'. OCDE/GD(94)103. Paris. <www.eldis.org/static/DOC4037.htm> (December 2005).

Organisation for Economic Co-operation and Development (2002). *Creditor Reporting System on Aid Activities: Aid Targeting the Objectives of the Rio Conventions 1998/2000*. Organisation for Economic Co-operation and Development, Paris.

Organisation for Economic Co-operation and Development (2003). 'Agricultural Trade and Poverty: Making Policy Analysis Count'. Paris.

Organization of African Unity (1985). 'Declarations and Resolutions as Adopted by the Twenty-First Ordinary Session of the Assembly of Heads of State and Government'. 18–20 July. Addis Ababa. <www.africa-union.org/Official_documents/Heads%20of%20State%20Summits/hog/uHoGAssembly1985.pdf> (December 2005).

Otterman, Joseph (1974). 'Baring High-Albedo Soils by Overgrazing: A Hypothesized Desertification Mechanism'. *Science* vol. 186, no. 4163, pp. 531–533.

Oxfam International (2002a). 'Rigged Rules and Double Standards: Trade, Globalisation, and the Fight Against Poverty'. <www.maketradefair.com/en/index.php?file=03042002121618.htm> (December 2005).

Oxfam International (2002b). 'Boxing Match in Agricultural Trade: Will WTO Negotiations Knock Out the World's Poorest Farmers?' Oxfam Briefing Paper No. 32. <www.oxfam.org.uk/what_we_do/issues/trade/bp32_agric_trade.htm> (December 2005).

Øygard, Ragnar, Trond Vedeld, and Jens Aune (1999). 'Good Practices in Drylands Management'. Paper prepared for the World Bank Dryland Program. Noragric Agricultural University of Norway. <www.eldis.org/static/DOC8322.htm> (December 2005).

Paoletto, Glen and Heike Schroeder (1997). 'Enhancing Participation of NGOs in the UNFCCC Process'. Global Environment Information Centre. <www.geic.or.jp/ngo-bonn.html> (December 2005).

Parry, John T. (1996). 'Land Degradation in Tropical Drylands'. In M.J. Eden and J.T. Parry, eds., *Land Degradation in the Tropics: Environmental and Policy Issues*. Pinter, London.

Penman, H.L. (1948). 'Natural Evaporation from Open Water, Bare Soil, and Grass'. *Proceedings of the Royal Society of London, Series A* vol. 193, no. 1032, pp. 120–145.

People's Republic of Bangladesh (2002). 'Second National Report on Implementation of United Nations Convention to Combat Desertification: Final Draft'. Report prepared for the First Session of the Committee for the Review of the Implementation, 11–22 November. Ministry of Environment and Forests, Dhaka. <www.unccd.int/cop/reports/asia/national/2002/bangladesh-eng.pdf> (December 2005).

Permanent Inter-State Committee on Drought Control in the Sahel and Club de Sahel (1994). 'Report on the CILSS/Club de Sahel Regional Conference on Land Tenure and Decentralisation'. 20–25 June. Praia, Cape Verde.

Pomareda, Carlos and Carlos Murillo (2003). 'The Relationship between Trade and Sustainable Development of Agriculture in Central America'. International Institute for Sustainable Development, Winnipeg. <www.iisd.org/publications/pub.aspx?pno=565> (December 2005).

Porter, Gareth, Janet Welsh Brown, and Pamela S. Chasek, eds. (2000). *Global Environmental Politics*. Westview Press, Boulder CO.

Poulsen, Lene (2003). 'Promoting the Role of Women in the Implementation of the UNCCD'. Introductory Presentation to the Global Biodiversity Forum on the Ecosystem Approach to Dryland Management: Integrating Biodiversity Conservation and Livelihood Security, Sixth Meeting of the Conference of the Parties to the UN Convention to Combat Desertification and Mitigate the Impacts of Drought, 30–31 August. Havana. <www.gbf.ch/Session_Administration/upload/paper%20Lene%20Poulsen.doc> (December 2005).

Pretty, Jules and Robert Chambers (1994). 'Towards a Learning Paradigm: New Professionalism and Institutions for Sustainable Agriculture'. In I. Scoones and J. Thompson, eds., *Beyond Farmer First: Rural People's Knowledge, Agricultural Research, and Extension Practice*. Intermediate Technology Publishing, London.

Prince, Stephen D., E. Brown de Colstoun, and L.L. Kravitz (1998). 'Evidence from Rain-Use Efficiencies Does Not Indicate Extensive Sahelian Desertification'. *Global Change Biology* vol. 4, pp. 359–374.

Projeto Áridas (1995). 'A Strategy for Sustainable Development in Brazil's Northeast'. Ministry of Planning and the Budget, Brasilia.

Reardon, Thomas and Stephen A. Vosti (1996). 'Link between Rural Poverty and the Environment in Developing Countries'. Centro Latinoamericano para el Desarrolo Rural. <www.rimisp.org/webpage.php?webid=117> (Deecember 2005).

Republic of Algeria (2004). 'Rapport national de l'Algériel sur la mise en oeuvre de la Convention des Nations Unies de Lutte Contre la Désertification'. Report prepared for the Third Session of the Committee for the Review of the Implementation, 2–11 May. Ministère de l'Agriculture et du Développement rural, Alger. <www.unccd.int/cop/reports/africa/national/2004/algeria-fre.pdf> (December 2005).

Republic of Argentina (1997). 'Programa de Acción Nacional de Lucha contra la Desertificatión: Documento de Base'. Ministerio de Desarrolo Social, Buenos

Aires. <www.unccd.int/actionprogrammes/lac/national/1997/argentina-spa.pdf> (December 2005).
Republic of Belarus (2002). 'National Report of the Republic of Belarus on Implementation of the UN Convention to Combat Desertficiation'. Report prepared for the First Session of the Committee for the Review of the Implementation, 11–22 November. Ministry of Natural Resources and Environmental Protection, Minsk. <www.unccd.int/cop/reports/centraleu/national/2002/belarus-eng.pdf> (December 2005).
Republic of Cape Verde (2004). 'Mis en oeuvre de la Convention des Nations Unies de Lutte contre la Désertification'. Report prepared for the Third Session of the Committee for the Review of the Implementation, 2–11 May. Ministère de l'Environnement, de l'Agriculture et des Pêches, Praia. <www.unccd.int/cop/reports/africa/national/2004/cape_verde-fre.pdf> (December 2005).
Republic of Djibouti (2004). 'Troisième rapport national sur la mise en oeuvre de la Convention des Nations Unies sur la lutte contre la désertification'. Report prepared for the Third Session of the Committee for the Review of the Implementation, 2–11 May. Ministere de l'Agriculture de l'élevage et de la mer chargé des ressources hydrauliques de Djibouti. <www.unccd.int/cop/reports/africa/national/2004/djibouti-fre.pdf> (December 2005).
Republic of Haïti (2004). 'Rapport national de la République d'Haïti sur la mise en oeuvre de la Convention sur la Lutte contre la Désertification'. Report prepared for the Third Session of the Committee for the Review of the Implementation, 2–11 May. Ministère de l'Environnement, Port-au-Prince. <www.unccd.int/cop/reports/lac/national/2002/haiti-fre.pdf> (December 2005).
Republic of Liberia (2002). 'First National Report of the Implementation of the United Nations Convention to Combat Desertification'. Report prepared for the First Session of the Committee for the Review of the Implementation, 11–22 November. Liberian National Coordinating Committee to Combat Desertification, Monrovia. <www.unccd.int/cop/reports/africa/national/2002/liberia-eng.pdf> (December 2005).
Republic of Mali (1998). 'Résumé du PNAE/PAN-CID'. Ministère de l'environnement, Secretariat permanent du PNAE-CID. Bamako. <www.gm-unccd.org/FIELD/Countries/Mali/nap.htm> (December 2005).
Republic of Mali (2004). 'Rapport national du Mali'. Report prepared for the Third Session of the Committee for the Review of the Implementation, 2–11 May. Ministère de l'Environnement et de l'Assainissement, Bamako. <www.unccd.int/cop/reports/africa/national/2004/mali-fre.pdf> (December 2005).
Republic of Poland (2004). 'The National Report of the Republic of Poland on the Implementation of the United Nations Convention to Combat Desertification'. Report prepared for the Third Session of the Committee for the Review of the Implementation, 2–11 May. Warszawa. <www.unccd.int/cop/reports/africa/developed/2004/poland-eng.pdf> (December 2005).
Republic of Senegal (2004). 'Troisième rapport national sur la mise en oeuvre de la Convention des Nations Unies sur la Lutte contre la Désertification'.

Report prepared for the Third Session of the Committee for the Review of the Implementation, 2–11 May. Ministère de l'Environnement et de la Protection de la nature, Dakkar. <www.unccd.int/cop/reports/africa/national/2004/senegal-fre.pdf> (December 2005).

Republic of Tajikistan (2002). 'National Report of the Republic of Tajikistan to Combat Desertification'. Report prepared for the First Session of the Committee for the Review of the Implementation, 11–22 November. State Committee of the Republic of Tajikistan for Land Management, Dushanbe. <www.unccd.int/cop/reports/asia/national/2002/tajikistan-eng.pdf> (December 2005).

Republic of Uganda (2004). 'Third National Report to the Conference of the Parties on the Implementation of the United Nations Convention to Combat Desertification in Uganda'. Report prepared for the Third Session of the Committee for the Review of the Implementation, 2–11 May. Ministry of Agriculture, Animal Industry, and Fisheries, Entebbe. <www.unccd.int/cop/reports/africa/national/2004/uganda-eng.pdf> (December 2005).

Reynolds, J.F. and D.M. Stafford Smith (2002). 'Do Humans Cause Deserts?' In J.F. Reynolds and D.M. Stafford Smith, eds., *Global Desertification: Do Humans Cause Deserts?* Dahlem University Press, Berlin.

Robbins, Peter (2003). *Stolen Fruit: The Tropical Commodities Disaster*. Zed Books, London.

Rosenfeld, Daniel, Yinon Rudich, and Ronen Lahav (2001). 'Desert Dust Suppressing Precipitation: A Possible Desertification Feedback Loop'. *Proceedings of the National Academy of Sciences of the United States of America* vol. 98, no. 11, pp. 5975–5980.

Sabine, Henry, Bruno Schoumaker, and Cris Beauchemin (2004). 'The Impact of Rainfall on the First Out-Migration: A Multi-level Event-History Analysis in Burkino Faso'. *Population and Environment* vol. 25, no. 5, pp. 423–460.

Sachs, Jeffrey, John W. McArthur, Guido Schmidt-Traub, et al. (2004). 'Ending Africa's Poverty Trap'. Brookings Papers on Economic Activity 1. <www.earthinstitute.columbia.edu/about/director/documents/bpea0104.pdf> (December 2005).

Sahel Club (1984). *Environmental Change in the West African Sahel ((Transformation de l'environnement dans le Sahel ouest africain))*. Club de Sahel, Paris.

Sanders, Thomas G. (1990-91). 'Northeast Brazilian Environmental Refugees: Part I — Why They Leave'. Field Staff Reports No. 20. National Heritage Institute.

Sandford, Rosemary (1992). 'Secretariats and International Environmental Negotiations: Two New Models'. In L. Susskind, E.J. Dolin and J.W. Breslin, eds., *International Environmental Treaty Making*. Harvard Law School, Cambridge MA.

Sandford, Rosemary (1994). 'International Environmental Treaty Secretariats: Stage-Hands or Actors?' In H.O. Bergesen and G. Parmann, eds., *Green Globe Yearbook of International Co-operation on Environment and Development 1994*. Oxford University Press, Oxford.

Scherl, Lea M. (1996). 'Relationships and Partnerships among Governments, NGOs, CBOs, and Indigenous Groups in the Context of the Convention to Combat

Desertification and Drought: An Analysis of Progress'. Environment Liaison Centre International.

Scherr, Sara (1999). 'Dryland Degradation and Poverty'. pp. 69–77, in 'Drylands, Poverty, and Development', Proceedings of the World Bank Round Table, 15–16 June, Washington DC.

Schlesinger, William H., James F. Reynolds, Gary L. Cunningham, et al. (1990). 'Biological Feedbacks in Global Desertification'. *Science* vol. 247, no. 4946, pp. 1043–1048.

Schlesinger, William H. (2004). Communication at the American Association for the Advancement of Science. Seattle, 12–16 February.

Schram Stokke, Olav and Oystein B. Thommessen, eds. (2003). *Yearbook of International Co-operation on Environment and Development 2003/2004*. Earthscan, London.

Scoones, Ian (1994). *Living with Uncertainty: New Directions in Pastoral Development in Africa*. Intermediate Technology Publishing, London.

Scoones, Ian (1996). 'Politics, Polemics, and Pastures: Range Management Science and Policy in Southern Africa'. In M. Leach and R. Mearns, eds., *The Lie of the Land: Challenging Received Wisdom on the African Environment*. International African Institute and Heinemann, Oxford.

Seck, Emmanuel S. (1996). 'Désertification: effets, lutte et convention'. ENDA TM, Dakar.

Sharma, Anju (1999). 'Rio's Stepchild'. *Down to Earth* vol. 7, no. 17, pp. 24–25.

Sharma, Manohar, Marito Garcia, Aamir Quershi, et al. (1996). 'Overcoming Malnutrition: Is There an Ecoregional Dimension?' Food, Agriculture, and the Environment Discussion Paper 10. International Food Policy Research Institute, Washington DC.

Sharma, Ravi (1993). 'Desertification: NGOs Reject Cautious Approach'. *Eco* vol. 2005, no. 85, pp. 10 (27 August). <www.climatenetwork.org/eco/INCs/Eco10_0893.html> (December 2005).

Shestakov, Alexander and Vladimir Streletsky (1998). *Mapping of Risk Areas of Environmentally Induced Migration in the Commonwealth of Independent States*. United Nations High Commissioner for Refugees, International Organization for Migration, and Refugee Policy Group, Geneva.

Siebenhüner, Bernd (forthcoming). 'Lean Shark in a Regulatory Jungle? The Case of the CBD Secretariat'. In F. Biermann and B. Siebenhüner, eds., *Managers of Global Change: Effectiveness and Learning of International Environmental Organizations*.

Sokona, Youba and Emmanuel S. Seck (2000). 'Report on the Consideration of Local Knowledge by the Action Programmes, Networks, and Mechanisms Set Up by the CCD Secretariat to Promote Programmes Combatting Desertification on the Regional and National Scales'. ENDA TM. <www.enda.sn/energie/desertif/local-knowl.htm> (December 2005).

Solomon Islands (2002). 'National Report to the United Nations Convention to Combat Desertification'. Report prepared for the First Session of the Committee

for the Review of the Implementation, 11–22 November. <www.unccd.int/cop/reports/asia/national/2002/solomon_islands-eng.pdf> (December 2005).
Stiles, Daniel (1997). 'Linkages between Dryland Degradation and Migration: A Methodology'. *Desertification Control Bulletin* no. 30, pp. 9–18.
Strutt, Anna (1998). 'Trade Liberalization and Land Degradation in Indonesia'. ACIAR Indonesia Research Project Working Paper 98.06. <www.adelaide.edu.au/cies/papers/iwp9806.pdf> (December 2005).
Suhrke, Astri (1993). 'Pressure Points: Environmental Degradation, Migration, and Conflict'. American Academy of Art and Science. Cambridge. <www.cmi.no/publications/1993%5Cpressure_points.pdf> (December 2005).
Swaziland (2000). 'National Action Programme'. <www.unccd.int/actionprogrammes/africa/national/2000/swaziland-eng.pdf> (December 2005).
Sweden (2002). 'Implementation of the UN Convention to Combat Desertification: Report by Sweden'. Report prepared for the First Session of the Committee for the Review of the Implementation, 11–22 November. Swedish International Development Agency, Stockholm. <www.unccd.int/cop/reports/developed/2002/sweden-eng.pdf> (December 2005).
Sweden (2004). 'Implementation of the UN Convention to Combat Desertification'. Report prepared for the Third Session of the Committee for the Review of the Implementation, 2–11 May. Swedish International Development Agency, Stockholm. <www.unccd.int/cop/reports/africa/developed/2004/sweden-eng.pdf> (December 2005).
Tamondong-Helin, Susan and William Helin (1991). 'Migration and the Environment: Interrelationships in Sub-Saharan Africa'. Field Staff Reports No. 22. UFSI Inc. and National Heritage Institute.
Thomas, David S.G. and Nick Middleton (1994). *Desertification: Exploding the Myth*. Wiley, Chichester.
Thornthwaite, C.W. (1948). 'An Approach toward a Rational Classification of Climate'. *Geographical Review* vol. 38, pp. 55–94.
Tiffen, Mary (1993). 'Productivity and Environmental Conservation under Rapid Population Growth: A Case Study of Machakos District'. *Journal of International Development* vol. 5, no. 2, pp. 207–223.
Tiffen, Mary, Francis Gichuki, and Michael J. Mortimore (1994). *More People, Less Erosion: Environmental Recovery in Kenya*. J. Wiley, Chichester.
Tiffen, Mary and Michael J. Mortimore (2002). 'Questioning Desertification in dryland sub-Saharan Africa'. *Natural Resources Forum* vol. 26, pp. 218–233.
Toulmin, Camilla (1994). 'Combating Desertification: Encouraging Local Action within a Global Framework'. In Fridtjof Nansen-stiftelsen på Polhøgda, ed., *Green Globe Yearbook of International Co-operation on Environment and Development*, pp. 79–88. Oxford University Press, Oxford.
Toulmin, Camilla (2001). 'Lessons from the Theatre: Should This Be the Final Curtain Call for the Convention to Combat Desertification'. World Summit on Sustainable Development Briefing Papers. International Institute for Environment

and Development, London. <www.iied.org/pubs/pdf/full/11017IIED.pdf> (December 2005).

Trolldalen, John Martin, Nina Birkeland, Jan Borgen, et al. (1992). 'Environmental Refugees: A Discussion Paper'. Paper prepared for the United Nations Conference on Environment and Development, 3–14 June. World Foundation for Enviornment and Development and Norwegian Refugee Council.

Tucker, Compton J. and S.E. Nicholson (1999). 'Variation in the Size of the Sahara Desert from 1980 to 1997'. *Ambio* vol. 28, pp. 587–591.

Union Mondiale pour la Nature and Bureau Régional de l'Afrique de l'Ouest (2004). 'Rapport d'activités du projet: «Appui à la mise en oeuvre de la CCD en Afrique de l'Ouest»'. January. <www.iucn.org/brao/articles/0402/appuiCCD.pdf> (December 2005).

United Kingdom (2004). 'Report by the United Kingdom of Great Britain and Northern Ireland on Measures Taken to Support the Implementation of the United Nations Convention to Combat Desertification with a Particular Focus on Affected Developing Country Parties in Africa'. Report prepared for the Third Session of the Committee for the Review of the Implementation, 2–11 May. Department for International Development, London. <www.unccd.int/cop/reports/developed/2004/united_kingdom_of_great_britain_and_northern_ireland-eng.pdf> (December 2005).

United Nations (1995). 'Report of the Fourth World Conference on Women'. A/CONF.177/20, 17 October. Beijing. <www.un.org/documents/ga/conf177/aconf177-20en.htm> (December 2005).

United Nations (2002). 'Plan of Implementation of the World Summit on Sustainable Development'. Johannesburg. <www.un.org/esa/sustdev/documents/WSSD_POI_PD/English/WSSD_PlanImpl.pdf> (December 2005).

United Nations (2002). 'Report of the World Summit on Sustainable Development'. A/CONF.199/20 and A/CONF.199/20/Corr.1, 26 August–4 September. Johannesburg. <www.unctad.org/en/docs/aconf199d20&c1_en.pdf> (December 2005).

United Nations (2002). 'Report of the International Conference on Financing for Development'. A/CONF.198/11, 18–22 March. Monterrey. <www.un.org/esa/ffd/aconf198-11.pdf> (December 2005).

United Nations (2002). 'Johannesburg Declaration on Sustainable Development: From Our Origins to the Future'. World Summit on Sustainable Development, 4 September. <www.un.org/esa/sustdev/documents/WSSD_POI_PD/English/POI_PD.htm> (December 2005).

United Nations (2003). '"Desertification Is Both a Cause and a Consequence of Poverty", Secretary General Says in Message for International Day'. Press Release SG/SM/8750 OBV/355, 12 June. <www.un.org/News/Press/docs/2003/sgsm8750.doc.htm> (December 2005).

United Nations Conference on Environment and Development (1992). 'Agenda 21'. 14 June. Rio. <www.un.org/esa/sustdev/documents/agenda21/english/agenda21toc.htm> (December 2005).

United Nations Conference on Trade and Development (2004). 'The Least Developed Countries Report, 2004: Linking International Trade with Poverty Reduction'. Geneva.

United Nations Conference on Trade and Development (2004). 'World Investment Report 2004: The Shift Towards Services'. Geneva.

United Nations Convention to Combat Desertification (1993). 'Elaboration of an International Convention to Combat Desertification in Countries Experiencing Serious Drought and/or Desertification, Particularly in Africa'. A/AC.241/12, 23 August. <www.unccd.int/cop/officialdocs/incd/pdf/24112eng.pdf> (December 2005).

United Nations Convention to Combat Desertification (1994). 'Text of the United Nations Convention to Combat Desertification'. <www.unccd.int/convention/text/convention.php> (December 2005).

United Nations Convention to Combat Desertification (1995). 'Down to Earth: A Simplified Guide to the Convention to Combat Desertification, Why It Is Necessary, and What Is Important and Different About It'. <www.unccd.int/publicinfo/downtoearth/downtoearth-eng.pdf> (December 2005).

United Nations Convention to Combat Desertification (1997). 'Report of the Conference of the Parties on Its First Session, Held in Rome from 29 September to 10 October 1997: Part Two — Actions Taken by the Conference of the Parties at Its First Session'. ICCD/COP(1)/11, 29 December. <www.unccd.int/cop/officialdocs/cop1/pdf/11add1eng.pdf> (2005 December).

United Nations Convention to Combat Desertification (1999). 'Report of the Conference of the Parties on Its Third Session, Held in Recife from 15 to 26 November 1999'. ICCD/COP(3)/20/Add.1, 31 December 1999. <www.unccd.int/cop/officialdocs/cop3/pdf/20add1eng.pdf> (December 2005).

United Nations Convention to Combat Desertification (1999). 'Report of the Conference of the Parties on Its Second Session, Held in Dakar from 30 November to 11 December 1998: Part Two — Action Taken by the Conference of the Parties at Its Second Session'. ICCD/COP(2)/14/Add.1, 5 February. <www.unccd.int/cop/officialdocs/cop2/pdf/14add1eng.pdf> (December 2005).

United Nations Convention to Combat Desertification (2000). 'Sub-regional Action Programme (SRAP) to Combat Desertification and Drought in West Asia'. Meeting on the Endorsement of the Sub-regional Action Program to Combat Desertification in West Asia, 16 February. Dubai. <www.unccd.int//actionprogrammes/asia/subregional/2000/westernasia-eng.pdf> (December 2005).

United Nations Convention to Combat Desertification (2001). 'Report of the Conference of the Parties on Its Fifth Session, Held in Geneva from 1 to 12 October 2001: Part Two — Action Taken by the Conference of the Parties on Its Fifth Session'. ICCD/COP(5)/11/Add.1, 7 November. <www.gm-unccd.org/English/DOCS/COP5Dec.pdf> (December 2005).

United Nations Convention to Combat Desertification (2001). 'Global Mechanism: Review, Pursuant to Article 21, Paragraph 5(d), of the Convention of the Report on the Activities of the Global Mechanism, and the Provision of Guidance to It'.

ICCD/COP(5)/4, 11 September. <www.unccd.int/cop/officialdocs/cop5/pdf/4eng.pdf> (December 2005).

United Nations Convention to Combat Desertification (2001). 'Implementation of the Convention: Conference of the Parties Fifth Session, Geneva, 1–12 October'. ICCD/COP(5)/3/Add.1, 29 August. <www.unccd.int/cop/officialdocs/cop5/pdf/3eng.pdf> (December 2005).

United Nations Convention to Combat Desertification (2001). 'Report of the Conference of the Parties on Its Fourth Session, Held in Bonn from 11 to 22 December 2000: Part Two — Action Taken by the Conference of the Parties at Its Fourth Session'. ICCD/COP(4)/11/Add.1, 25 January 2001. <www.unccd.int/cop/officialdocs/cop4/pdf/11add1eng.pdf> (December 2005).

United Nations Convention to Combat Desertification (2002). 'Statement by Mr. Arba Diallo, Executive Secretary of the UNCCD'. World Summit on Sustainable Development, 30 August. Johannesburg. <www.un.org/events/wssd/statements/unccdE.htm> (December 2005).

United Nations Convention to Combat Desertification (2003). 'Programme and Budget'. ICCD/COP(g)/2/Add.1, 23 May 2003. <www.unccd.int/cop/officialdocs/cop6/pdf/2add1eng.pdf> (December 2005).

United Nations Convention to Combat Desertification (2003). 'Review of the Implementation of the Convention and of Its Institutional Arrangements, Pursuant to Article 22, Paragraphs 2(a) and (b), and Article 26 of the Convention: Review of the Report on Enhanced Implementation of the Obligations of the Convention'. ICCD/CRIC(2)/3, 2 July. <www.unccd.int/cop/officialdocs/cric2/pdf/3eng.pdf> (December 2005).

United Nations Convention to Combat Desertification (2003). 'Report of the Committee on Its First Session, Held in Rome from 11 to 22 November 2002'. ICCD/CRIC(1)/10, 17 January. <www.unccd.int/cop/officialdocs/cric1/pdf/10eng.pdf> (December 2005).

United Nations Convention to Combat Desertification (2003). 'National Reporting Process of Affected Country Parties: Explanatory Note and Help Guide'. ICCD/CRIC(3)/INF.3, 19 November. <www.unccd.int/cop/officialdocs/cric3/pdf/inf3eng.pdf> (December 2005).

United Nations Convention to Combat Desertification (2003). 'Report of the Conference of the Parties on Its Sixth Session, Held in Havana from 25 August to 5 September 2003: Part One — Proceedings'. ICCD/COP(6)/11, 3 November 2003. <www.unccd.int/cop/officialdocs/cop6/pdf/11eng.pdf> (December 2005).

United Nations Convention to Combat Desertification (2003). 'Report of the Conference of the Parties on Its Sixth Session, Held in Havana from 25 August to 5 September 2003: Part Two — Action Taken by the Conference of the Parties at Its Sixth Session'. ICCD/COP(6)/11/Add.1, 7 November. <www.unccd.int/cop/officialdocs/cop6/pdf/11add1eng.pdf> (December 2005).

United Nations Convention to Combat Desertification (2003). 'Independent Evaluation of the Global Mechanism'. Final Report to the World Bank

Development Grant Facility, ICCD/COP(6)/MISC.1, 24 August. <www.unccd.int/cop/officialdocs/cop6/pdf/misc1eng.pdf> (December 2005).
United Nations Convention to Combat Desertification (2003). 'Global Mechanism: Review, Pursuant to Article 21, Paragraph 7, of the Convention of the Policies, Operational Modaliites and Activities of the Global Mechanism, and the Provision of Guidance to It'. ICCD/CRIC(2)/5, 20 June. <www.unccd.int/cop/officialdocs/cric2/pdf/5eng.pdf> (December 2005).
United Nations Convention to Combat Desertification (2005). 'Combating Desertification in Africa'. <www.unccd.int/publicinfo/factsheets/showFS.php?number=11> (December 2005).
United Nations Convention to Combat Desertification (2005). 'UN Maps New Ways to Mainstream Desertification'. 12 May. Bonn. <www.unccd.int/publicinfo/pressrel/showpressrel.php?pr=press12_05_05> (December 2005).
United Nations Convention to Combat Desertification (2005). 'Status of Ratification and Entry into Force'. <www.unccd.int/convention/ratif/doeif.php> (December 2005).
United Nations Convention to Combat Desertification (2005). 'Mainstreaming of National Action Programmes and Their Contribution to Overall Poverty Eradication'. ICCD/CRIC(3)/Misc.1, 2–11 May. Bonn. <www.unccd.int/cop/officialdocs/cric3/pdf/misc1eng.pdf> (December 2005).
United Nations Convention to Combat Desertification (2005). 'Reports Submitted by Africa'. <www.unccd.int/cop/reports/africa/africa.php> (December 2005).
United Nations Convention to Combat Desertification (2005). 'Glossary'. <www.unccd.int//knowledge/glossary-eng.html> (December 2005).
United Nations Convention to Combat Desertification (2005). 'The Bottom-Up Approach'. <www.unccd.int/ngo/menu.php> (December 2005).
United Nations Convention to Combat Desertification (2005). 'Explanatory Leaflet'. <www.unccd.int/convention/text/leaflet.php> (December 2005).
United Nations Development Programme (2001). 'Human Development Report 2003: Millennium Development Goals: A Compact among Nations to End Poverty'. United Nations Development Programme, New York. <hdr.undp.org/reports/global/2003> (December 2005).
United Nations Development Programme (2003). 'Human Development Report 2003: Millennium Development Goals — A Compact among Nations to End Human Poverty'. New York. <hdr.undp.org/reports/global/2003/> (December 2005).
United Nations Development Programme's Office to Combat Desertification and Drought (UNSO) (1994). 'Poverty Alleviation and Land Degradation in the Drylands: Issues and Action Areas for the International Convention on Desertification'. Paper produced by the United Nations Sudano-Sahelian Office, Food Studies Group, University of Oxford. <www.undp.org/seed/unso/text/public/pov-eng.htm> (December 2005).
United Nations Development Programme's Office to Combat Desertification and Drought (UNSO) (1999). 'A Preliminary Overview of National Action Programme Processes of the United Nations Convention to Combat Desertification and

Drought'. Report prepared as a contribution to the deliberations of COP-2. <www.undp.org/seed/unso/lessons/trad/documemts/nap-eng.pdf> (December 2005).
United Nations Economic and Social Council (1998). 'Measures to Improve the Situation and Ensure the Human Rights and Dignity of All Migrant Workers'. Report of the Working Group of Intergovernmental Experts on Human Rights of Migrants E/CN.4/1998/76, 10 March. <www.unhchr.ch/Huridocda/Huridoca.nsf/TestFrame/e696166bf66373f3c12566180046b9c6> (December 2005).
United Nations Economic and Social Council (1999). 'The Right to Adequate Food (Article 11)'. E/C.12/1999/5, Committee on Economic, Social, and Cultural Rights, 5 December. Geneva.
United Nations Economic Commission for Europe (1998). 'Convention on Access to information, Public Participation in Decision-Making and Access to Justice in Environmental Matters'. Aahrus Convention, 25 June. Aahrus. <www.unece.org/env/pp/documents/cep43e.pdf> (December 2005).
United Nations Education, Science, and Cultural Organization. (1979). 'Map of the World Distribution of Arid Regions'. Man and the Biosphere Technical Notes 7. Paris.
United Nations Education, Science, and Cultural Organization. (2005). 'Migrant/Migration'. <www.unesco.org/shs/migration/glossary> (December 2005).
United Nations Environmental Programme (1991). 'Status of Desertification and Implementation of the United Nations Plan of Action to Combat Desertification'. Nairobi. <www.na.unep.net/des/uncedtoc.php3> (December 2005).
United Nations Environmental Programme (2002). 'Integrated Assessment of Trade Liberalization and Trade-Related Policies — Round II: A Country Study on the Ecuador Banana Sector'. Synthesis Report. <www.unep.ch/etu/publications/Synth_Ecuador.PDF> (December 2005).
United Nations Environmental Programme (2002). 'Integrated Assessment of Trade Liberalization and Trade-Related Policies — Round II: A Country Study on the Export Crop Sector in Nigeria'. Synthesis Report. <www.unep.ch/etu/publications/Synth_Nigeria.PDF> (December 2005).
United Nations Environmental Programme (2003). 'Global Environmental Outlook 3'. <www.unep.org/geo/geo3/english> (December 2005).
United Nations Environmental Programme (2003). 'An Indicator Model for Dryland Ecosystems in Latin America'. Final report submitted to the Global Environment Facility. GF/1040-00-10.
United Nations Environmental Programme (2004). 'UNEP's Strategy on Land Use Management and Soil Conservation: A Strengthened Functional Approach'. UNEP/GC.22/INF/25, 4 December. <www.unep.org/GC/GC22/Document/k0263473.pdf> (December 2005).
United Nations General Assembly (1977). 'Plan of Action to Combat Desertification'. Resolution 32/172, 19 December.
United Nations General Assembly (1992). 'Report of the United Nations Conference on Environment and Development'. 3–14 June. Rio de Janeiro. <www.un.org/documents/ga/conf151/aconf15126-2.htm> (December 2005).

United Nations General Assembly (1992). 'Establishment of an Intergovernmental Negotiating Committee for the Elaboration of an International Convention to Combat Desertification in Those Countries Experiencing Serious Drought and/or Desertification, particularly in Africa'. A/RES/47/188, 22 December. <www.un.org/documents/ga/res/47/a47r188.htm> (December 2005).

United Nations General Assembly (1996). 'Organization of Scientific and Technological Cooperation'. Intergovernmental Negotiating Committee for the Elaboration of an International Convention to Combat Desertification in Those Countries Experiencing Serious Drought and/or Desertification, Particularly in Africa, Ninth Session, 3–13 September, A/AC.241/57. <www.unccd.int/cop/officialdocs/incd/pdf/24157eng.pdf> (December 2005).

United Nations General Assembly (2000). 'Resolution Adopted by the General Assembly: 55/2. United Nations Millennium Declaration'. A/RES/55/2, 8 September. <www.un.org/millennium/declaration/ares552e.htm> (December 2005).

United Nations General Assembly (2001). 'Road Map towards the Implementation of the United Nations Millennium Declaration: Report of the Secretary General'. A/56/150, 6 September. <www.un.org/documents/ga/docs/56/a56326.pdf> (December 2005).

United Nations Human Settlements Programme (2002). 'International Legal Instruments Addressing Good Governance'. Nairobi. <www.unhabitat.org/campaigns/governance/documents/Intl%20legal%20instruments%20addressing%20good%20gov.pdf> (December 2005).

United Nations Millennium Project (2004). 'Halving Hunger by 2015: A Framework for Action'. Task Force 2 on Hunger, New York. <www.unmillenniumproject.org/documents/tftwointerim.pdf> (December 2005).

United Nations Millennium Project (2005). 'Halving Hunger: It Can Be Done'. New York. <www.unmillenniumproject.org/reports/tf_hunger.htm> (December 2005).

United Nations Millennium Project (2005). 'About the Goals'. <www.unmillenniumproject.org/goals> (December 2005).

United Nations Millennium Project (2005). 'Investing in Development: A Practical Plan to Achieve the Millennium Development Goals'. <www.unmillenniumproject.org/reports/index.htm> (December 2005).

United Nations Office for the Coordination of Humanitarian Affairs (2002). 'Nigeria: Dozens Reported Dead in Clashes between Farmers, Herders'. IRINews, 8 January. <www.irinnews.org/print.asp?ReportID=18545> (December 2005).

United Nations Office for the Coordination of Humanitarian Affairs (2003). 'Agreement Reached Allowing Humanitarian Access to Darfur Region of Sudan'. Press release AFR/701 IHA/795, 17 September. <www.un.org/News/Press/docs/2003/afr701.doc.htm> (December 2005).

United States (2002). 'United States Activities in Support of the UN Convention to Combat Desertification'. Report prepared for the First Session of the Committee for the Review of the Implementation, 11–22 November. Washington DC.

<www.unccd.int/cop/reports/africa/developed/2002/united_states_of_america-eng.pdf> (December 2005).
United States Geological Survey (1997). 'Distribution of Non-polar Arid Land'. <pubs.usgs.gov/gip/deserts/what/world.html> (December 2005).
Vitalis, Vangelis (2004). 'Trade, Agriculture, the Environment, and Development: Reaping the Benefits of a Win-Win-Win'. Paper presented at a Strategic Dialogue on Agriculture, Trade Negotiations, Poverty, and Sustainability, 14–16 July. Windsor UK. <www.ictsd.org/dlogue/2004-07-14/Vangelis.pdf> (December 2005).
Vorley, Bill (2002). 'Sustaining Agriculture: Policy, Governance, and the Future of Family-Based Farming'. International Institute for Environment and Development, London.
Walker, B.H., D. Ludwig, C.S. Holling, et al. (1981). 'Stability of Semi-Arid Savannah Grazing Systems'. *Journal of Ecology* vol. 69, pp. 473–498.
Weiss, Thomas G. (1982). 'International Bureaucracy: The Myth and Reality of the International civil Service'. *International Affairs* vol. 58, no. 2, pp. 287–306.
Werth, Alexander (2002). 'Agri-Environment and Rural Development in the Doha Round'. International Institute for Sustainable Development, Winnipeg. <www.tradeknowledgenetwork.net/pdf/tkn_ruraldev_doha.pdf> (December 2005).
Westing, Arthur (1994). 'Findings regarding the Ethiopian-Somali War of 1977–78'. Presented at the International Symposium on Desertification and Migrations, 9–11 February. Almeria, Spain.
Westoby, M., B.H. Walker, and Imanuel Noy-Meir (1989). 'Opportunistic Management for Rangelands Not at Equilibrium'. *Journal of Range Management* vol. 42, pp. 266–274.
Wettestad, Jorgen (2001). 'Designing Effective Environmental Regimes: The Conditional Keys'. *Global Governance* vol. 7, no. 317–341.
White, Gilbert F., ed. (1956). *The Future of Arid Lands: Papers and Recommendations from the International Arid Lands Meetings*. Washington DC.
White, Robin P. and Janet Nackoney (2003). 'Drylands, People, and Ecosystem Goods and Services: A Web-Based Geospatial Analysis'. World Resources Institute. <biodiv.wri.org/drylandsgis-pub-3813.html> (December 2005).
Whitford, Walter G. (2002). *Ecology of Desert Systems*. Academic Press, San Diego.
Wiens, John A. (1984). 'On Understanding a Nonequilibrium World: Myth and Reality in community Patterns and Processes'. In D.R. Strong, D. Simberloff, L.G. Abele et al., eds., *Ecological Communites: Conceptual Issues and the Evidence*, pp. 439–457. Princeton University Press, Princeton.
Wilhite, Donald A. and Michael H. Glantz (1985). 'Understanding the Drought Phenomenon: The Role of Definitions'. *Water International* vol. 10, no. 3, pp. 111–120.
Wilhite, Donald A., ed. (2000). *Drought: A Global Assessment*. Routledge, London.
Wood, William B. (2001). 'Ecomigration: Linkages between Environmental Change and Migration'. In A.R. Zolberg and P. Benda, eds., *Global Migrants, Global Refugees: Problems and Solutions*, vol. 42–61. Berghahn Books, New York.

World Bank (2001). 'Making Sustainable Commitments: An Environment Strategy for the World Bank'. Washington.

World Bank Group (2004). 'About the Goals'. <www.developmentgoals.org> (2005 December).

World Commission on Environment and Development (1987). *Our Common Future* (Brundtland Report). Oxford University Press, Oxford.

World Food Programme (2005). 'Keep Niger on the Map' — Renewed Support Urged as World Attention Wanes'. Niamey, 26 August. <www.wfp.org/english/?ModuleID=137&Key=1806> (December 2005).

World Trade Organization (2001). 'Ministerial Declaration'. WT/MIN(01)/DEC/1, 14 November. Doha. <www.wto.org/english/thewto_e/minist_e/min01_e/mindecl_e.htm> (December 2005).

World Wildlife Fund (2003). 'Sustainability Assessment of Export-Led Growth in Soy Production: Full Report'. <www.panda.org/downloads/policy/soylongeng.pdf> (December 2005).

World Wildlife Fund, World Bank, International Centre for Trade and Sustainable Development, et al. (2004). 'From Negotiations to Global Adjustment: Preparatory Phase Summary Report'. 14 June. Paris. <www.panda.org/downloads/policy/ntoasummary.pdf> (December 2005).

Yin, Yan, Zev Levin, Tamir G. Reisin, et al. (2000). 'The Effects of Giant Cloud Condensation Nuclei on the Development of Precipitation in Convective Clouds'. *Atmospheric Research* vol. 53, pp. 91–116.

Young, Oran R., Mark A. Levy, and Gail Osherenko, eds. (1999). *Effectiveness of International Environmental Regimes: Causal Connections and Behavioral Mechanisms*. MIT Press, Cambridge MA.

Zürn, Michael (1998). 'The Rise of International Environmental Poliitcs'. *World Politics* vol. 50, no. 4, pp. 617–649.

Zweifler, Mark O., Michael A. Gold, and Robert N. Thomas (1994). 'Land Use Evolution in Hill Regions of the Domincan Republic'. *Professional Geographer* vol. 46, pp. 39–53.

Index

Aarhus Convention 121–2
Africa *see also* Algeria; Burkina Faso; Chad; Egypt; Eritrea; Ethiopia; Guinea; Kenya; Malawi; Mali; Morocco; Namibia; Niger; Nigeria; Senegal; Sudan; Swaziland; Tunisia; Uganda; West Africa; Zimbabwe
    droughts 45
    drylands 28–9
    Intergovernmental Negotiating Committee on Desertification (INCD) 64
    migration 44–6
    overpopulation 33
African Union 52
agricultural commodity prices 181
agricultural development financing 133
agricultural productivity 2–3, 182
agricultural trade
    desertification 178–9, 186–9
    development 179–82
    liberalisation 178–84
        drylands 184–6
    poverty 179–84
albedo hypothesis 20
Algeria 121
Almeria Declaration on Migration and Desertification 50–51
Americas, migration 46–7
Anuak tribes 49
Aral Sea 47–8
Argentina 47, 119–20
arid environments *see* drylands
Arid Zone Research Programme (UNESCO) 11
aridity
    causes of 15
    measurement 14
    nature of 14–15
Asia, migration 47–8
atmospheric dust, drought 20–21

Bangladesh 48
Belarus 121
Bolivia 47
Brazil 47, 64, 182, 185
Burkina Faso 45, 120, 121, 122, 151, 182, 185

carbon sequestration
    drylands 141
    land degradation 140
carrying capacity 31–2
CBD *see* Convention on Biological Diversity
CDM (Clean Development Mechanism) 135
CETP *see* Community Exchange and Training Programme
Chad 29
China 136–7
civil servants, international 74
civil society 154 *see also* nongovernmental organisations (NGOs)
    definition 89
    National Action Programmes (NAPs) 100–101
    United Nations 89–90
Clean Development Mechanism (CDM) 135
Community Exchange and Training Programme (CETP) 101, 165–75
    knowledge exchange 170–75
    pilot project 165–9
Convention on Biological Diversity (CBD) 3–4
    nongovernmental organisations (NGOs) 92
    secretariat 75

Darfur 29, 49
decentralisation 157–8
deforestation 19, 32–3, 62–3, 140
desertification 17–19 *see also* land degradation
    agricultural trade 178–9, 186–9
    as an international policy issue 60–61

conflict 48–9
definition 4, 17, 199
discourse 78–9
global warming 21
intervention areas 156–7
migration 43–52
usage of term 12, 79, 199
women 90, 94–5
Doha Round (World Trade Organization) 177–8
drought 16–17
albedo hypothesis 20
atmospheric dust 20–21
migration 45
drylands 1–2
Africa 28–9
agricultural trade liberalisation 184–6
aridity measurement 14
carbon sequestration 141
classification 12
definition 12
development financing 132–5
as equilibrium systems 13
human impact 18–19
hunger 28–9
land privatisation 35
land use 31–2
marginalisation 36
moisture index 12
as non-equilibrium systems 13
physical characteristics 12–14
population 11
poverty 28–9
private financing 134–5
rainfall variability 15–16
research 11–12
resilience 13

Earth Summit *see* Rio Conference on Environment and Development (UNCED)
economic migrants 43–4
Egypt 121
El Niño Southern Oscillation (ENSO) 15–16
EMG (Environmental Monitoring Group (South Africa)) 170–71
environment, ministries of 157–8
environmental information 121–2

Environmental Monitoring Group (South Africa) 170–71
environmental refugees 43–4
environmental treaties, secretariats 73–5
equilibrium systems 13
Eritrea 119
Ethiopia 45, 48–9, 122
European Union 62–3

farmers 182–4, 186–9
Financial Information Engine on Land Degradation (FIELD) 137
foreign direct investment (FDI) 134

G77 62–6, 69, 70–71, 197
GEF *see* Global Environment Facility
Georgia 122
Global Commission on International Migration 52
Global Environment Facility (GEF) 52, 76, 118–19, 133–4, 196
land degradation 134
United Nations Convention to Combat Desertification (UNCCD) 134
secretariat 79
Global Mechanism *see* United Nations Convention to Combat Desertification (UNCCD), Global Mechanism
global warming 21
good governance
accountability 115, 122
basic elements 110–11
coherence 114, 120–21
definition 110
definition of terms 115–16
effectiveness 114, 120–21
efficiency 114, 122
equality 113, 120
equity 113
framework 116–17
information 115, 121
legitimacy 115
openness 113
participation 112, 120, 121
predictability 114
regional annexes 115
responsiveness 113–14
transparency 113, 121

Index

United Nations Convention to Combat Desertification (UNCCD) 111–24
governance 199–201 *see also* good governance
governments, land degradation 33–4
Group of 77 plus China *see* G77
Guinea 32

Havana Declaration of Heads of States and Governments 82
hunger 2
   drylands 28–9

INCD *see* Intergovernmental Negotiating Committee on Desertification
India 48
Indonesia 186
institutions for sustainable resource use 164
Intergovernmental Negotiating Committee on Desertification (INCD) 50–51, 61, 64–7
   Africa 64
   funding 65–6
   Geneva Session (1993) 64–5
   Geneva Session (1994) 66
   global mechanism 67–8
   interim phase 67–8
   Nairobi session 64
   New York session 65–6
   nongovernmental organisations (NGOs) 89–90, 109
   decision making 91–2
   Paris session 66–7
international civil servants 74
International Symposium on Migration and Desertification 50–51

Kazakhstan 48
Kenya 33
knowledge exchange 170–75

land *see also* land degradation
   carrying capacity 31–2
   low potential 29
   privatisation 35
   rights 35–6
   tenure 35

land degradation 5, 17–19, 187–9 *see also* desertification
   carbon sequestration 140
   causes of 2
   costs 2
   definition 4–5, 17
   Global Environment Facility (GEF) 134
   governments 33–4
   land rights 35–6
   land tenure 35
   monocropping 33–4
   overpopulation 33
   poverty 27–8, 30–37
least developed countries (LDCs) 180–81
livestock management 32

Malawi 33–4
Mali 45, 49, 100, 121
marginalisation of drylands 36
MDGs (Millennium Development Goals) 3, 37–9, 117
Mexico 47–8, 184
Middle East, migration 47–8
migrants 43–4
migration
   Africa 44–6
   Americas 46–7
   Asia 47–8
   conflict 48–9
   desertification 43–52
   droughts 45
   Middle East 47–8
   United Nations Convention to Combat Desertification (UNCCD) 49–51
Millennium Development Goals (MDGs) 3, 37–9, 117
Millennium Project 187
   Task Force on Hunger 38–9
monocropping 33–4, 187
Morocco 46

NAFTA (North American Free Trade Association) 184, 186
Namibia 121
National Action Programmes (NAPs) 5–6, 51, 81, 112, 119–21, 139, 147–59, 188–9, 196
   civil society 100–101

decentralisation 157–8
integration 151–3
intervention areas 156–7
investment 148–55
ministries of environment 157–8
National Environmental Action Plans (NEAPs) 120–21
nongovernmental organisations (NGOs) 99–101
operational guidelines 149–50
'shopping list' 150
strategic 150–51
weaknesses 154–5
West Africa 147–59
women 120
National Co-ordinating Bodies (NCBs) 152
National Environmental Action Plans (NEAPs) 120–21
NGOs *see* nongovernmental organisations
Niger 1, 8
Nigeria 49
nomadic pastoralism 32
non-equilibrium systems 13
nongovernmental organisations (NGOs)
capacity building 101
co-ordination 98–9
Convention on Biological Diversity (CBD) 92
definition 89
Intergovernmental Negotiating Committee on Desertification (INCD) 89–90, 109
decision making 91–2
National Action Programmes (NAPs) 99–101
network 93
north-south partnerships 101–2
Réseau international des ONG pour la désertification (RIOD) 93, 98–9
Rio Conference on Environment and Development (UNCED) 89–90
traditional knowledge 96–7
United Nations Convention to Combat Desertification (UNCCD) 196
Committee on Science and Technology (CST) 96–7
Conferences of the Parties (COPs) 95–6
implementation 93–4
negotiations 90

United Nations Framework Convention on Climate Change (UNFCCC) 92
North American Free Trade Association (NAFTA) 184, 186

official development assistance (ODA) 117, 118, 133
overcultivation 18
overgrazing 18, 31–2
overpopulation 33

PACD (Plan of Action to Combat Desertification) 60–61
Pacific Decadal Oscillation (PDO) 15–16
Pakistan 185, 187
Plan of Action to Combat Desertification (PACD) 60–61
poverty
agricultural trade 179–84
alleviation 3
desertification 2
drylands 28–9
land degradation 27–8, 30–37
poverty-environment nexus 27–8, 31–4
privatisation of land 35
productivity 17–18

rainfall, variability 15–16, 29
refugees 43–4
Réseau international des ONG pour la désertification (RIOD) 93, 98–9
Rio Conference on Environment and Development (UNCED) 50, 61
nongovernmental organisations (NGOs) 89–90
Preparatory Committee 62–3
RIOD *see* Réseau international des ONG pour la désertification
rural development financing 133
rural livelihoods, agricultural trade liberalisation 182–4

SADC (South African Development Community) 168–73
SCS (Solidarité Canada Sahel) 99
sea surface temperature (SST) anomalies 15–16
semi-arid environments *see* drylands
Senegal 45–6, 52, 121

Solidarité Canada Sahel (SCS) 99
Solomon Islands 121
South Africa, Environmental Monitoring Group (EMG) 170–71
South African Development Community (SADC) 168–73
Sudan 29, 49
sustainable development 1, 3, 199–202
    United Nations Convention to Combat Desertification (UNCCD) 76
sustainable resource institutions 164
Swaziland 119
Syria 48

Tajikistan 121
Task Force on Hunger (Millennium Project) 38–9
traditional knowledge 96–7
Tuareg 49
Tunisia 136

Uganda 120, 122–3
UNCCD see United Nations Convention to Combat Desertification
UNCED see Rio Conference on Environment and Development
UNCOD (United Nations Conference on Desertification) 60–61
UNEP (United Nations Environment Programme) 60
UNESCO (United Nations Education, Science and Cultural Organization) 11
UNFCCC see United Nations Framework Convention on Climate Change
United Nations, civil society 89–90
United Nations Conference on Desertification (UNCOD), Plan of Action to Combat Desertification (PACD) 60–61
United Nations Convention to Combat Desertification (UNCCD) 3–6, 206–56
    Action Facilitation Co-ordinators 78, 80
    action programmes 5–6, 112
    Ad Hoc Working Group (AHWG) 148–9
    agricultural trade 186–9
    collaborations 4, 51–2, 81
    Committee for the Review of the Implementation of the Convention (CRIC) 61, 80, 148–9, 189
    Committee on Science and Technology (CST)
        composition 96
        nongovernmental organisations (NGOs) 96–7
    Community Exchange and Training Programme (CETP) 101, 165–75
        knowledge exchange 170–75
        pilot project 165–9
    community participation 153–4
    Conferences of the Parties (COPs) 61, 68, 77, 82
        nongovernmental organisations (NGOs) 95–6
    as the 'convention of endangered populations' 201–2
    developed countries 197–8
    development governance 199–201
    as developmental convention 138, 197
    as environmental convention 138, 197
    financing 6, 76, 131–43, 197–8
        constraints 141–3
        effectiveness 137–41
        National Action Programmes (NAPs) 139
        opportunities 141–3
        structure 138–40
    focus 197
    Global Environment Facility (GEF) see Global Environment Facility (GEF)
    Global Mechanism 76, 131–43
        catalysing funds 136–7
        constraints 141–3
        direct funding 135–6
        effectiveness 137–41
        Financial Information Engine on Land Degradation (FIELD) 137
        functions of 132
        governance 132
        investment funding 135–6
        opportunities 141–3
        process funding 135–6
        structure 138–41
    good governance 111–24
        affected countries 119–23
        definition of terms 115–16
        donors' response to 117–19
        framework 116–17

governance 6, 199–201 *see also* good governance
Havana Declaration of Heads of States and Governments 82
information base 198–9
institutionalisation 79–80
Intergovernmental Negotiating Committee on Desertification (INCD) 50–51, 61
International Symposium on Migration and Desertification 50–51
liaison officers 78
mainstreaming 200–201
membership 188–9
migration 49–51
Millennium Development Goals (MDGs) 37–9, 117
National Action Programmes (NAPs) 5–6, 51, 81, 112, 119–21, 147–59, 188–9, 196
 decentralisation 157–8
 environment, ministries of 157–8
 financing 139
 integration 151–3
 intervention areas 156–7
 investment 148–55
 ministries of environment 157–8
 National Environmental Action Plans (NEAPs) 120–21
 nongovernmental organisations (NGOs) 99–101
 operational guidelines 149–50
 'shopping list' 150
 strategic 150–51
 weaknesses 154–5
 women 120
National Co-ordinating Bodies (NCBs) 152
negotiations 59–71
 nongovernmental organisations (NGOs) 90
nongovernmental organisations (NGOs) 89–90, 196
 implementation 93–4
nongovernmental stakeholders 76
participatory approach 5
regional annexes 115
Regional Co-ordination Units (RCUs) 80

revision of 201
as a Rio Convention 75–6
scientific base 198–9
scope 197
secretariat 75–84
 capacity development 80–81
 desertification discourse 78–9
 form 76–8
 functions 76–8
 Global Environment Facility (GEF) 79
 political conduct 81–2
surprises 70
sustainable development 76
text of 206–56
United Nations Framework Convention on Climate Change (UNFCCC) 140–41
women 113
United Nations Education, Science and Cultural Organization (UNESCO) 11
United Nations Environment Programme (UNEP) 60
United Nations Framework Convention on Climate Change (UNFCCC) 3–4
 Clean Development Mechanism (CDM) 135
 nongovernmental organisations (NGOs) 92
United Nations Convention to Combat Desertification (UNCCD) 140–41

West Africa
 decentralisation 155
 National Action Programmes (NAPs) 147–59
 National Co-ordinating Bodies (NCBs) 152
women
 desertification 90, 94–5
 United Nations Convention to Combat Desertification (UNCCD) 113
 National Action Programmes (NAPs) 120
Women and Desertification Caucus 95
World Trade Organization (WTO) 177–8

Zimbabwe 29